BEFORE CONSCIOUSNESS

In Search of the
Fundamentals of Mind

Edited by
Zdravko Radman

imprint-academic.com

Copyright © Imprint Academic, 2017

Individual contributions © the respective authors 2017

The chapter titled 'Auditory Verbal Hallucinations and Inner Speech'
by Wilkinson and Fernyhough is made freely available via a CC-BY licence.

The moral rights of the authors have been asserted.
No part of this publication may be reproduced in any form
without permission, except for the quotation of brief passages
in criticism and discussion.

Published in the UK by
Imprint Academic, PO Box 200, Exeter EX5 5YX, UK

Distributed in the USA by
Ingram Book Company,
One Ingram Blvd., La Vergne, TN 37086, USA

ISBN 9781845409203

A CIP catalogue record for this book is available from the
British Library and US Library of Congress

Contents

Foreword — v
    Zdravko Radman

Introductory Essay
    Consciousness in the Light of the Non-Conscious Brain — ix
    Joseph LeDoux

Consciousness Meets the Unconscious

Elegant Actors for a Smooth Unconscious to Conscious Transition — 1
    Alain Berthoz

There is No Pure Consciousness and No Innocent Unconscious — 26
    Zdravko Radman

The Interdependence between Conscious and Unconscious Processes — 50
    Donish Cushing, Reza D. Ghafur and Ezequiel Morsella

As Above, So Below:
    Tangled Loops between Consciousness and the Unconscious — 83
    Axel Cleeremans

Higher-Order Awareness, Misrepresentation, and Function — 106
    David Rosenthal

Doing Complex Cognitive Tasks in a Non-Conscious Way

Unconscious Perception and the Function of Consciousness — 142
    Jesse Prinz

Prenoetic Effects on Perception and Judgment — 163
    Shaun Gallagher

The Anatomy of an (Unconscious) Decision — 174
    Ben R. Newell

Memory, Consciousness, and the Hippocampus — 192
    Thomas Reber

Can We Think Unconsciously Via Analogy? 206
  Penka Hristova

The Unconscious Sophistication of Skills

Habits and the Intergration of Conscious
  and Non-Conscious Human Actions 225
  Javier Bernacer and Jose Ignacio Murillo

Flow, Choke, Skill:
  The Role of the Non-Conscious in Sport Performance 246
  Massimiliano Cappuccio

Auditory Verbal Hallucinations and Inner Speech:
  A Predictive Processing Perspective 284
  Sam Wilkinson and Charles Fernyhough

'Efforts of the Cultivated Mind':
  Neurological Impairment, Intention, and Attention to the Body 304
  Jonathan Cole

An Epilogue 319
  Chris D. Frith

Notes on Contributors 329

Index 337

Zdravko Radman

Foreword

Without consciousness we would not have the experientially flavoured world we have, but without the non-conscious we would not have it at all; for we would not be able to breathe, eat, move, walk, feel, mimic, gesture, laugh, etc., and even see, talk, remember, understand, think, imagine, and make myriad spontaneous decisions as we continuously do in all life situations, from trivial to existential ones.

Without consciousness we would not be the kind of creatures we are, but what makes us really unique is our specific non-conscious constellation—a basis from which all mentality germinates and which is irreducible, that is, not representable or in any way simulable. Much effort has been put in the philosophy of mind to emphasize the irreducible nature of consciousness and to prove that the qualitative cannot be adequately transcribed. What is really irreducible, however, is our 'silent' self, acting in the background, yet powerful enough to become 'audible' in the conscious mode. The long-lasting conviction that what makes us different is our thoughts, ideas, views, etc. is now ripe for a revision: what makes each of us specific is not so much a matter of intellect but has to do with the fact that our individual non-conscious 'physiognomies' are unique. These differences, in turn, are a result of social and cultural happenings that leave their trace on the flesh. That is why my cat is differently unconscious than myself and why my unconscious 'it' does not match the one of an infant or a savage.

The non-conscious apparatus is the mind's vital organ that is never at rest and from which the deliberative 'self' can never retreat or find shelter. As there are no mind-lids, there is no way we can shut off its ongoing fabrication of 'mute' (but effective) inputs, signalling, and suggesting. The non-conscious intervenes in virtually all acts of mind and it is a steady companion to agency; not even experience is exempt from this. Experience is thus not to be seen as autonomous (and neither, I think, can we study it as an isolated phenomenon, as the mainstream currently does). The qualitative does not originate out of the blue—the 'blueness' emerges powered by the non-conscious mechanisms that give it growth.

Further, the non-conscious is neither passive nor static for it undergoes permanent reconfiguration so that it is justified to claim that it has its own growth and evolution. There would be no development in the matters of mind if the non-conscious were not capable of keeping in mind (or 'keeping in body') all that is no longer actuality; there would be no progress in our cognitive competence if the non-conscious were not empowered to acquire whatever the conscious agent is capable of accomplishing. If it were not so, for instance, memory and learning would not be able to fulfil their functions, for the frame of the attentive mind is small and the processes underlying it slow. It simply means that the potentiality for any sort of action, motor or mental, lies outside of that frame.

We become masters of crafts and arts as skills required for their performance get acquired by the body and become routine actions that are exercised best if there is no intrusion of an attentive subject. There would hardly be any expertise of any kind if the unconscious mind and body were not able to absorb and store even the most complex motor and cognitive tasks. Because it is the case, the non-conscious has the potential of increasing its cognitive capacity. How else can we explain the highly skilled and competent coping of which we are capable without engaging in conscious awareness?

Minds are unconscious long before they are conscious—as Joseph LeDoux states straightforwardly in his introductory essay to this volume. Yet the unconscious does not cease to exist with the emergence of consciousness, for it permanently exercises its impact on whatever we do. In addition, we can no longer ignore the scientific fact that most of what constitutes mind by far is non-conscious in nature (from the neuroscientific perspective as much as 95% to 98% of all neural processing is not conscious). The unconscious mind is ubiquitous and always at stake. Nothing we mentally perform can bypass the unconscious dynamics at some level or at some point.

There are sufficient philosophical reasons to want to devise methodological strategies to account for mind's 'extendedness' not only in the outward direction (as suggested by the originators of the hypothesis, Andy Clark and David Chalmers), but also to awaken sensitivity for the alternative approach according to which minds are 'extended' also inwardly, towards the sphere of the mental below the threshold of awareness from which it all originates.

It is somewhat awkward that the most prominent view from 'within' that has ambitions to capture the subjective basis of our being seldom goes further than the description of phenomenal experience with emphasis on the qualitative. Such attempts are in that respect weak, as they more often than not leave out the genuine 'within-ness' associated with the fundamentals of mind not under the surveillance of an attentive subject. Any scientifically founded insight into the mental world *within* should then open toward (or at least not ignore) the 'silent' rooms of the mind.

Our current philosophies of mind are almost exclusively philosophies of *conscious* mind and are in that sense discriminatory of all that lies outside of its confines. Textbooks and introductions to philosophy of mind habitually define the mental in terms of 'beliefs', 'desires', 'wishes', 'plans', etc., and in such a way create the conviction that these are not only representative of the mind but that they also exhaust the scope of mentality. The truth is there is a lot of mental activity before believing, desiring, wishing, planning, etc. become present to the conscious mind. Nowadays we have sufficient scientific reasons to claim that such approaches are partial and inadequate for they leave out a huge sphere of mentality which by its nature, however, cannot speak for itself.

Inhibitions and preselections, motivations and moods, affinities and aversions, inclinations and intuitions, prejudices and dislikings, as well as endless variants of emotionality, all pre-shape mental events before they become conscious acts. Actually, consciousness comes as a late (but eloquent and vivid) witness of the processes already taking place at some more fundamental level. To think that when it comes to complex mental processes the non-conscious loses its import is largely unfounded; in all high-level cognition the non-conscious is at stake and serves as a tool without which cognition seems impossible. For instance, *perception* is shaped in such a way; no less is this the case with *memory*, which mutates with the ever-changing actual mental landscape, with 'self' having no authority over this very process; *learning* is unimaginable without the possibility of storing in the background and 'forgetting' the learned; *unconscious thinking* is no oxymoron, as thought is just in its most complex version inspired by and impacted by the processes of which we are largely unaware; *decision making*, as well as *judgments*, have their roots in the sphere preceding the emergence of consciousness. Its stamp is also present in all of *intentionality*. Before the intentional bond to the world gets expressed in consciousness, 'aboutness' is already (pre-)shaped by means of non-conscious modulations.

One of the conclusions we reach in this collective attempt to better understand the mind in its entirety is that by the metonymic taking of consciousness to represent the entire mind we neglect, or refuse to accept, that the mental is instantiated in incomparably many more ways than is conceived by intellectualist approaches, and those mechanisms need no consciousness for the mind to matter and be relevant.

A further lesson following from this project could probably be that *the mind has many forms* and that we have for too long been blinded by the intellectualist dogma that narrowed the scope of the mind and impoverished it, ignoring the function and import of the huge sphere that's not accessible to consciousness and not in command of volition. I guess we are now in a good position to recognize the relevance of the claim that 'mind is more than consciousness' (John Dewey) and provide support to the credo 'the whole man counts' (William James). Indeed, the mind itself is not discriminatory in regard to the processes and features that do not conform to dictates of the conscious 'self'; our theories, however, obviously are.

It will probably still take a while until the science of the brain and the study of mind feel mature (and courageous) enough to proclaim a *new paradigm* that would mark a shift away from conscious-centredness toward recognizing the fundamentality of the non-conscious, though the scientifically supported conviction grows that we have already advanced enough to inaugurate it.

Acknowledgment

I am indebted to Joseph LeDoux and Chris Frith who kindly accepted my invitation to write the introductory and the postscript to this volume. I also want to thank the participants of the conference 'The Power of the Non-Conscious', that took place at the Interuniversity Centre Dubrovnik, 3–5 October 2014, who gave that meeting a stimulating intellectual energy that spilled over into this volume.

I appreciate enormously Shaun Gallagher's and Jesse Prinz's steady support (not only) to this project. David Rosenthal has provided generous encouragement throughout various stages of the production of this book and I want to thank him for that.

Special thanks go to Graham Horswell from Imprint Academic for recognizing the importance of the theme the volume is devoted to and also for being an agile and thorough (also patient) copy editor.

Zdravko Radman

Joseph LeDoux

Introductory Essay
Consciousness in Light of the Non-Conscious Brain

Scientific psychology emerged in the late nineteenth century when problems about the mind, which had been the province of philosophers, began to be addressed using the scientific experimental methods of physics and physiology (Boring, 1950). The number one topic of interest to early psychologists was the nature of conscious experience. Wilhelm Wundt developed the method of introspection in which the psychologist was both the experimenter and subject (Wundt, 1874). This approach was rejected by John Watson and fellow behaviourists who argued that, to be scientific, psychology had to be objective, and this could be best achieved by using observable behaviour as the gold-standard of measurement in research (Watson, 1925; Skinnner, 1938). Behaviourists felt no need to go inside the 'black box', as behaviour could be accounted for in terms of relations between observable stimuli and responses, and neither mental nor neural explanations were needed.

Near the middle of the twentieth century, the behaviourist rein began to weaken when the emergence of computing devices compelled some to note similarities between the way the human mind and electronic machines process information (Turing, 1950; Newell and Simon, 1972; Miller et al., 1960). The result was the field of cognitive science (Neisser, 1967; Gardner, 1987). But the mind that the behaviourist got rid of (the introspecting mind) was not the mind that the cognitive scientists were promoting. The cognitive mind was not a place where thoughts and feelings occurred but was instead a set of processes that create internal representations that are used to guide action. Cognitive scientists were less interested in the introspectively accessible conscious content that can result from such processing than in the underlying processes themselves (Neisser, 1976). The cognitive mind was thus, in many ways, a set of non-conscious processes that came to be known as the cognitive unconscious (Kihlstrom, 1987); some of these sometimes give rise to conscious experiences, but they do not necessarily do so.

The cognitive unconscious differed from the Freudian unconscious (Freud, 1915), a place where troubling thoughts and memories were banished in an effort to prevent anxiety. But Freud also proposed the existence of pre-conscious information, which referred to content that is not in consciousness at the moment but that can be readily retrieved into consciousness from memory when needed. And the cognitive unconscious included processes that bore strong similarity to Freud's pre-conscious. Explicit or declarative memory is like this—when not being retrieved you are not conscious of the information, but, barring a retrieval failure, you can bring the information into conscious awareness when needed. The cognitive unconscious also included information that is unconscious because it is, by virtue of brain connectivity, not accessible by the cognitive systems that make consciousness possible. So-called implicit or non-declarative memory is like this—you do not have conscious access to the specific memories that allow you to ride a bike or play an instrument, or increase your heart rate when you encounter a stimulus that caused you harm in the past.

Initially, cognitive scientists sought 'functional' explanations of mind, and had little interest in brain research. One common view was that, just as computer software is independent of the hardware, so too is the mind independent of the brain (Putman, 1960; Fodor, 1975). Nevertheless, the emergence of cognitive science changed the way that psychological processes were studied in patients with brain damage, transforming neuropsychology from a field concerned with characterizing the clinical picture of patients (e.g. Geschwind, 1965) to a field, cognitive neuroscience, that used cognitive methods to study the brain, including brain pathology, for the purpose of understanding the normal brain and its cognitive and behavioural functions (Gazzaniga and LeDoux, 1978; Gazzaniga, 1984).

The advent of functional imaging was a tremendous boon to cognitive neuroscience. An impressive amount of data has been acquired, and has, for the first time, allowed us to understand in some detail what goes on in the brain when people can give introspective reports of stimuli (usually visual stimuli) and when they cannot due to subliminal stimulus presentations or brain damage. This work shows that, when people can report on seeing a stimulus, both the visual cortex (necessary for representing the visual stimulus) and prefrontal (including lateral and medial areas) and parietal cortex (involved in such processes as attention and working memory) are activated. However, when people are not aware, only the visual cortex is active. The conclusion from these data is that prefrontal/parietal circuits constitute a general network of consciousness that make possible introspectively accessible experiences (Rees and Frith, 2007; Frith *et al.*, 1999; Lau and Passingham, 2006; Dehaene and Naccache, 2001; Dehaene and Changeux, 2011; Gaillard *et al.*, 2009; Dehaene *et al.*, 2006; Dehaene, 2014; Naccache *et al.*, 2002; LeDoux, 2015).

There are debates about the extent to which awareness is prevented with subliminal methods, with some methods doing a better job than others (Peters and Lau, 2015; Szczepanowski and Pessoa, 2007; Pessoa, 2005; Mitchell and Greening, 2012; Block, 2015; Overgaard, 2012). Some claim this shows that the non-conscious mind is greatly limited. But such studies are hindered by the necessity to use degraded stimulus exposures in order to prevent conscious awareness. The findings thus say as much if not more about methods used to study non-conscious processing as they do about the limits of non-conscious processing. Studies of blindsight patients with damage to the right visual cortex who are able to respond to stimuli that they cannot verbally report on also provide strong evidence that the brain can detect and respond to unreportable stimuli (Weiskrantz *et al.*, 1974; Weiskrantz, 1986; Cowey and Stoerig, 1991). It is important to note that because degraded stimuli do not have to be used in such studies, these may be more revealing about the limits of non-consious processing. Also, studies of healthy subjects show that stimuli that are fully present, but unnoticed, affect subsequent behaviour (Hassin *et al.*, 2005).

I got interested in the question of non-conscious processing through my PhD work with Michael Gazzaniga at SUNY Stony Brook in the mid-1970s. I explored how the left hemisphere dealt with behaviours produced by the right hemisphere. From the point of view of the left, anything done by the right was done non-consciously. We were fascinated with how the left hemisphere, when observing the responses of the right, confabulated an explanation of why those responses occurred. If the right caused the person to laugh or stand up, the left explained the response as being due to the humour of the situation (for the laugh) or the need to stretch (in the case of standing up). We drew upon the emerging field of cognitive social science, explaining the left hemisphere's confabulations as efforts to reduce cognitive conflict by attributing cause to non-consciously controlled processes. Gazzaniga and I summarized this work in our 1978 book, *The Integrated Mind*, in which we argued that the interpretation of experience in such a way as to maintain a coherent sense of self was an important role of conscious experience.

My career has, ever since, been dedicated to understanding how systems that operate non-consciously provide fodder for the construction of conscious experiences. I was especially interested in how so-called emotion systems generate signals that are consciously interpreted as feelings. There was little work on emotion and the brain at the time, and this seemed to be a topic begging to be researched. But I decided that the way to go was the non-obvious one of studying emotion in humans. There were no good techniques for exploring the detailed mechanisms of the human brain at the time, so I turned to rats to make progress on how the emotion systems work.

My work implicated the amygdala as a key brain region that connects external threats with behavioural responses that cope with the impending danger (for a summary of my early work see LeDoux, 1996). True to my split-brain work, I have long maintained that the conscious experience of fear, the feeling of being afraid, was a product of a different processing channel in the brain that involved cognitive systems in the neocortex (see LeDoux, 1984; 1996; 2002; 2008), systems like attention and working memory, that are now firmly grounded in prefrontal and parietal circuits (see above). The amygdala, I argued, was a non-conscious processor. Nevertheless, I came to be associated with the idea that the amygdala is the seat of fear in the brain. I am probably somewhat to blame, as I discussed the amygdala as processing non-conscious aspects of fear.

In my recent book, *Anxious*, and several recent papers (LeDoux, 2012; 2014; 2015), I clear the air and argue that fear should only be used to describe the conscious feelings we have when in danger, and not the non-conscious processes that detect and respond to threats. The non-conscious processes contribute indirectly to fear. Non-conscious threat processing is an evolutionarily old function, very old, in fact—in order to survive and give rise to the history of life, the earliest microbes had to detect and respond to danger. Conscious feelings of fear, I argue, did not appear until organisms had brains, and only in brains that were sufficiently sophisticated as to be able to be aware of their own activities. This is undoubtedly the case in humans, and is unknown, and possibly unknowable, in other organisms.

Other organisms may have inner experiences but if they do they are likely to be different from those made possible by the human brain. This is not a claim for human exceptionalism but an argument for recognizing that claims about conscious feelings in other animals are based on assumptions, not on data that directly address the question. It has been said that to assess whether a behaviour depends on consciousness a researcher should both test the hypothesis that the behaviour is plausibly accounted for in terms of conscious experience and also the hypothesis that the behaviour can be accounted for by non-conscious processes (Heyes, 2008). Seldom are both tests performed. In humans we can get some leverage on whether a nonverbal response reflects conscious processes by using verbal report. In other animals, though, verbal reports are not available to distinguish between conscious and non-conscious explanations of nonverbal behavior (LeDoux, 2015).

This book on the unconscious is a welcome addition to the literature. It covers many important topics and raises relevant questions that will define the scope of future work aimed at advancing our understanding of the nature, limits, and extent of unconscious processing in the human brain.

Much is made about the wonders of consciousness and how weak and limited the unconscious is. But, unless we assume that consciousness has always existed, we have to acknowledge that brains were unconscious long before they were conscious. It is unfortunate that the primordial condition of the nervous system (non-conscious) as named by the negation of the newer condition (consciousness). One could argue that it is consciousness (i.e. the non-non-conscious) that needs defending—that explanations of behavioural or psychological processes in terms of consciousness should be refrained from unless there is compelling evidence that consciousness is involved.

References

Block, N. (2015) The Anna Karenina principle and skepticism about unconscious perception, *Philosophy and Phenomenological Research*, doi: 10.1111/phpr.12258.

Boring, E.G. (1950) *A History of Experimental Psychology*, New York: Appleton-Century-Crofts.

Cowey, A. & Stoerig, P. (1991) The neurobiology of blindsight, *Trends in Neuroscience*, **14**, pp. 140–145.

Dehaene, S. (2014) *Consciousness and the Brain: Deciphering How the Brain Codes Our Thoughts*, New York: Penguin Books.

Dehaene, S. & Changeux, J.P. (2011) Experimental and theoretical approaches to conscious processing, *Neuron*, **70**, pp. 200–227.

Dehaene, S., Changeux, J.P., Naccache, L., Sackur, J. & Sergent, C. (2006) Conscious, preconscious, and subliminal processing: A testable taxonomy, *Trends in Cognitive Science*, **10**, pp. 204–211.

Dehaene, S. & Naccache, L. (2001) Towards a cognitive neuroscience of consciousness: Basic evidence and a workspace framework, *Cognition*, **79**, pp. 1–37.

Fodor, J. (1975) *The Language of Thought*, Cambridge, MA: Harvard University Press.

Freud, S. (1915) The unconscious, in Strachey, J. (ed.) *The Standard Edition of the Complete Psychological Works of Sigmund Freud*, vol. 14, pp. 161–215, London: The Hogarth Press.

Frith, C., Perry, R. & Lumer, E. (1999) The neural correlates of conscious experience: An experimental framework, *Trends in Cognitive Sciences*, **3**, pp. 105–114.

Gaillard, R., Dehaene, S., Adam, C., Clemenceau, S., Hasboun, D., Baulac, M., Cohen, L. & Naccache, L. (2009) Converging intracranial markers of conscious access, *PLoS Biology*, **7**, e61.

Gardner, H. (1987) *The Mind's New Science: A History of the Cognitive Revolution*, New York: Basic Books.

Gazzaniga, M.S. (ed.) (1984) *Handbook of Cognitive Neuroscience*, New York: Plenum.

Gazzaniga, M.S. & LeDoux, J.E. (1978) *The Integrated Mind*, New York: Plenum.

Geschwind, N. (1965) The disconnexion syndromes in animals and man, Part I, *Brain*, **88**, pp. 237–294.

Hassin, R.R., Uleman, J.S. & Bargh, J.A. (eds.) (2005) *The New Unconscious*, New York: Oxford University Press.

Heyes, C. (2008) Beast machines? Questions of animal consciousness, in Weiskrantz, L. & Davies, M. (eds.) *Frontiers of Consciousness: Chichelle Lectures*, Oxford: Oxford University Press.

Kihlstrom, J.F. (1987) The cognitive unconscious, *Science*, **237**, pp. 1445–1452.

Lau, H.C. & Passingham, R.E. (2006) Relative blindsight in normal observers and the neural correlate of visual consciousness, *Proceedings of the National Academy of Sciences USA*, **103**, pp. 18763–18768.

LeDoux, J.E. (2012) Rethinking the emotional brain, *Neuron*, **73**, pp. 653–676.

LeDoux, J.E. (1984) Cognition and emotion: Processing functions and brain systems, in Gazzaniga, M.S. (ed.) *Handbook of Cognitive Neuroscience*, New York: Plenum.

LeDoux, J.E. (1996) *The Emotional Brain*, New York: Simon and Schuster.

LeDoux, J.E. (2002) *Synaptic Self: How Our Brains Become Who We Are*, New York: Viking.

LeDoux, J.E. (2008) Emotional colouration of consciousness: How feelings come about, in Weiskrantz, L. & Davies, M. (eds.) *Frontiers of Consciousness: Chichele Lectures*, Oxford: Oxford University Press.

LeDoux, J.E. (2014) Coming to terms with fear, *Proceedings of the National Academy of Sciences USA*, **111**, pp. 2871–2878.

LeDoux, J.E. (2015) *Anxious: Using the Brain to Understand and Treat Fear and Anxiety*, New York: Viking.

Miller, G.A., Glanter, E.G. & Pribram, K.H. (1960) *Plans and the Structure of Behavior*, New York: Holt, Rinehart and Winston.

Mitchell, D.G. & Greening, S.G. (2012) Conscious perception of emotional stimuli: Brain mechanisms, *The Neuroscientist*, **18**, pp. 386–398.

Naccache, L., Blandin, E. & Dehaene, S. (2002) Unconscious masked priming depends on temporal attention, *Psychological Science*, **13**, pp. 416–424.

Neisser, U. (1967) *Cognitive Psychology*, Englewood Cliffs, NJ: Prentice Hall.

Neisser, U. (1976) *Cognition and Reality: Principles and Implications of Cognitive Psychology*, New York: Freeman.

Newell, A. & Simon, H. (1972) *Human Problem Solving*, Boston, MA: Little, Brown and Company.

Overgaard, M. (2012) Blindsight: Recent and historical controversies on the blindness of blindsight, *Wiley Interdisciplinary Review of Cognitive Science*, **3**, pp. 607–614.

Pessoa, L. (2005) To what extent are emotional visual stimuli processed without attention and awareness?, *Current Opinion in Neurobiology*, **15**, pp. 188–196.

Peters, M.A.K. & Lau, H. (2015) Human observers have optimal introspective access to perceptual processes even for visually masked stimuli, *eLife*, **4**, e09651.

Putnam, H. (1960) Minds and machines, in Hook, S. (ed.) *Dimensions of Mind*, New York: Collier Books.

Rees, G. & Frith, C. (2007) Methodologies for identifying the neural correlates of consciousness, in Velmans, M. & Schneider, S. (eds.) *A Companion to Consciousness*, Oxford: Blackwell.

Skinner, B.F. (1938) *The Behavior of Organisms: An Experimental Analysis*, New York: Appleton-Century-Crofts.

Szczepanowski, R. & Pessoa, L. (2007) Fear perception: Can objective and subjective awareness measures be dissociated?, *Journal of Vision*, **7**, p. 10.

Turing, A.M. (1950) Computing machinery and intelligence, *Mind*, **59**, pp. 433–460.

Watson, J.B. (1925) *Behaviorism*, New York: W.W. Norton.

Weiskrantz, L. (1986) *Blindsight: A Case Study and Implications*, Oxford: Clarendon Press.

Weiskrantz, L., Warrington, E.K., Sanders, M. & Marshall, J. (1974) Visual capacity in the hemianopic field following a restricted occipital ablation, *Brain*, **97**, pp. 709–728.

Wundt, W. (1874) *Principles of Physiological Psychology*, Leipzig: Engelmann.

Alain Berthoz

Elegant Actors for a Smooth Unconscious to Conscious Transition

1. Prologue: A Few Thoughts as a Starter

When I was challenged by Zdravko Radman to write an essay on the relation between un-conscious and conscious processes, I first thought about the influence of emotion on decision making. But I have already expressed some views on the theme (Berthoz, 2006). In addition, many recent studies have been devoted to decision making and its relation with emotion. However I was stimulated[1] by the goal statement proposed by Zdravko Radman[2] that 'it is justified to say that the non-conscious is not cognitively inferior, but is informed, skilled, competent, smart... Because it is so it can convert into consciousness with ease and in an instance (no "translation" needed)'.

Modern neuroscience studies confirm that most operations of our brain are not only pre-conscious but also completely unconscious. Notions such as 'awareness'[3] indicate also some processes which are, like many *attentional* processes, at the border between conscious and unconscious.[4] They also show that unconscious operations are not only what was called automatic, or reflexes. They require a high level of sophistication and cognitive meaning. The recently documented 'default system' (Raichle, 2015; Raichle and Snyder, 2007) is a beautiful

---

[1] Some of the ideas in this chapter have been developed in my books published first in French by O. Jacob and translated: *The Brain's Sense of Movement* (1996a); *Simplexity: Simplifying Principles for a Complex World* (2012).
[2] In an email.
[3] It is interesting that there is no equivalent of 'awareness' in French, which only uses 'la conscience', i.e. consciousness. I believe that this is because of the strong Cartesian abstract foundation of the French tradition and the very axiomatic and theoretical, disembodied orientation of our relation with nature.
[4] An interesting proposal was made by Graziano and Webb (2015) of an 'attentional schema' similar to the body schema but for attentional processes.

example of the rich operations occurring during rest or sleep without us being aware of them. For example, the transfer of memories from the hippocampus to the cortex during sleep (Buzsáki, 2015),[5] or rest, which is performed by a re-coding of the hippocampal memorized information into bursts of activities transmitted to the cortex for long-term storage. Another example is that the cortical contribution to vision has completely led to us forgetting the major subcortical role of the superior colliculus (SC) in vision. You cannot consciously catch a service ball in tennis with your cortex, it is too fast, or save a penalty that is taken at a football game. A very clever anticipation is made by the superior colliculus, which is specialized in anticipatory movement detection, and we have shown (Senot *et al.*, 2008) that MT (medio-temporal nucleus) is activated before V1 when catching a ball probably through rapid pathways feeding MT from the SC. (See also the chapter by Prinz in this volume.) This is probably due to direct connections from the thalamus to MT (Gaglianese *et al.*, 2012).

The intuition of Freud, in spite of the perversion of his ideas by some of his followers, was seminal. He is often quoted as the one who discovered the unconscious brain. However, for him, in addition to a conscious and an unconscious level there is a *pre-conscious* level which he viewed as a translator from the unconscious world to consciousness (Freud, 1991, p. 289). By contrast, I propose that many unconscious processes are already 'coded' in a way which facilitates the transition to consciousness *without any need for 'translation'*. This may be possible thanks to an elegant set of mental models or activities, coding principles, at the fringe of consciousness but not requiring full conscious access (William James had wonderful intuitions about this). I will propose that *simplex* mechanisms are fundamental for this.

Modern neuroscience has shown that all singular concepts, like 'decision, attention, memory, motivation', are in fact each hiding a multiplicity of processes which appeared throughout evolution (we have many kinds of memory systems, we have many kinds of attentional processes, etc.). I agree with the proposal of Semir Zeki (2008) when he talks about the 'multiplicity of conscious processes'. I believe it is not useful or correct to speak about '**La** conscience', or '**Le**

---

[5] Sharp wave ripples (SPW-Rs) are rapid high frequency bursts of firing of hippocampal neurons which occur during 'offline' states of the brain, associated with consummatory behaviours and non-REM sleep, and are influenced by numerous neurotransmitters and neuromodulators. Their firing patterns replay fragments of waking neuronal sequences in a compressed format. They contribute to the transfer of hippocampal representation to distributed circuits to support memory consolidation. This replay of recent events and episodes may influence decisions and even creative thoughts (Maingret *et al.*, 2016).

code de la conscience' (Dehaene, 2014).[6] For example, we know that there is a contrast between current theories of consciousness such as those of (a) Rodolfo Llinàs (2002; 2008[7]), who suggests that consciousness arises from a combination of two thalamo-cortical loops involved respectively in *context* and *content*, and which is also related with the geometrical properties of the brain; (b) Jean-Pierre Changeux and Stanislas Dehaene (following the theory of Baars) that consciousness arises from cortico-cortical global activity involving long axon connections between various specialized modules; or (c) that consciousness has a relation to the time synchronization of oscillators (Buzsáki *et al.*, 1994, p. 303) in the brain (see for instance the studies of Wolf Singer and the proposals of the Giulio Tonini and Jerry Edelman groups) which may be instrumental in the 'binding' of information of different categories to create a dynamic ever-changing unity of perception. I am not at all a specialist of the brain mechanisms of consciousness, but my intuition is that there is not a unique global definition of what the neural substrates of conscious, unconscious, or pre-conscious are. In addition, the diversity of brain configurations for a single function — that I have named 'vicariousness' (Berthoz, 2016) — is remarkable.

We have to reformulate the whole problem of analysing traditional concepts in light of modern science. This is going to be a difficult task and none of us has the answer. This chapter does not have the answer either — it will only modestly propose some facts and ideas. In this essay I have chosen to deal with the *concept of self* which, I believe, is at the root of many aspects of the problem of the relation between conscious and unconscious processes. It tries to establish a dialogue with Zdravko Radman (this volume) who writes: 'If cognitive competence is not limited to the thinking self, but also (always) realizable through embodied coping, then why not conceive the possibility of an *organismic self* capable of complex behaviour without having to engage in awareness.'

2. The Concept of Self:
General Considerations

Many recent theories have dealt with the concept of self. This polysemic notion is interesting to discuss within the frame of a book on

---

[6] For a crticial review see Salti *et al.* (2015).
[7] In this paper they examine the matching between an internal space or functional spaces endowed with stochastic metric tensor properties, allowing a dynamic correspondence between events in the external world and their specification in the internal space. They suggest a role of this dynamic geometry in brain function modelling and the neuronal basis of consciousness.

unconscious processes. This concept has challenged philosophers, scientists, and clinicians, including psychoanalysts like Freud. It also has a very deep meaning in immunology. In the cognitive domain recent attempts have been made, for instance by Pascale Piolino and Francis Eustache (Leblond *et al.*, 2016; Duval *et al.*, 2009), to suggest that there are two major components in the construction of self: an *episodic* component and a *semantic* one. These two components combine to give the rich notion of self. The *episodic* component, defined initially by Tulving (1985), concerns what we may call the 'phenomenological self'. It includes remembrances which deal with *where* the episode happened, *what* happened, *who* experiences it, and details. This requires a *binding* of these characteristics, and constitutes an auto-noetic conscience with a 'mental time travel' element and the reminiscence of the context of the encoding of the memory (the proponents of this idea deal with this for humans but we should remember that even rats have this capability at a more elementary level). It induces a feeling of meaning and continuity in humans. The *semantic* component as described by Conway (2001) is a conceptual self which deals with beliefs, personal goals, decision rules, and concerns, not experience but knowledge, and is related to the question of identity. Clinical evidence for this dichotomy has been found in cases of developmental amnesia, which destroys episodic memories but not semantic ones. However, other theories have been proposed for this notion. For instance the philosopher Paul Ricoeur (1990) has suggested that there are at least three types of self: the first is based upon a permanent identity ('mêmeté', sameness); a second ('ipséité') is fundamentally likened to alterity and justifies the title of his book, *Soi-même comme un autre*, or 'Oneself as Another' — here the self is embedded in the relation with others; and the third type of self is a narrative self. Shaun Gallagher (2000; 2013) has discussed the possibility of a distributed self with several components.

Another reason for which a discussion about the notion of self is important is that there are arguments for a deficit in the construction of the self in neurological or psychiatric pathologies like autism, but also in many other pathologies including so-called 'sensory-motor' pathologies like cerebral palsy. In addition, the dissociations of willed actions in schizophrenia, for instance, which lead a person to have the feeling that he is not the author of his actions induces a destruction of the unity of self and self-responsibility of one's actions and decisions. This is why so much attention is given today to the question of the initiation of action in view of the Benjamin Libet (2002) paradigm, for instance. Recent work by Patrick Haggard and his team (e.g. Conway, 2001), for instance, is a good example of this approach.

2.1. Consciousness requires a unity of the perceived body

Whatever the mechanisms underlying conscious processes are, they all require that the brain creates a **unified construction of the body**. Let me make the use of the term 'body' clear. I do not mean only the physical combination of muscles and bones which constitute our physical body. I mean what the philosopher Maurice Merleau-Ponty named 'le corps en acte',[8] that is the physical body as a server of the brain's goals, intentions, desires, and at the same time as the master of its possibilities, though today man has duplicated these possibilities with technology, extending the notion of the body to the immense number of artefacts which occupy our modern world. Merleau-Ponty was interested in self-awareness via the notions of '*corps propre*' (I have discussed this with Jean-Luc Petit in our book *Phenomenology and Physiology of Action*, 2008)[9] and of 'body schema'. He discussed the way Freud treated this question and the difference between 'me' and 'self'. Merleau-Ponty suggests that there is no need to have any transformation between unconscious and conscious processes. He writes that the body is not 'in' the world, it is '*au monde*'.[10]

The construction of self requires that all information concerning the body, whether physical or neural, can be combined. Another way to say this is that the content of both the conscious and unconscious mind may be similar, but something is added in consciousness, and the

---

[8] Andrieu and Berthoz (2009).

[9] Jean-Luc Petit suggested that the following paragraph from our book could be inserted in the present chapter to show the importance of the acting body for the 'direct' transmission of ideas: '*Even if it might be taken as a caricature, the behaviour of that one of us two (Alain Berthoz) who is a physiologist could be cited as an example of the importance of the act of live thinking. When explaining facts and theories, he gets up from his chair and, in order to explain locomotion, orientation, change of point of view etc., makes himself understood by acting. As if he felt that in remaining seated and inviting his audience to take account of objects of purely theoretical import he would be instilling in himself, and in his auditors, forgetfulness of their common posture. At one stroke, by moving his body, he becomes once again the agent he never should have stopped being, the agent who moves and who, in moving, moves himself and also his audience, an audience composed of agents just like himself.*' Jean-Luc Petit has discussed these questions in several publications (see Petit, 2010a,b,c,d).

[10] He wrote, 'En tant que j'ai un corps et que j'agis à travers lui dans le monde, l'espace et le temps ne sont pas pour moi une somme de points juxtaposés, pas d'avantage d'ailleurs une infinité de relations dont ma conscience opèrerait la synthèse et où elle impliquerait mon corps. Je ne suis pas **dans** l'espace et le temps; je suis **à** l'espace et au temps, mon corps s'applique à eux et les embrasse. Mon corps a son monde ou comprend son monde sans avoir à passer par des «représentations», sans se subordonner à une «fonctions symbolique» où «objectivante»' (Merleau-Ponty, 1945, p. 164).

question is what this is. Let us first discuss briefly the question of the construction of a coherent perception of one's body as the basis of self. This is not a simple process for living creatures because, for example, all senses work in *different reference frames*. Vision is in the retinotopic frame, tactile sense is in skin space, audition is in world space modified by the processes of acoustic perception, etc. Stein and Meredith (1993) have elegantly shown how the brain solves the problem of mixing acoustic and visual space unconsciously in the colliculus, a structure fundamental for our perception and largely ignored by the visual community, which devotes most of its effort to understanding the operations of the visual cortex.

It just so happens that throughout evolution one particular sense has served as a common reference frame — that is, the *vestibular system*. It is remarkable that in all world literature people keep speaking about the five senses, *ignoring the most fundamental sense for the construction of the self and probably also for our relation with others*, namely the vestibular system and the proprioceptive system from the body itself, subserved by several sensory systems (muscle receptors, tendon organs, etc.). The reason for this universal neglect is probably precisely because the messages sent by these receptors are generally unconscious; they become conscious in the case of lesion or conflict (vertigo, for example). Let us describe the main functions of these two sets of receptors.

Muscular proprioception is an unconscious system of receptors for measuring several variables of muscle contraction and movement. Neuromuscular spindles measure muscle length and the velocity of elongation. They are under the control of the brain through intrafusal muscles which regulate their static and dynamic sensitivities (Perret and Berthoz, 1973). They contribute to local spinal reflexes and motor regulation but they also have access to the brainstem and to the cortex. The cognitive role of spindles has been shown by the illusion that is induced by vibrations of the muscles: if a muscle of the arm is stimulated mechanically at frequencies of around 50 to 100 Hz, a conscious illusion of displacement of the arm is induced. Some illusions of this kind are also induced by local block anaesthesia. This reveals that we have an unconscious knowledge of the position of our arm through the complex treatment of signals from the spindles. Information from the spindles is combined with the measures made by receptors in the joints, which measure joint angle, and by Golgi tendon organs in series with the spindles, which contribute to the measurement of force. In fact the 'sense of effort' is a complex perception, which is also in part unconscious. Basically the information given by muscular and joint proprioception can be considered as unconscious, but it does contribute to the elaboration of the *'body schema'*.

The *vestibular organs* measure head movements. The three semi-circular canals measure angular acceleration, and the otoliths (utricle and saccule) measure linear acceleration and also provide a measure of head tilt with respect to gravity. All this information is essentially unconscious. It is used to stabilize the eye in the socket through the vestibulo-ocular reflex, providing a fundamental way to stabilize the visual image on the retina during head movements. It also induces postural reactions to stabilize the body, when equilibrium is disturbed, through vestibulo-spinal reflexes which are combined and sometimes in competition with other postural reflexes triggered by proprioception or tactile inputs.

All these mechanisms are fundamentally unconscious. But the measurements of head orientation and movements are also sent to several stations of the cerebral cortex and provide cognitive evaluation for several conscious operations: first, spatial orientation with respect to gravity and with respect to the environment. This is done in cooperation with vision through interactions which occur at different levels from the vestibular nuclei in the brainstem to the higher centres of the cortex and even through a direct interaction in visual areas. But it also plays an important role in spatial memory and in navigation via projection to the hippocampus, for instance (O'Mara *et al.*, 1994; Berthoz *et al.*, 1995; Vitte *et al.*, 1996). Even the vestibulo-ocular system is accessible to consciousness. We have shown that with mental imagery (Berthoz and Melville-Jones, 1985, p. 386; Melville-Jones *et al.*, 1984) we can modify the gain of the VOR in a similar way to the wearing of prisms, i.e. a sensory conflict.

Lastly, vestibular signals influence the emotional system and are also responsible for many interoceptive regulations in part through pathways implying the cerebellum. My proposal is that the vestibular organ is used as a fundamental dynamic 3D reference frame, not only due to the geometrical properties of the canals and otoliths but also their dynamic properties (Dimiccoli *et al.*, 2013).

But this is not enough. Dealing with multiple kinds of sensory information is a time consuming task. Evolution has found elegant solutions to reduce the complexity of the task. The brain is not a machine which receives sensory information and transforms it into action. I have suggested the opposite view in *The Sense of Movement* (Berthoz, 2000): *the brain is essentially a predictor and imposes upon the world its hypotheses and interpretations*. The theories of von Uexküell (1934/1965) are, in this respect, to be considered very seriously. According to him, the design of living organisms is not 'bottom-up' but 'top-down', in the sense that an initial schema is prevalent from which all parts and bits are combined. His example of a watch is useful: the watchmaker does not assemble the pieces of the watch to make a watch; he has a general plan of the principles which will allow the machine to indicate time, and from them deduces the organization of the parts.

There are many examples of these 'projective' properties of the brain. Even at the level of the thalamus, the only door from the world to the brain, signals coming from the outer world are modified according to influences coming from the cortex and other parts of the brain. *In other words, the brain is a comparator between its expectations and the incoming information. This is a permanent process, not dependent upon a conscious examination of the world but with a profound influence on consciousness.*

We should therefore re-examine the problem of 'multisensory integration'. Many recent studies have provided evidence that the brain has *'internal models'* which are not conscious but are in fact neural networks which have properties similar to the body and even of the laws of physics (MacIntyre *et al.*, 2001). These internal models allow an internal simulation of action without the outside world, as in dream states. Roboticists have long used the concept of 'internal models' in order to account for the capacity of their machines to adapt and produce actions in an anticipatory way. There is *no multisensory integration stricto sensu*: the brain takes the signals coming from the world, compares them with predictions, and reformulates them according to *a priori* schema depending upon its mood, intentions, emotions, goals, etc. We have to elaborate on this. The Bayesian approach favoured today by many groups is a first attempt in this direction (e.g. Llinás and Roy, 2009; see also the review by Friston, 2010). It introduces the fact that we live in an uncertain world with ambiguities.

But constructing a coherent and dynamic self requires more than these sensory-motor and cognitive mechanisms. Two other components are necessary, namely *the emotional and the vegetative* (also named interoceptive). The capacity to combine emotion and cognition in a unified manner and thus contribute to the unity of the self is in part implemented in the *insula*, which is a kind of hub for the integration of external and internal knowledge and emotions in the self. A recent study (Nguyen *et al.*, 2016) showed interoceptive states represented in the posterior insula are integrated with exteroceptive representations by the anterior insula to highlight emotionally salient moments. In addition, emotional states involve a large network. We know also that the orbito-frontal cortex is crucial for blending value information from the limbic system (amygdala) and the cognitive system. Capgras syndrome is a beautiful example of a deficit in this cooperation. The patient recognizes a familiar person (their father, for instance), but denies that this is the original (i.e. claiming it is an imposter). The modern explanation is that the *conscious cognitive identification* of the father made by the temporal cortex is not combined with the *unconscious affective evaluation* made by the limbic system, due to a functional lesion between the cognitive and the emotional systems. This is a nice example of the necessity of blending emotion and cognition for the construction of a conscious shared 'social self'.

2.2. *The virtual body, a self for simulation and emulation*

In the preceding section we discussed how the brain can build a unified dynamic schema of the body: the *body schema*. This virtual body duplicates our physical body.[11] It is not conscious when we are functioning normally. But it has remarkable properties that make it far from being a simple 'representation' of the physical body. It is a set of neural structures which allows us to simulate the physical body as if it were the real one. It can be consciously perceived, e.g. when we lose an arm and experience the illusory perception of phantom limbs. I also think that dreams are a particular form of consciousness. I know that this could be debated, but I have nightmares every day and I do 'live' them as if I were completely 'conscious' of what happens (I understand, of course, that this is not a hard scientific proof!).

The Canadian neurologist Wilder Penfield made pioneering discoveries via electrical stimulation of the brains of epileptic patients and recording their conscious perceptions. In particular he stimulated the *temporo-parietal region* and concluded that this area is involved in '*body awareness and spatial relationships*'. It is interesting that he wrote the word 'awareness' and not consciousness. As mentioned earlier, in French the word awareness has no equivalent; we only use consciousness. Philosophers may relate this to the tendency of the French language to insist on the abstract description of reality, although our British colleagues stay closer to nature with its implicit meanings. Penfield's discovery has recently been confirmed by several sets of studies. For example, we have, with Philippe Kahane and colleagues (Kahane *et al.*, 2003; Kolev *et al.*, 2014), described the importance of this brain area for inducing illusions of body rotations and spatial disorientations very different from vertigo. And Olaf Blanke (Park *et al.*, 2016) and his groups have documented the fact that this area is indeed involved in the body schema because its stimulation or paroxistic activities in patients induce 'out of body experiences' (Blanke *et al.*, 2005). They also documented the importance of the vestibular system in

---

[11] Antonio Damasio has developed important ideas concerning what he calls 'as if body loops', 'proto-self', etc. and one may like to refer to his books. For instance see Damasio (2012) *Self Comes to Mind: Constructing the Conscious Brain*.

self-awareness and consciousness. Recently they have shown that self-consciousness is related to interoceptive functions.

The remarkable fact is that the parieto-insular region is also the top cortical station of the vestibular system. This was shown first by Joachim Grüsser (Guldin and Grüsser, 1998) in the monkey. He actually described what he called an 'inner vestibular circuit' involving several areas in the cortex. My colleagues and I have since been the first to establish this in humans (Lobel et al., 1999) and were followed by a number of other studies from the London (Sedda et al., 2016) and Munich (Thomas Brandt and Marian Dietrich) groups. Recent work by Francesco Lacquaniti and colleagues (2015) has also suggested that there is a neural network in this area which allows the brain to implement the laws of gravity. Recent data also confirm the implication of this area in self-awareness (Ganesh et al., 2015).

The virtual body is not a visual image, or representation, or homunculus of some sort. It is really an emulation of our living body. Although the temporo-parietal area is crucial for this operation, it is part of a network, or maybe several networks, which use information from other parts of the brain (somatosensory areas, parietal areas, etc.) to build the coherent and potentially animate body schema. For instance, the posterior cingulate cortex may be involved in the relation between body awareness and self-location, which are important for consciousness of our situation in the world (Guterstam et al., 2015).

The great physiologist Victor Gurfinkel, himself a pupil of Nicolas Bernstein,[12] insisted throughout his life that the fundamental role of this inner body schema was for controlling actions, adapting the physical body state to intentions, context, etc. This is a highly dynamic neural network, which can adapt and is very flexible (Ivanenko et al., 2011). Experiments like the rubber-hand illusion, or its whole body version or our own work on the virtual hand (Olivé and Berthoz, 2012), have provided recent evidence for the capacity we have to not only have a flexible body schema but also to project this schema into the world as if it were our real body (see also the work of the Ehrsson group: e.g. Gutersam and Ehrsson, 2012). Using the example of a second illusory body it was shown recently (Canzoneri et al., 2016) that *conceptual processing about the self can be located where the illusory body is located and not in the physical body*. The 'translocation' of the perceived body is therefore a powerful physiological mechanism.

The main lesson we can learn from what precedes is that there cannot be a conscious self, whether episodic or semantic, centralized or

---

[12] I am referring to the famous book of Nicolai Bernstein (1967), *The Co-ordination and Regulation of Movements*.

distributed, narrative or intersubjective, if there is not what I will call a 'coherent dynamic adaptable model' of the body in our brain. I believe that one of the reasons for which evolution built this inner model of our living, feeling, acting, and perceiving physical body is that there cannot be survival without anticipation. It is already too late if you have to become conscious that a dinosaur wants to eat you. If you wait for consciousness to tell you what is happening, you will be eaten. You have to anticipate and make a bet!

The beauty of the inner body schema is that with it we can simulate action without execution. The capacity to simulate the action of others is not just a property of the 'mirror system'. It could also be that the old distinction of neurologists between the body schema and body image corresponds to the respective functions of the parieto-insular cortex, and of the precuneus (Fuster named it the 'mind's eye'), which has been shown to be involved in conscious *mental imagery*. Many brain imaging studies have shown that to imagine an action activates similar area as to execute it. For example, if you hit the table repetitively, or just imagine doing it, you will activate similar areas in your brain. However, the term imagery has given rise to many confusions (Pearson and Kosslyn, 2015) and we shall not enter into the debate here, although it is relevant for the question of the relation between unconscious and conscious processes.

2.3. Simplexity and natural laws: an elegant proto-conscious set of principles?

The functional properties that I have described above indicate that evolution has built a unified body schema which can be used to simulate actions and emulate worlds. Evolution has allowed the brain to solve complex problems with solutions that are unconscious, but very elegant. This is the essence of what I have called *simplexity* (Berthoz, 2012). By simplex I mean that evolution has found *simplifying principles* which allow the brain to avoid complex computations at the price of some apparent complexities, detours, redundancy, inhibition, modularity, etc. For example, if you draw an ellipse, the trajectory kinematics obeys a law called the $1/3$ power law, which relates the curvature and the tangential velocity. In short, although you want to draw at constant velocity you will in fact produce a movement which will be rapid when the curvature is low and slow when the curvature is great. The remarkable thing is that the same law is valid for finger or arm movement, but also for a locomotor trajectory. This is due to the fact that some geometrical properties (non-Euclidian) subserve all movements. This is of course a remarkable simplification. The fact that this control is performed by non-Euclidian geometries (Bennequin et al., 20109) is an apparent complexity which allows the brain to perform this unification and this task very efficiently.

These mechanisms are completely unconscious but they also have another remarkable property: they also *rule perception*. If I draw an ellipse and I ask you if my movement is at a constant velocity, you will only answer yes if the movement in fact obeys the above law and is NOT at constant velocity. In other words, *the laws of production of movement are the same as the laws of perception.* This is a beautiful unconscious matching between self-production of action and perception of the other, which allows an unconscious understanding of the action of others. The importance of these natural laws for an easy translation between unconscious and conscious is illustrated by the literature on biological motion: if we represent a person only via a small number of dots on a screen and if we give these dots movement corresponding to the natural laws, we consciously immediately perceive a person moving.

One of the most important functions in living organisms and especially in us humans is walking or navigation in the environment and finding our way. Way-finding requires a hierarchy of physiological and cognitive mechanisms which link the most basic unconscious functions of locomotion to the highest cognitive performances for spatial orientation, navigation, and identification of landmarks, etc. I have elsewhere (Berthoz, 2012) already described some of the simplifying principles which allow us to move without consciously having to control the complex coordination of muscles. For walking they imply a hierarchy of 'laws',[13] engaging spinal, brainstem, cerebellar, basal ganglia, and cortical networks involved in walking and generating locomotor trajectories. They use head stabilization by the vestibular system to create a stabilized reference frame (Berthoz and Pozzo, 1988), co-planar variation and various other coordinations called 'primitives' (Dominici *et al.*, 2011) to simplify descending commands, trajectory optimizing principles (Arechaveleta *et al.*, 2007; Hicheur *et al.*, 2007), anticipations, etc. *These elegant principles allow the brain to control the multiple degrees of freedom involved without explicitly calling on conscious control. But their simplex properties allow conscious control to be called upon easily and rapidly when necessary.* My hypothesis is that they are what in the title of this essay I call *elegant actors* of the transition from unconscious to conscious.

---

[13] For instance the law relating curvature and tangential velocity during natural movements that we have generalized to 3D movements in Maoz, Berthoz and Flash (2009).

Elegant and very smart unconscious *simplex* and *vicariant* mechanisms (Berthoz, 2017) are also involved in the higher aspects of spatial memory during navigation. We can walk to the post office and return home in a complex city with very little conscious control of our trajectory. The path has been memorized by very elegant combinations of cognitive strategies. We can use *egocentric* sequences of records of our path (I walk straight, turn right at the red light, go by the post office, etc.), combining landmarks and movements and episodes. The vestibular access to the hippocampus has a role in this ability to blend movements and spatial location memory. The brain can also use *allocentric* strategies,[14] that is, a coding independent from first-person perspective body movements and episodic memories. Even these may be unconsciously called upon when, for instance, we make an error in the egocentric planning of the path, and realize our error. This perceptive change often induces a cognitive episode of course. Humans seem to be able to build cognitive allocentric maps of their environment as they walk along. Many animals of course do it.

2.4. Inhibition – a fundamental tool in transition from unconcious to conscious

In the previous section we saw that we possess elegant, smart mechanisms, which, I suppose, may allow a smooth, effortless transition between unconscious and conscious processes. But we have no evidence yet as to *how* this can happen.

Often access to consciousness requires a drastic switch to a new cognitive approach, which requires a second type of transition. It requires inhibition of one strategy and vicarious replacement by another. This section will briefly deal with this idea.

Among the simplex principles created by evolution *inhibition* is a fundamental one which underlies all processes in the brain. My hypothesis is that *consciousness very often involves inhibition for action selection, or for inhibiting non-conscious processes, or for changing cognitive strategy requiring inhibition of a primary strategy, or perspective change*. For example, conscious exploration of the world often involves gaze to the consciously examined target. For this we need to select among many possible targets the one of interest. But when we explore the world with vision we sample a very small part of the world through rapid ballistic eye-movements, saccades — movements of the eyes at a very

---

[14] The brain mechanisms underlying these strategies have been widely studied in the last ten years. Poirel *et al.* (2011), Galati *et al.* (2010), Lambrey *et al.* (2012), Bastin *et al.* (2013), Landgraf *et al.* (2010), Iglói *et al.* (2009; 2010), Bullens *et al.* (2010), Meilinger, Berthoz and Wiener (2011).

high speed (up to 500 degrees per second). The selection of the target we aim to is in fact due to a cascade of inhibitory mechanisms (Berthoz, 1996b). The frontal eye field activates a specific group of neurons according to the desired target. These neurons are located in a sub-cortical area—the superior colliculus. This internal retina is a map of the visual world in log-polar coordinates and each of its neurons in the intermediate layers is connected to the brainstem and may trigger a saccade to a target in space. But the array is also under complete inhibitory control from the basal ganglia through the substantia nigra. Thus, in order for the frontal eye field orders to look somewhere to be executed, the inhibition (from the substantia nigra) has to be stopped. A cascade of inhibitory signals regulates our exploration of the visual space. The beauty of inhibition is that at each stage of these inhibitory cascades' learning, plasticity can be introduced. All these mechanisms are completely unconscious although they are the expression of higher-order cognitive factors. They are not perceived explicitly, but my hypothesis is that they are the embodied part of the conscious process. They express the agency of my intentions to explore consciously.

Other examples have been offered by Patrick Haggard's group on the decision and agency of action (Parkinson and Haggard, 2015). It is also known that during development children have to inhibit cognitive strategies that appear early, in order to substitute them with more elaborate strategies. I could refer the reader to the work of Adèle Diamond, but I have chosen here to cite a very nice example from Olivier Houdé's group in Paris.[15] They submitted subjects to a logical task on a visual array that consisted of shapes (squares, triangles, etc.) of different colours. This logical task could be completed using perception, and indeed the subjects started by adopting this strategy and activated their parietal areas in addition to the visual areas. But they made errors. Then they *inhibited the first strategy* and adopted a strategy using the frontal and prefrontal areas involved in logical thinking. Errors disappeared. Obviously this is a good example of a shift between two strategies, which have a varying degree of consciousness. We could say that the first was accompanied by awareness and the second by conscious thinking. This is very close to what happens when we inhibit an egocentric strategy during navigation and replace it with an allocentric strategy, or when we take another person's point of view (Aïte *et al.*, 2016).

These inhibitory processes, which occur also throughout the child-hood development, are very elegant and cannot really be called

---

[15] A recent paper summarizes results from this group: Borst, Aïte and Houdé (2015).

unconscious because they are in fact the substrate of decisions made by the brain according to context, meaning, goals, values, etc. They also belong to the category of processes which imply a very close cooperation between perception and action, memory and anticipation. We should remember that all the main structures in the brain involved with higher cognitive processes (cerebellum, basal ganglia, prefrontal cortex) are in fact inhibitory. Inhibition, I believe, is the most important discovery of evolution because, every time we have the involvement of an inhibitory synapse, the brain can introduce decision, flexibility, substitution, learning, etc.

2.5. Changing perspective, a fundamental mechanism of consciousness: the example of sympathy and empathy

Finally, I have the intuition that the passage from unconscious to conscious requires changes in 'reference frames'. To be conscious is always to be conscious of something or someone. I would like to propose the following hypothesis: *the object of conscious experience is perceived as having an existence independent of our egocentric perception of the world*. The unconscious to conscious transition therefore requires a change in reference frames based upon both a solid coherent awareness of the body and, at the same time, a change in perspective. I have to inhibit my egocentric, self-centred perception and change perspective in order to place the object of consciousness in its own world, and in a sense adopt its 'point of view' (Belmonti *et al.*, 2015). The advantage of this definition is that consciousness is not a top-down artefact, linked to a magical appearance of language, but can be studied as a progressive emergence of competence during evolution. In childhood, it corresponds to the progressive transition of the private self into an independent allocentric, or heterocentric, perspective change, which occurs around the age of four in children, with maturation not until after adolescence (*ibid.*).

A very clear example of a transition between two states of the brain, with a graded level of consciousness, based on an embodied perception of self, is provided by the differentiation of sympathy and empathy.

The German philosophers[16] distinguished clearly between at least two different ways of interacting with others. They called the first way *sympathy* (*Mitfühlung*), which means 'to feel *with*', a spatial and temporal process by which individuals experience the same thing as the other is experiencing and at the same time. It is a unitary self–other experience, a subjective and private sensory and/or emotional experi-

---

[16] *Über das optische Formgefühl*, Robert Vischer (1872); see also Lipps (1913); Husserl (Hua XII–XV, XVI).

ence. It does not require that we change place and reference frame. It is akin to emotional contagion, mirror neurons, and the behavioural resonance which appears very early in childhood. It is a remarkable physiological function, which allows us to be aware of the actions and, to a certain degree, the intentions of others. It does not require a conscious effort. It is not unconscious as we feel in our body a mirror simulation of the actions and emotions of others. In other words, it corresponds to what we were looking for at the beginning of this chapter: an unconscious process which can be immediately called to consciousness without any complex transformation. I call it 'proto-conscious'.

Empathy is very different. It was defined as *Einfühlung* (to feel *into*), a spatial process by which individuals mentally project themselves, their 'self', into the other's body. It is a *conscious* complex dynamic process of competing and cooperating networks, because it requires that we change our point of view and feel the emotion of the other from their own perspective, that is, as if we were in their body. It is therefore an embodied change of perspective. But it can also include a sympathetic component. However, to be empathic we have eventually to *inhibit* the sympathetic process. For instance, if somebody in front of us is suffering, but smiles, we may misinterpret their smile for an absence of suffering. We therefore have to inhibit the sympathetic contagion of the smile to really put ourselves in the body of the other and consciously feel their pain.

We have proposed (Berthoz and Jorland, 2005, p. 308) that empathy implies at least three processes: (a) a spatial perspective change which requires *inhibition of self-centred frames of reference* and *switching to allo- or heterocentric frames of reference* (Sulpizio et al., 2015); (b) the use of our virtual body and the capacity of this virtual body to carry with it/him some of our capacity to feel emotions (see Sedda *et al.*, 2016; Lacquaniti *et al.*, 2015); and (c) inhibition of the sympathetic reaction.

In order to clearly show the spatial reference frame switch between sympathy and empathy we have used a behavioural model with virtual reality (Thirioux *et al.*, 2009; 2010; 2014a; Berthoz and Thirioux, 2010; Cleret de Langavant *et al.*, 2012). The subject was standing in front of a tightrope walker and had to imitate the walker's lateral body tilts by tilting the body to the left or to the right. Two types of behavioural responses were possible in a mirrored fashion (if the virtual creature leaned to the right the subject leaned to the left) or with a perspective change (the subject leaned to the right performing a mental spatial perspective change). The recordings of brain activity with EEG showed that two different networks are implied in these two strategies, and they also correlate with neurological descriptions of the difference between autoscopy (no change in perspective) and heautoscopy (change of perspective). We suggested that the latter requires a disembodied jump which, I suppose, also requires a conscious event.

We were also able to show the moment in the brain between the quasi-automatic mirroring behaviour and the switch to what we believe is a proper empathic response (Thirioux et al., 2014b). Schizophrenic patients have difficulty performing this change. We are presently studying these processes in normal and pathological children.

3. Epilogue

In this essay I have proposed, around the theme of the self, a few examples of what I call 'actors' that contribute to making consciousness easily accessible. The body schema, simplex principles linking perception and action, simulation, imagery, and emulation, inhibition, and manipulation of spatial reference frames, are amongst the many that evolution has given us. But the above examples confirm that a strict dichotomy between conscious and unconscious processes is unwarranted.[17] These categories are marked by the scholastic tradition that prevailed in our Western countries for several centuries. They correspond to a desire to classify with axiomatic concepts what is in fact a remarkable unity of the flow of life. They correspond also sometimes to a hidden dualist view separating reason from emotion, thoughts from the body, and creating an abstract world which is undoubtedly a powerful tool for mankind that has led to its domination of the world. They neglect the multiplicity of what is called in French 'Les états de conscience', which should actually be 'Les états de consciences'.

I would like to finish by quoting some thoughts of the sociologist Pierre Bourdieu in order to stimulate the discussion raised by this multidisciplinary volume. In his book on the painter Manet (Bourdieu, 2013) (which brings together his lectures at the College de France shortly before his premature death), Pierre Bourdieu discusses the aesthetic revolution created by the paintings of Manet and particularly by the 'Déjeuner sur l'herbe', which generated a scandal not only because it featured a nude woman but because of the composition and context in which she was placed, as well as the general setting (although it was in fact inspired by an illustration by Raimondi).

Bourdieu in this text discusses the difficulty that societies have to face when trying to radically change their frame of thought to adopt

---

[17] See for instance the essay of neurologist Lionel Naccache (2006) 'Le nouvel inconscient: Freud, Christophe Colomb des Neurosciences'.

new views (*ibid.*, pp. 104, 105).[18] This is, of course, well known. The reason for which the proposals of Bourdieu seem interesting to me framed in Zdravko Radman's project is that it goes along with the general trend of ideas which, today, places the act (not only action) and the 'savoir-faire', the acting body, the learning habits and skills, at the origin and the centre of the way we perceive the world. Jean-Paul Sartre wrote that consciousness is always consciousness of something; Bourdieu believes that consciousness is consciousness of what we know *and can do* in our interactions with the world (habitus). This is at least how I interpret his concept of *'disposition'*. This approach is close to the one of Emanuelle Danblon in her (2013) book *L'homme réthorique*. She aims at a completely new foundation of rhetoric, switching from a Platonic view to a more Aristotelian view based on *'doing'* rather than on the abstract concepts of rational thinking. It is interesting today to see that in many disciplines a deep effort is being made to understand higher cognitive processes. The few ideas proposed in this essay are only small pebbles on a long and difficult path, but the challenge is worth it!

References

Aïte, A., Berthoz, A., Vidal, J., Roëll, M., Zaoui, M., Houdé, O. & Borst, G. (2016) Taking a third-person perspective requires inhibitory control: Evidence from a developmental negative priming study, *Child Development*, 9 June, doi: 10.1111/cdev.12558 [Epub ahead of print].

Andrieu, B. & Berthoz, A. (eds.) (2009) *Le corps en acte. A l'occasion du centenaire de la naissance de Maurice Merleau-Ponty (1908–1961)*, (*Epistémologie du corps*), Nancy: Presses Universitaires de Nancy.

Arechaveleta, G., Laumond, J.P., Hicheur, H. & Berthoz, A. (2007) An optimality principle governing human walking trajectories, *IEEE Transactions on Robotics*, **24** (1), pp. 5–14.

---

[18] Bourdieu writes about his theory of 'dispositions': 'A *dispositionalist theory of action goes in an opposite direction to all the cultural occidental traditions, of all the philosophy of consciousness of the intentional subject, of the philosophy we have received from Descartes or Kant, or even the soft version of the Christian philosophy. We all are subdued by a philosophy of the intentional subject. If one asks who painted the Déjeuner sur l'herbe one will get the answer: a subject (Manet) painted this picture. I will answer Manet, but from the sociological point of view this individual who painted the picture is not the traditional subject of the occidental tradition. It is a habitus inserted in a field. A habitus that is a biological being socialized and owner of dispositions. These dispositions have been socially and historically constituted and it is in the relation between the habitus and this field that a new way of thinking is invented*' (my translation with a few modifications to shorten the text).

Bastin, J., Committeri, G., Kahane, P., Galati, G., Minotti, L., Lachaux, J.P. & Berthoz, A. (2013) Timing of posterior parahippocampal gyrus activity reveals multiple scene processing stages, *Human Brain Mapping*, **34** (6), pp. 1357–1370.

Belmonti, V., Fiori, S., Guzzetta, A., Cioni, G. & Berthoz, A. (2015) Cognitive strategies for locomotor navigation in normal development and cerebral palsy, *Developmental Medicine & Child Neurology*, **57** (Suppl. 2), pp. 31–36.

Bennequin, D., Fuchs, R., Berthoz, A. & Flash T. (2009) Movement timing and invariance arise from several geometries, *PLoS Computational Biology*, **5** (7), e1000426.

Bernstein, N. (1967) *The Co-ordination and Regulation of Movements*, New York/Oxford: Pergamon Press.

Berthoz, A. (1996a) *The Brain's Sense of Movement*, Cambridge, MA: Harvard University Press.

Berthoz, A. (1996b) The role of inhibition in the hierarchical gating of executed and imagined movement, *Cognitive Brain Research*, **3**, pp. 101–113.

Berthoz, A. (2000) *The Brain's Sense of Movement*, Cambridge, MA: Harvard Univeristy Press. (Translated from French, Odile Jacob.)

Berthoz, A. (2006) *Emotion and Reason: The Cognitive Foundations of Decision Making*, Oxford: Oxford University Press. (Translated from French, Odile Jacob.)

Berthoz, A. (2012) *Simplexity: Simplifying Principles for a Complex World* (An Editions Odile Jacob Book), New Haven, CT: Yale University Press.

Berthoz, A. (2016) *The Vicarious Brain*, Cambridge, MA: Harvard University Press. (Translated from French, Odile Jacob.)

Berthoz, A. (2017) *The Vicarious Brain: Creator of Worlds*, Weiss, G. (trans.), Cambridge, MA: Harvard University Press.

Berthoz, A. & Melville-Jones, G. (1985) *Adaptive Mechanisms in Gaze Control*, Amsterdam: Elsevier.

Berthoz, A. & Pozzo, T. (1988) Intermittent head stabilisation during postural and locomotory tasks in humans, in Amblard, B., Berthoz, A. & Clarac, F. (eds.) *Posture and Gait: Development, Adaptation and Modulation*, pp. 189–198, Amsterdam: Elsevier.

Berthoz, A., *et al.* (1995) Spatial memory of body linear displacement: What is being stored?, *Science*, **269**, pp. 95–98.

Berthoz, A. & Jorland, G. (2005) *L'empathie*, Paris: Odile Jacob.

Berthoz, A. & Petit, J.-L. (2008) *The Physiology and Phenomenology of Action*, Oxford: Oxford University Press. (Translated from French, Odile Jacob.)

Berthoz, A. & Thirioux, B. (2010) A spatial and perspective change theory of the difference between sympathy and empathy, *Paragrana*, **19** (1).

Blanke, O., Mohr, C., Michel, C.M., Pascual-Leone, A., Brugger, P., Seeck, M., Landis, T. & Thut, G. (2005) Linking out-of-body experience and self-processing to mental own-body imagery at the temporo-parietal junction, *Journal of Neuroscience*, **25** (3), pp. 550–557.

Borst, G., Aïte, A. & Houdé, O. (2015) Inhibition of misleading heuristics as a core mechanism for typical cognitive development: Evidence from behavioral and brain-imaging studies, *Developmental Medicine & Child Neurology*, **57** (Suppl. 2), pp. 21–25.

Bourdieu, P. (2013) *Manet. Une révolution symbolique*, Paris: Seuil.

Bullens, J., Iglói, K., Berthoz, A., Postma, A. & Rondi-Reig, L. (2010) Developmental time course of the acquisition of sequential egocentric and allocentric navigation strategies, *Journal of Experimental Child Psychology*, **107** (3), pp. 337–350.

Buzsáki, G. (2015) Hippocampal sharp wave-ripple: A cognitive biomarker for episodic memory and planning, *Hippocampus*, **25** (10), pp. 1073–1088.

Buzsáki, G., Llinás, R., Singer, W., Berthoz, A. & Christen, Y. (eds.) (1994) *Temporal Coding in the Brain*, Berlin: Springer Verla.

Canzoneri, E., di Pellegrino, G., Herbelin, B., Blanke, O. & Serino, A. (2016) Conceptual processing is referenced to the experienced location of the self, not to the location of the physical body, *Cognition*, **154**, pp. 182–192.

Cleret de Langavant, L., Trinkler, I., Remy, P., Thirioux, B., McIntyre, J., Berthoz, A., Dupoux, E. & Bachoud-Lévi, A.C. (2012) Viewing another person's body as a target object: A behavioural and PET study of pointing, *Neuropsychologia*, **50** (8), pp. 1801–1813.

Conway, M.A. (2001) Sensory-perceptual episodic memory and its context: Autobiographical memory, *Philosophical Translations of the Royal Society of London B*, **356** (1413), pp. 1375–1384.

Damasio, A. (2012) *Self Comes to Mind: Constructing the Conscious Brain*, London: Penguin Random House.

Dehaene, S. (2014) *Le code de la Conscience*, Paris: Odile Jacob.

Dimiccoli, M., Girard, B., Berthoz, A. & Bennequin, D. (2013) Striola magica: A functional explanation of otolith geometry, *Journal of Computational Neuroscience*, **35** (2), pp. 125–154.

Dominici, N., Ivanenko, Y.P., Cappellini, G., d'Avella, A., Mondì, V., Cicchese, M., Fabiano, A., Silei, T., Di Paolo, A., Giannini, C., Poppele, R.E. & Lacquaniti, F. (2011) Locomotor primitives in newborn babies and their development, *Science*, **334** (6058), pp. 997–999.

Duval, C., Desgranges, B., Eustache, F. & Piolino, P. (2009) Looking at the self under the microscope of cognitive neurosciences: From self-consciousness to consciousness of others, *Geriatrie et Psychologie Neuropsychiatrie du Vieillissement*, **7** (1), pp. 7–19.

Freud, S. (1991) Il est facile d'apercevoir que le moi est la partie du ça modifiée sous influence directe du monde extérieur par l'intermédiaire du Pc-Cs, en quelque sorte une continuation de la différentiation de surface, *Œuvres completes*, Vol. XVI PUF.

Friston, K. (2010) The free-energy principle: A unified brain theory?, *Nature Review of Neuroscience*, **11**, pp. 127–138.

Gaglianese, A., Costagli, M., Bernardi, G., Ricciardi, E. & Pietrini, P. (2012) Evidence of a direct influence between the thalamus and hMT+ independent of V1 in the human brain as measured by fMRI, *Neuroimage*, **60** (2), pp. 1440–1447.

Galati, G., Pelle, G., Berthoz, A. & Committeri, G. (2010) Multiple reference frames used by the human brain for spatial perception and memory, *Experimental Brain Research*, **206** (2), pp. 109–120.

Gallagher, S. (2000) Philosophical conceptions of the self: Implications for cognitive science, *Trends in Cognitive Sciences*, **4** (1), pp. 14–21.

Gallagher, S. (2013) A pattern theory of self, *Frontiers in Human Neuroscience*, **7**, pp. 1–7.

Ganesh, S., van Schie, H.T., Cross, E.S., de Lange, F.P. & Wigboldus, D.H. (2015) Disentangling neural processes of egocentric and allocentric mental spatial transformations using whole-body photos of self and other, *Neuroimage*, **116**, pp. 30–39.

Graziano, M.S. & Webb, T.W. (2015) The attention schema theory: A mechanistic account of subjective awareness, *Frontiers of Psychology*, **6**, art. 500.

Guldin, W.O. & Grüsser, O.J. (1998) Is there a vestibular cortex?, *Trends in Neuroscience*, **6**, pp. 254–259.

Guterstam, A. & Ehrsson, H.H. (2012) Disowning one's seen real body during an out-of-body illusion, *Consciousness & Cognition*, **21** (2), pp. 1037–1042.

Guterstam, A., Björnsdotter, M., Gentile, G. & Ehrsson, H.H. (2015) Posterior cingulate cortex integrates the senses of self-location and body ownership, *Current Biology*, **25** (11), pp. 1416–1425.

Hicheur, H., Pham, Q.C., Arechaveleta, G., Laumond, J.P. & Berthoz, A. (2007) The formation of trajectories during goal-oriented locomotion in humans. I. A stereotyped behavior, *European Journal of Neuroscience*, **26** (8), pp. 2376–2390.

Iglói, K., Zaoui, M., Berthoz, A. & Rondi-Reig, L. (2009) Sequential egocentric strategy is acquired as early as allocentric strategy:

Parallel acquisition of these two navigation strategies, *Hippocampus*, **19** (12), pp. 1199-1211.

Iglói, K., Doeller, C.F., Berthoz, A., Rondi-Reig, L. & Burgess, N. (2010) Lateralized human hippocampal activity predicts navigation based on sequence or place memory, *Proceedings of the National Academy of Sciences USA*, **107** (32), pp. 14466-14471.

Ivanenko, Y.P., Dominici, N., Daprati, E., Nico, D., Cappellini, G. & Lacquaniti F. (2011) Locomotor body scheme, *Human Movement Science*, **2**, pp. 341-351.

Kahane, P., Hoffmann, D., Minotti, L. & Berthoz, A. (2003) Re-appraisal of the human vestibular cortex by cortical electrical stimulation study, *Annals of Neurology*, **54** (5), pp. 615-624.

Kolev, O.I., Georgieva-Zhostova, S.O. & Berthoz, A. (2014) Anxiety changes depersonalization and derealization symptoms in vestibular patients, *Behavioural Neurology*, Article ID 847054.

Lacquaniti, F., Bosco, G., Gravano, S., Indovina, I., La Scaleia, B., Maffei, V. & Zago, M. (2015) Gravity in the brain as a reference for space and time perception, *Multisensory Research*, **28** (5-6), pp. 397-426.

Lambrey, S., Doeller, C., Berthoz, A. & Burgess, N. (2012) Imagining being somewhere else: Neural basis of changing perspective in space, *Cerebral Cortex*, **22** (1), pp. 166-174.

Landgraf, S., Krebs, M.O., Olié, J.P., Committeri, G., van der Meer, E., Berthoz, A. & Amado I. (2010) Real world referencing and schizophrenia: Are we experiencing the same reality?, *Neuropsychologia*, **48** (10), pp. 2922-2930.

Leblond, M., Laisney, M., Lamidey, V., Egret, S., de La Sayette, V., Chételat, G, Piolino, P., Rauchs, G., Desgranges, B. & Eustache, F. (2016) Self-reference effect on memory in healthy aging, mild cognitive impairment and Alzeimer's disease, *Cortex*, **74**, pp. 177-190.

Libet, B. (2002) The timing of mental events: Libet's experimental findings and their implications, *Consciousness & Cognition*, **2**, pp. 291-299; discussion, 304-333.

Llinás, R. (2002) *I of the Vortex: From Neuron to Self*, Cambridge, MA: MIT Press.

Llinás, R. (2008) Dynamic geometry, brain function modeling, and consciousness, *Progress in Brain Research*, **168**, pp. 133-144.

Llinás, R. & Roy, S. (2009) The 'prediction imperative' as the basis for self-awareness, *Philosophical Transactions of Royal Society London B*, **364** (1521), pp. 1301-1307.

Lobel, E., Kleine, J.F., Leroy-Willig, A., Van de Moortele, P.F., Le Bihan, D., Grüsser, O.J. & Berthoz, A. (1999) Cortical areas activated by bilateral galvanic vestibular stimulation, *Annals of the New York Academy of Sciences*, **871**, pp. 313-323.

Maingret, N., Girardeau, G., Todorova, R., Goutierre, M. & Zugaro, M. (2016) Hippocampo-cortical coupling mediates memory consolidation during sleep, *Nature Neuroscience*, **7**, pp. 959-964.

Maoz, U., Berthoz, A. & Flash, T. (2009) Complex unconstrained three-dimensional hand movement and constant equi-affine speed, *Journal of Neurophysiology*, **101** (2), pp. 1002-10015.

McIntyre, J., Zago, M., Berthoz, A. & Lacquaniti, F. (2001) Does the brain model Newton's laws?, *Nature Neuroscience*, **4**, pp. 693-694.

Meilinger, T., Berthoz, A. & Wiener, J.M. (2011) The integration of spatial information across different viewpoints, *Memory & Cognition*, **39** (6), pp. 1042-1054.

Melville-Jones, G., Berthoz, A. & Segal, B. (19984) Adaptive modification of the vestibulo-ocular reflex by mental effort in darkness, *Experimental Brain Research*, **56**, pp. 149-153.

Merleau-Ponty, M. (1945) *Phénoménologie de la perception*, Paris: Gallimard.

Naccache, L. (2006) *Le nouvel inconscient. Freud, Christophe Colomb des Neurosciences*, Paris: Odile Jacob.

Nguyen, V.T., Breakspear, M., Hu, X. & Guo, C.C. (2016) The integration of the internal and external milieu in the insula during dynamic emotional experiences, *Neuroimage*, **124**, pp. 455-463.

O'Mara, S.M., Rolls, E., Berthoz, A. & Keshner, R.P. (1994) Neurons responding to whole body motion in the primate hippocampus, *Journal of Neuroscience*, **14**, pp. 6511-6523.

Olivé, I. & Berthoz, A. (2012) Combined induction of rubber-hand illusion and out-of-body experiences, *Frontiers of Psychology*, **3**, art. 128.

Park, H.D., Bernasconi, F., Bello-Ruiz, J., Pfeiffer, C., Salomon, R. & Blanke, O. (2016) Transient modulations of neural responses to heartbeats covary with bodily self-consciousness, *Journal of Neuroscience*, **36** (32), pp. 8453-8460.

Parkinson, J. & Haggard, P. (2015) Choosing to stop: Responses evoked by externally triggered and internally generated inhibition identify a neural mechanism of will, *Journal of Cognitive Neuroscience*, **27** (10), pp. 1948-1956.

Pearson, J. & Kosslyn, S.M. (2015) The heterogeneity of mental representation: Ending the imagery debate, *Proceedings of the National Academy of Science USA*, **112** (33), pp. 10089-10092.

Perret, C. & Berthoz, A. (1973) Evidence of static and dynamic fusimotor actions on the spindle response to sinusoidal stretch during locomotor activity in the cat, *Experimental Brain Research*, **18**, pp. 178-188.

Petit, J.-L. (2010a) Intention in phenomenology and neuroscience: Intentionalizing kinesthesia as an operator of constitution, in Grammont, F., Legrand, D. & Livet, P. (eds.) *Naturalizing Intention in Action*, pp. 269–292, Cambridge, MA: MIT Press.

Petit, J.-L. (2010b) A Husserlian, neurophenomenologic approach to embodiment, in Gallagher, S. & Schmicking, D. (eds.) *Handbook of Phenomenology and Cognitive Science*, pp. 201–216, Dordrecht: Springer.

Petit, J.-L. (2010c) The brain, the person and the world, in D'Agostino, M., Giorello, G., Laudisa, F., Pievani, T. & Sinigaglia, C. (eds.) *SILFS Vol. 1, New Essays in Logic and Philosophy of Science*, pp. 585–599, London: College Publications.

Petit, J.-L. (2010d) Corps propre, schéma corporel et cartes somatotopiques, in Berthoz, A. & Andrieu, B. (eds.) *Le corps en acte. Centenaire Maurice Merleau-Ponty*, pp. 41–59, Nancy: Presses Universitaires de Nancy.

Poirel, N., Vidal, M., Pineau, A., Lanoë, C., Leroux, G., Lubin, A., Turbelin, M.R., Berthoz, A. & Houdé, O. (2011) Evidence of different developmental trajectories for length estimation according to egocentric and allocentric viewpoints in children and adults, *Experimental Psychology*, **58** (2), pp. 142–146.

Raichle, M.E. (2015) The brain's default mode network, *Annual Review of Neuroscience*, **38**, pp. 433–447.

Raichle, M.E. & Snyder, A.Z. (2007) A default mode of brain function: A brief history of an evolving idea, *Neuroimage*, **37** (4), pp. 1083–1090; discussion 1097–1099.

Ricoeur, P. (1990) *Soi-même comme un autre*, Paris: Le Seuil.

Salti, M., Monto, S., Charles, L., King, J.R., Parkkonen, L. & Dehaene, S. (2015) Distinct cortical codes and temporal dynamics for conscious and unconscious percept, *Elife*, **4**.

Sedda, A., Tonin, D., Salvato, G., Gandola, M. & Bottini, G. (2016) Left caloric vestibular stimulation as a tool to reveal implicit and explicit parameters of body representation, *Consciousness & Cognition*, **41**, pp. 1–9.

Senot, P., Baillet, S., Renault, B. & Berthoz, A. (2008) Cortical dynamics of anticipatory mechanisms in interception: A neuromagnetic study, *Journal of Cognitive Neuroscience*, **20** (10), pp. 1827–1838.

Stein, B. & Meredith, A. (1993) *The Merging of the Senses*, Cambridge, MA: MIT Press.

Sulpizio, V., Committeri, G., Metta, E., Lambrey, S., Berthoz, A. & Galati, G. (2015) Visuospatial transformations and personality: Evidence of a relationship between visuospatial perspective taking and self-reported emotional empathy, *Experimental Brain Research*, **233** (7), pp. 2091–2102.

Thirioux, B., Mercier, M.R., Jorland, G., Berthoz, A. & Blanke, O. (2010) Mental imagery of self-location during spontaneous and active self-other interactions: An electrical neuroimaging study, *Journal of Neuroscience*, **30** (21), pp. 7202–7214.

Thirioux, B., Jorland, G., Bret, M., Tramus, M.H. & Berthoz, A. (2009) Walking on a line: A motor paradigm using rotation and reflection symmetry to study mental body transformations, *Brain Cognition*, **70** (2), pp. 191–200.

Thirioux, B., Tandonnet, L., Jaafari, N. & Berthoz, A. (2014a) Disturbances of spontaneous empathic processing relate with the severity of the negative symptoms in patients with schizophrenia: A behavioural pilot-study using virtual reality technology, *Brain Cognition*, **90**, pp. 87–99.

Thirioux, B., Mercier, M.R., Blanke, O. & Berthoz, A. (2014b) The cognitive and neural time course of empathy and sympathy: A neuroimaging study on self-other interaction, *Neuroscience*, **267**, pp. 286–306.

Tulving, E. (1985) *Elements of Episodic Memory*, Oxford Psychology Series, Oxford: Oxford University Press.

Vitte, E., Derosier, C., Caritu, Y. & Berthoz, A. (1996) Activation of the hippocampal formation by vestibular stimulation: A functional magnetic resonance imaging study, *Experimental Brain Research*, **112**, pp. 523–526.

von Uexküll, J. (1934/1965) *Mondes animaux et monde humain suivi de: La théorie de la signification*, Paris: Denoël.

Zeki, S. (2008) The disunity of consciousness, *Progress in Brain Research*, **168**, pp. 11–18.

Zdravko Radman

There is No Pure Consciousness and No Innocent Unconscious

1. Introduction: Can the Unconscious Solve (or Soften) the 'Mystery of the Mind'?

In spite of the general agreement that what we most intimately know is our own conscious experience, and in spite of the fact that contemporary science has gained significant new insights into the functioning of the human brain underlying mental processing, mind — and specifically consciousness — is still considered a mystery. How is it that something that feels so close and immediately experienced as consciousness (of which we tend to think in terms of 'directness',' 'transparency', 'privileged access'; Agassi, 1969; see also Crane, 2002) appears to be so difficult, if not impossible, to be adequately represented and scientifically explained? How is it that in spite of the huge progress in neuroscience and brain research, as well as cognitive and computer science, the 'mystery of the mind' has become a frequent companion of many authors researching and publishing in the field (e.g. Penfield, 1975; Searle, 1997; McGinn, 1999; etc.; see also Radman, 2007)? Illustrative of this is the opening sentence of Gerald Edelman and Giulio Tononi's book (2000), which reads: 'Consciousness has been seen as both a mystery and a source of mystery.' Even David Chalmers' seminal work humbly states at the very outset: 'Consciousness is the biggest mystery' (1996, p. vi), and further adds: 'Conscious experience is at once the most familiar thing in the world and the most mysterious' (*ibid.*, p. 3). Colin McGinn confirms the general feeling that we are still very far from deciphering the riddle, saying that 'consciousness is indeed a deep mystery, a phenomenon of nature on which we have virtually no theoretical grip. The reason for this mystery... is that our intelligence is wrongly designed for understanding consciousness. Some aspects of nature are suited to our mode of intelligence, and science is the result;

but others are not of the right form for our intelligence to get its teeth into, and then mystery is the result' (McGinn, 1999, p. xi). Even the most recent sources cannot but confirm that 'how neurons engender the conscious field remains a mystery' (Cushing *et al.*, this volume; see also Hameroff, 2006).

Views, however, differ on whether we will ever be in the position to fully decipher 'the mystery' and get a scientifically exhaustive grasp of the conscious mind. Some are in this respect more optimistic and claim that with the advance of science we will gain enough expert knowledge, e.g. in neurobiology and related fields, which would suffice for a full scientific explanation of the nature of consciousness (e.g. Searle, 1997), while others deny that it will ever happen (e.g. McGinn, 1999; Pinker, 2007; Nagel, 1974) since, as they propagate, our mentality is not tuned to theoretically cope with something as unique and peculiar as consciousness. Owen Flanagan (1991) refers to this sort of attitude as 'mysterianism' — a position that holds that our intellectual capacities are limited by biology, and the impossibility to scientifically decipher consciousness is not an issue that can be solved with time, for the problem is *in principle* not solvable. The 'hard problem' (Chalmers, 1995; 1996; see also Shear, 1997; Gray, 2004) additionally fuels the conviction that the irreducibility of consciousness implies the impossibility of its accountability by available investigatory means.

Because consciousness seems to resist reliable scientific methodologies, various theoretical constructs, hypotheses, (imaginative and sometimes bizarre) thought experiments, and speculations have been devised to provoke thoughts on this still enigmatic phenomenon; for instance, 'microtubuli' (Penrose, 1989); orchestrated objective reduction (Hameroff and Penrose, 2014a,b); 'ego tunnels' (Metzinger, 2010); 'global workspace theory' (Baars, 1988; 1997); extended (and 'super-sized') mind (Clark and Chalmers, 1998; Clark, 2007a); Chinese Room argument (Searle, 1989; 2009; Preston and Bishop, 2002); 'intuition pumps' (Dennett, 1991); Doppelgänger and zombie argument (Kirk, 1974; Polger, 2000); knowledge argument (Jackson, 1986; 1995); speculation about what is it like to be a bat (Nagel, 1974) or could a Martian feel the same way we do (Feigel, 1958; Lewis, 1980); twin earth (Putnam, 1973; 1985); panpsychism (Fechner, 1848; 1906; Drake, 1925; Strawson, 2006); quantum theory (Koch and Hepp, 2006; Litt *et al.*, 2006); and many more. The issue intrigues even the wider audience (e.g. Burkeman, 2015; Pinker, 2007) and gets generally recognized as *the* scientific problem of the twenty-first century.

The 'mysterianist' credo with its defeatist position and the 'hard problem' challenge have, in an important way, enlightened the uniqueness of consciousness, but at the same time evidently brought us to a blind alley from which no view opens toward alternatives. The result is a lasting stagnation in the discussion, with endless recycling of the arguments following from the (in principle) impossibility of a founded insight. It seems that insisting on the authenticity of the phenomenon terminated in the conclusion that consciousness, because it is irreducible, is also for that reason unrelated to, i.e. detached from, the rest of mentality. Such a position has prevented us from seeing the possibility that no matter how specific conscious (particularly phenomenal) experience might be, it does not follow that it must exist in isolation and, for that matter, remain inaccessible. It turned out that the radicalism of the irreducibility thesis has significantly deprived us of the courage to outline alternative views. For, as previously stated, there is no evidence that irreducibility necessarily implies isolationism; that specificity must mean segregation; and that the qualitative must not only be seen as distilled but also detached from other traits of the mental.

If we manage to free our methodological strategies from biases, it may allow us to open the vista and gain a more integrative perspective on consciousness in the broader context of the mind. In the first instance it may make us be theoretically open to the presence and import of the non-conscious—the vast sphere of mentality (Gazzaniga, 1989;[1] Mlodinow, 2012[2]) that both precedes and follows every conscious mental act. The view presented here is that consciousness, no matter how unique and important, is not a sole player. Mental 'games' (to make use of Wittgensteinian terminology) are played also with unconscious players and, as more recent findings suggest, they may be considered playmakers.

I want to believe that such an approach has the potential to, if not solve, at least soften the radicalism of the message from mysteriousness and help us gain a more inclusivist view on the mind; this will enable a shift away from treating consciousness in its exclusiveness and also free it from the theoretical niche in which it is placed.

Placing consciousness in the (much wider) world of the unconscious generates a number of questions, such as the following: can the unconscious in any significant way help us understand consciousness? Is there a way to see both as interdependent? Is there enough support for the assumption that the unconscious is fundamental and underlying

---

[1] 'Ninety-eight percent of what the brain does is outside of conscious awareness' (Gazzaniga, 1998, p. 21).

[2] 'Some scientists estimate that we are conscious of only about 5 percent of our cognitive function. The other 95 percent goes on beyond our awareness and exerts a huge influence on our lives—beginnning with making our lives possible' (Mlodinow, 2012, p. 34).

other forms of mentality? Do we have plausible arguments to see consciousness and the unconscious as being co-constituted? My response to all of these is yes. It is also in accord with the general claim of this chapter: that consciousness and the non-conscious are so interrelated and interdependent that attempts to explain one without the other are futile.

In order to make the idea convincing that inquiries into the nature of the non-conscious are crucial for understanding the phenomenon of consciousness, and vice versa, I first have to briefly critique the dual schema as exemplified in the dichotomy that splits consciousness and the non-conscious apart and presents them as counterpointed entities.

2. The Disaster of Dichotomizing

Our desire for making discriminations is ubiquitous and our inclination for radicalization quite strong: 'In order to explain, we discriminate; in order to discriminate, we delineate; delineations lead to oppositions; oppositions are understood as polarities; polarizations imply exclusiveness; exclusions lead to irreducibility' (Radman, 2013a, p. 19). And so what starts as benign, as a mode of helpful description, ends up as radical dichotomy that strongly patterns our way of conceptualization. It basically means that whenever we try to describe an element, it seems unavoidable to point out what makes it distinctive and what it is opposite of. For instance, in the philosophy of mind and theory of consciousness we contrast organic and inorganic, matter and mind, 'it' (brain) and 'I' (self), motor and mental, nature and nurture, genes and culture, etc. In the string of further polarization, reason is dissociated from emotion, rationality from intuition, logic from imagination, facticity from fiction. The demand for contrasting is then saturated with opposing science and art, exactness and expressiveness, truth and beauty.

The conscious/non-conscious dichotomy is but an outcome of this kind of conceptualizing. The prefixes 'non' and 'un' additionally strengthen the conceptual divide in that they suggest that what is characteristic of the non-conscious cannot also be associated with consciousness, and vice versa. In other words, attributes attached to one are not expected to be affiliated with the other. What this kind of reasoning implies is that the authenticity of each element is warranted only if there is no intrusion of traits from its supposed counterpart.

This is consistent with the general view according to which differences necessarily imply exclusion. That is, if A is different in some sense from B, then A and B are not seen as related, and that in turn means they cannot share properties. This kind of conceptualization is fatal for understanding the subject at stake for it is guilty of rigid exclusiveness according to which predicates applied to consciousness cannot also be applied to the non-conscious, and vice versa. In order to establish the specificity of a favoured option, we emphasize the difference by claiming that for each segment to be authentic means for it to be possibly devoid of traces of the other.

Disastrous above all is the consequence of such an attitude that unavoidably leads to the 'gapped' image of mentality and a conception of a 'divided man' (see, for example, Radman, 2013a). There are at least two reasons why such a schema is inadequate.

First (logical reason): the tendency to only make distinctions does not imply that the contrasting elements are necessarily exclusive and incommensurate, i.e. discrimination does not necessarily mean that the resulting differences imply mutual isolation and irreducibility.

Second (empirical reason): if applied to living beings, such dichotomies would make us profoundly dysfunctional. If we, in a sort of thought experiment, imagine beings that are a literal incarnation of mainstream theoretical constructs (with all the conceptual rigidity and divisions they imply), those creatures would have problems maintaining even the simplest life functions.

Fortunately, natural minds do not have such problems, and this leads us to the conclusion that, in some non-trivial sense, we are not the kind of creatures science has promoted us to; what this means is that at least some aspects of current theories are inadequate.

This chapter tries to tackle just this kind of stereotype, which has biased us for too long and has produced a conviction that there are kinds of mental processes (cognitive and other) that can be performed in one mode (e.g. consciousness) and not in the other (e.g. the non-conscious). It further aims to outline an alternative view according to which the elements standardly perceived as contrasted are seen as conjoined.

Once we accept as a possibility that the differentiation of consciousness and the unconscious is not a matter of content but of mode, then the road is open for the conclusion that all mental processes can be performed in both modes; and that consciousness is in no way privileged in claiming power over, say, high-level cognition (e.g. perception, language, thought) but also that the implicit, emotional, and intuitive is not to be seen as operated exclusively by unconscious mechanisms.

What we learn from contemporary neuroscience confirms the view that the unconscious is not ignorant or blind; on the contrary, it is cognitively as potent as the conscious mind. To the question, what are all these non-conscious processes?, Joseph LeDoux responds: 'Actually, they include almost everything the brain does, from standard body maintenance like regulating heart rate, breathing rhythm, stomach contractions, and posture, to controlling many aspects of seeing, smelling, behaving, feeling, speaking, thinking, evaluating, judging, believing, and imagining' (LeDoux, 2002, p. 11). And in a similar vein, Leonard Mlodinow says: '...to ensure our smooth functioning in both the physical and the social world, nature has dictated that many processes of perception, memory, attention, learning, and judgement are delegated to brain structures outside of conscious awareness' (Mlodinow, 2012, p. 18).

Chapters in this volume also document that major forms of more complex cognition, such as perception, memory, thinking, etc., are not an exclusive privilege of the conscious mind but can also be instantiated away from it.

Believing that dichotomization is an inadequate way of understanding the nature of the mental, I will outline two theses in favour of the idea of the undivided mind.

3. Affirming Continuity and Bi-directionality

In re-reading Marr's (1982) seminal work, which states that (computationally) conscious and unconscious processes are very similar (in a similar vein, Prinz, this volume, also says: 'Conscious and unconscious vision seem to be very much alike'), we may feel encouraged to start considering both as sharing more than is conventionally conceived. In order to do so, I want to outline a *continuity thesis* — an assumption that, just as there is an ungapped transition from low-level cognitive processing (e.g. motor reactions) to so-called high-level cognitive processing (e.g. speech), there is also an ungapped transition between unconscious and conscious mental processes. The continuity thesis is not meant to erase differences between various mental and behavioural phenomena; it rather suggests that there is a smooth transition between them.

The core of the idea can be found, for instance, in Popper and Eccles as they draw on this interrelatedness from the perspective of the empirical sciences: 'All experience is already interpreted by the nervous system a hundredfold — or a thousandfold, before it becomes conscious experience' (Popper and Eccles, 1977, p. 431). That is to say, all consciousness is 'contaminated' by unconscious elements and there is no way one can extricate the latter from the former. In addition, according to them, the unconscious appears to be primary: 'so in a sense, when you get the experience, you can say it is not primary. It is based upon all of this immense patterned development that is a necessary prelude to a conscious experience' (*ibid.*).

However, in addition to pointing out that there is a continuity between elementary and more complex forms of mentality (e.g. Sun *et al.*, 2001), it is also important to convincingly argue that the latter has not only an impact on the former but that it also has the power to reshape it. In order to account for this aspect of mutuality, I want to outline a notion of *bi-directionality*—a concept needed to explain the changing and evolving competence in skill and the increase in automaticity. For only if we allow for the possibility that higher mental processes can be acquired by the body are we in a position to explain how corporeal know-how evolves. Thus, both tendencies of the cognitive organism have to be recognized: the one in which conscious states tend to 'disappear' in the non-concious, and the inclination of the non-conscious to shape (conscious) experience. Consequently, the applied idea of bi-directionality may prove to be an explanatory mode of bridging *knowing that* and *knowing how* (Snowdon, 2004) as well as that of the explicit and implicit (Berry *et al.*, 1988). It may further help us realize how metacognition exercises its impact on, and fertilizes, lower cognitive processes (Rosenthal, this volume; Frith and Metzinger, 2016; also Cole, this volume); and I am prone to add that whatever we do consciously acts back on the non-conscious.

The argument from bi-directionality makes it clear that consciousness does not exist in isolation from what is occurring below the threshold of awareness, but also that the non-conscious is not cognitively innocent, that is, devoid of the cultural.

To put it in more straightforward terms: embodiment goes both ways (or as Chris Frith says in his epiloque to this volume: 'This interplay runs in both directions'). As advanced forms of knowledge and highly complex performances that require consciousness at some point become 'unknown' and 'incognizable', the unconscious generates processes that find their expression in conscious experience, too. In a somewhat Kantian terms,[3] one could then say consciousness without the non-conscious is empty, the non-conscious without consciousness is blind.

A consequence of the bi-directionality thesis is that there is no 'innocent' eye, naïve ear, neutral touch, or ignorant body. Even phenomenal experience does not come in its raw qualitativeness but is rather coloured by culture. As Heidegger famously observed, 'It requires a very artificial and complicated frame of mind to "hear" a "pure noise."' 'What we "first" hear is never noise or complex of sound, but the cracking wagon, the motorcycle. We hear the column on the march, the north wind, the woodpecker tapping, the fire cracking" (1926, p. 163).

---

[3] 'Thoughts without content are empty, intuitions without concepts are blind', Introducton to the *Critique of Pure Reason* (Kant, 1781–87/1933).

As cultured beings we are permanently seeking meanings, and we recognize them in aural, optical, tactile, and other patterns. The same kind of thought can be found in William James: 'No word in an understood sentence comes to consciousness as a mere noise' (1890, 1, p. 281). C.I. Lewis similarly observes: 'We do not see patches of color, but trees and houses; we hear not the indescribable sound, but voices and violins' (1929, p. 54). Seeing is in no way different from hearing. It is just as difficult, if not impossible, to see naïvely (for instance, it requires an effort of attentive awareness to see a tilted coin as an ellipse because it appears round to us), and it is questionable whether such an effort can ever succeed. Similarly, a plate persists in appearing round in spite of the form of its sensory imprint, which is mostly elliptical (Wartofsky, 1972). Actually, contrary to the standard interpretation that the phenomenal is a kind of subjective experience that is not necessarily faithful to the 'real', in the latter case the appearance conveys not the sensory given (ellipse) but the real shape of the plate (circle) (see, e.g., Thouless, 1931).

Not only is there no pure sound and pure form, there is generally no pure consciousness. The human mind is so profoundly encultured (see, e.g., Menary, 2015; 2013) that consciousness simply never comes in a mentally distilled form, with the result that even phenomenal experience is not freed from intrusions of the symbolic. A consequence from continuity and bi-directionality further confirms that none of the entities that constitute the mental exist in isolation but rather interact and mutually cross-fertilize so that the input from any sphere can generate reactions and modulations in virtually any other. A touch can induce a more complex sense of togetherness (Farmer and Tsakiris, 2013); a smell can attract us to another person or do the opposite (Freeman, 1995); a sound can stir fear just as a word can 'hurt'; hunger may not only influence moods but also impact cognitive processes (Piech *et al.*, 2010) or judgment (Gallagher, this volume); hearing can inform about what is going on in the surrounding world so that it is possible to say, 'I can *hear* what you are *doing*' (Iacoboni, 2008, p. 34; emphasis added);[4] vision informs hearing (McGurk and MacDonald, 1976); gaze can disclose another's intention and goal of action (Iacoboni, 2008); mimics can disclose how you feel (Cole, 1999); gesture can free mental energy for better results in mathematics (Goldin-Meadow *et al.*, 2001; Goldin-Meadow, 2003; see also Clark, 2013);

---

[4] 'I know a lot of things by simply listening. Clapping, tearing paper, typing, breaking peanuts — these are all actions that produce sounds and can be easily recognized by all of us. We don't give this ability a second thought' (Iacoboni, 2008, p. 35).

walking can enhance creative capacity (Oppezzo and Schwartz, 2014; also Gretchen, 2014); etc. In all of these cases, a simple, ephemeral, or even trivial (mostly unconscious) act may stir more complex forms of behaviour and define their course.

We can now better appreciate Eric Kandel's saying that 'every conversation changes the brain' (quoted in Dresler *et al.*, 2013, p. 538). Indeed, every act of pointing may miraculously make many things non-existent in an instant and focus instead on a detail that suddenly fills the entire field of attention; every colour changes the visible scene (e.g. 'redness' warns; 'greenness' signals the opposite, etc.); every smile softens attitudes, and laughter may make us feel different and change moods that may even impact our decision making;[5] every joke leaves a trace, as does bad news; every compliment has a pleasurable effect on the one it is directed to; every stroke of the brush changes not only the configuration of the canvas but the mental image, too; every piece of music that we hear refurnishes the entire internal mental architecture, not only the auditory, etc. Which aspect of the latter will be moved by initial stimulation and what effect will it induce is almost completely out of reach of consciousness.

Not only conversation, but a single word can alter mental setting. The use of the word 'good' (or 'bad') for a person significantly shapes our attitude toward him or her; a 'guilty' (or 'not guilty') defines the destiny of a person in court; 'and the winner is...' brings euphoria to one or a few and disappointment to all others who have hoped; a 'beautiful', 'valuable', 'great', or 'worthless' for a piece of art may make you discover traits and qualities not evident at first glance; a 'wrong', 'true', 'difficult', 'easy', 'dangerous', etc. has the power of moulding the mental in a profound way. We are certainly conscious of those words as we pronounce or read them, but we are unaware of the effects they have upon our minds. In other words, we are conscious users of labels but are unaware of what these labels do to us. That is, words have a power over us in that they mould our cognitive world by imposing meanings whose impact upon us we do not control.

If one kind of mental act (e.g. narrative) can influence a quite different one (e.g. emotions or judgments) then we may state once again that what constitutes mentality is not so neatly discriminated, sorted out, and separated as is the case with concepts as we describe them in our theories. Mental phenomena are not as strictly delineated as we find them in our conceptual schemas. Living creatures' minds are hardly what standard textbooks and lexical entries say they are. They are

---

5   '[F]ear, anxiety, and sadness all help us in our decision making' (Gazzaniga, 1998, p. 10; see also LeDoux, 2015).

much more variegated, flexible, interconnected, dynamic, and capable of generating processes that self-transform, all of which are done with disregard to what is conscious and what is not. The counterpointing of the conscious and non-conscious that obviously matters for theoretical minds is something that does not concern the living minds that theorists are investigating.

Such is the case, for instance, with our standard notions of thought, which we take to be the epitome of reflective behaviour and see as being instantiated in an exclusively conscious mode and through the engagement of high-level cognition. However, it becomes more and more evident that we have been deceived in that regard, too (see, e.g., Dijksterhuis, 2016). For thinking is not reducible to what intellectualists say it is (see, e.g., Noë, 2005). There are various ways that thoughts can be brought about and they are, more often than we want to believe, dictated by the unconscious.

If 'one can have thoughts without thinking' (Prinz, 2004, p. 49) and if 'different kinds of vehicles can fill the thinking role' (Prinz and Clark, 2004, p. 58) then we can also assume that thought need not be tailored out of its conscious cloth. Following Wolf Singer (2003), who claims that unconscious percepts can trigger thinking,[6] and Cushing et al. (this volume), who say that 'Conscious contents seldom arise from elaborate chains of thought', we may say that the unconscious is not alien to thought but is a dimension of it. John Searle goes even further: 'All of our mental life is conditioned by a set of presuppositions that are not part of our conscious awareness' (Searle, 1998, p. 6). Though many researchers are still hesitant to acknowledge it, unconscious processing of thoughts is ubiquitous.

Unconscious thought is not only much quicker (for the general issue of slowness of consciousness, see Libet, 1998; 2004), it also entails an indefinitely heterogeneous set of dispositions (see, e.g., Stanley and Williamson, 2001). According to Dijksterhuis and Nordgren (2006), it even has advantages over deliberative thought.[7] For instance, it is a better guide when approaching complex decisions: 'Rather than thinking much consciously, you can delegate the labor of thinking to the unconscious, and at some point you will intuitively "feel" what the best option is' (Dijksterhuis and Nordgren, 2006, p. 95). Furthermore, according to the same authors, it seems that conscious thought favours

---

[6] 'Ja, der Köper reagiert auf eine unbewusste Wahrnehmung und das kann einem Denkvorgang auslösen' (Singer, 2003, p. 122).

[7] Compare also Prinz (this volume), who says that 'unconscious vision can perceive much more'.

the familiar and habitual and so relies more on standard solutions, whereas in an unconscious mode people tend to stereotype less.

We are unaware of two basic things in regard to thinking. First, we are ignorant of the fact that most of what constitutes mind is not thought. Second, even the belief that thinking is exclusively realized by means of conscious cognitive effort is an illusion. Not having to process thought consciously means that it can 'just happen'. This is exactly how Galen Strawson interprets understanding: 'Certainly understanding is not something one does intentionally. In the normal case, it is something that *just happens*' (Strawson, 1994, p. 7, emphasis added). In previous work I have been (following Wittgenstein, 1953, and Searle, 1983) referring to this as 'just doing' (Radman, 2012). However, while we, as a rule, take 'just doing' as synonymous with the unconscious, interestingly enough Ezequiel Morsella (Cushing *et al.*, this volume) claims that it is also how consciousness manifests itself in mentality: 'In most cases, conscious contents "just happen".'

It is thus not only that simple motor acts of which we are unaware (like walking, eating, drinking, driving) 'just happen'; high-level cognitive tasks are also performed as unconscious routines (Bargh, 1994; Bargh and Chartrand, 1999). In the final instance we can say that the symbolic gets acquired by the knowing body, or, in Steven Pinker's words, for instance, that language is an instinct (Pinker, 1994). Obviously inspired by Darwin, Pinker claims that 'people know how to talk in more or less the sense that spiders know how to spin webs' (*ibid.*, p. 5). For him 'language is a product of well-engineered biological instinct' (*ibid.*, p. 6). He then concludes that 'language is an art, like brewing or baking' (*ibid.*, p. 6) — or, according to Alva Noë, it is like barking: '[M]uch talking is more like barking than it is anything like what the linguists have in mind' (Noë, 2009, p. 108). As he further observes: 'One of the very many false ideas about language is that its primary function is to express information or communicate thoughts. Speech has many functions, but surely a large part of it is more like the grooming behaviour of chimpanzees or the shepherding behaviour of dogs than it is like reasoned discourse among parliamentarians' (*ibid.*, p. 107).

Even mathematical cognition may be seen as an embodied skill that basically happens backstage (Lakoff and Núñez, 2000; see also Prinz, this volume[8]) and diagnosing can be seen as not much different from muscular know-how. According to Michael Polany, the 'medical diagnostician's skill is as much an art of doing as it is an art of knowing. The skill of testing and tasting is continuous with the more actively

---

[8] 'People unconsciously perform arithmetic.'

muscular skills, like swimming or riding a bicycle' (Polanyi, 1958, p. 54). In much the same way Andy Clark remarks: 'Making the expert medical judgment... has more in common with knowing how to ride a bicycle than with consulting a set of rules in a symbolic data-base' (2003, p. 312).

Doing surgery, playing piano, or conducting an orchestra alike are a matter of embodied acting which, at the expert level, are performative routines like eating, shaving, or dressing (e.g. Dreyfus, 1986). If the symbolic can be internalized and acquired by the body then one can easily agree with, and appreciate, the saying: 'Culture gets under the skin and skull' (Roepstorff et al., 2010, p. 1052). That may mean that, on the one hand, symbols can be accessed via non-conscious paths and, on the other hand, that the impact of the cultural is more immediate and affects even the basic levels of cognition than has been habitually conceived.

Enculturation is so overwhelming and profound that not even the most elementary mental processes are exempt from its impact. Realizing this may prompt us to conclude that pure experience is a myth, as is the idea that the non-conscious is by its nature ignorant and uneducated. It is thus justified to claim there is no virgin unconscious and no pure consciousness. Our concepts may be 'pure', but it is definitely not the case with what they apply to.

4. Towards an Organismic Self

Let me tackle the methodology of dichotomization once again, this time targeting the contrasting of 'it' (the brain/body) and 'self', which is habitually used to polarize the unconscious body and the Cartesian conscious surveyor, ignorance and knowing, intuition and reason. Taking into account all of the above, it is now clear that such exclusiveness is not tenable. If cognitive competence is not limited to the thinking self but also (always) realizable through embodied coping, then why not conceive the possibility of an *organismic self* capable of complex behaviour without having to engage in awareness.

Accepting this is in discord with Wittgenstein (1953) as he says, very much like Ryle (1949), that there is nothing going on when one understands something, and also that 'understanding is not a mental process' (Wittgenstein, 1953, §§149, 152, 154). The truth is, a lot is going on when one understands something even though it need not be witnessed by awareness. It would also be inadequate, and unscientific, to imply that what is not shaped in consciousness does not deserve the status of mentality and so does not belong to mind. Living minds in many respects do not match the models that mainstream modern cognitive scientists have designed to represent them. Much more theoretical effort is needed to reconceptualize many standard notions related to the nature of the mental. This chapter is an attempt to affirm the idea that whatever is achieved automatically and with ease (and thus need not be assisted by consciousness), as well as intentionality of which we are unaware (Bargh and Morsella, 2008; *cf.* also Searle, 1998[9]), cannot be ousted from mentality but rather must be integrated as a constitutive element of the mind.

This prompts us further to say that consciousness is never a sole player—but neither is the unconscious. Just as no conscious act can bypass the unconscious, it is difficult to imagine that the unconscious can be instructive for the cognitive organism unless it is moved and stimulated by conscious sources.

Perhaps the most far-reaching consequence following from the postulation of the existence of an organismic self and the bi-directionality thesis is that virtually everything that is processed as conscious activity can also be performed in the non-conscious mode (see also Prinz, this volume). Nowadays we have enough (and empirical) support for this kind of claim. Unlike the habitual preconception, according to which there are mental activities or forms of knowing specifically instantiated as conscious contents and those that exist only in non-conscious form, my assumption is that any mode of mentality or kinds of cognition can exist in both the conscious and non-conscious modes. That is to say (as already pointed out above), perception (Marcel, 1983; Schacter, 1987; Debner and Jacoby, 1994; Holender and Duscherer, 2004), memory (Fuchs, 2012; also T. Reber, this volume), language usage (Pinker, 1994; Noë, 2009), decision making (Gold and Shadlen, 2007; Dijksterhuis, 2004; also Newell, this volume), learning (S.A. Reber, 1993; Seger, 1994), and even thinking (Dijksterhuis and Nordgren, 2006) and calculating (Lakoff and Núñez, 2000; Prinz, this volume), forms of cognition we usually take as prototypes of conscious activity and effort, can also be performed in a non-conscious mode. However (and here we take the continuity and bi-directionality theses consequently), emotion, intuition, imagination, etc.—processes we almost exclusively affiliate with the non-conscious—are to be seen as inextricably connected with conscious forms of mentality. For instance, being in love is not necessarily only a matter of emotional 'blindness' and neither is science reducible to emotionless calculation; both entail traits from their 'opposites', so that the usage of phrases such as 'reasoned emotion' and 'felt', or aesthetic, reason are not out of place (Radman, 2004; 1997; see also Radman, 2013b). This allows us not only

---

[9] '...not all intentional states are conscious; and not all conscious states are intentional' (Searle, 1998, p. 6).

to talk about emotional intelligence (LeDoux, 1998; Goleman, 2005) but also of intelligent emotions, of embodied mind but also of 'enminded' body.

All this amounts to the conviction that *the unconscious is to be taken as a function of consciousness*. Yet the opposite is true, too: the unconscious does not reside in a mental enclave but is exposed to the impact of conscious happenings that permanently shape and reshape it.

Because the impact domain of the organismic self is broad and covers a wide range of mental events, it is appropriate to say that there are many grades (or shades) of the unconscious or, in other words, that it is scalar. Plainly, we are differently unconscious when we are in coma, sleep dreamlessly, drive the car, type, talk, or play the piano. Concepts we use (here exemplified through the contrastive basic terms 'consciousness' and 'the unconscious') have solid and clear-cut boundaries; the phenomena they are applied to, however, do not conform to this kind of stricture. Our conceptual tools are created to help us discriminate features of the mental, but they generally fail to account for the transitional forms and for the fluent and effortless conversion from one kind of mental state or event to another.

In accordance with the continuity and bi-directionality theses (to stress it once again), we are ill-advised to theoretically treat consciousness as a self-contained instance. However, to insist on the big picture of mentality is not to relativize the import of consciousness (for it is due to consciousness that we experience the world the way we do), but to learn more about what generates, impacts, and pre-shapes it. We further realize that cognitive competence is shared and both the (conscious) self and the (unconscious) organismic self[10] are at stake in the processes that will result in our behaviour, though some processes are typically performed by one or the other. For instance, the organismic self is in charge of most inhibition (see, e.g., Berthoz, this volume) and selection processes (Morsella *et al.*, 2015; Cusching *et al.*, this volume), while the 'self' gets articulated at the level of semantics.[11]

On the other hand, the organismic self actively searches for meaningfulness before semantics sets in; it defines the taste before the taster finds words for the gustatory quality; it prepares the terrain in decision making for which the deliberative mind later claims exclusive merit; it is at stake in judging before we are aware of the reasons and criteria

---

[10] Compare also Scheper-Hughes and Lock (1987), who talk about 'body-self' and distinguish it from 'social body' and 'body politic'.

[11] An interesting attempt is made by Bruner and Kalmar (1998), who integrate these two aspects: '…the Self is both outer and inner, public and private, innate and acquired, the product of evolution and the offspring of culturally shaped narrative' (p. 326).

used in making the judgment; it tells us what is right before the conscious self finds reasons to justify it; it grants us 'criteria' of beauty before they become an aesthetic preference we can report or reflect on; it pre-shapes what will in narration, and poetry, become 'love'; in science and other creative endeavours it may provide us with a sense of doing something right (or wrong) and even impact what will be felt as true before evidence or proof is provided.

Because our mental capacity to process actual stimuli is so limited, whatever happens on the conscious stage quickly fades into the background which, in turn, becomes a potential foreground for the next step in mental processing. So seen, conscious actuality is a brief moment between the past and forthcoming unconscious. It seems appropriate to say that consciousness can only function the way it does because there is a massive body of the unconscious temporally extended from the remembered to the expected and the anticipated. If it is theorists' and other scientists' ambition to provide a more encompassing picture of the mind, they are advised to consider the 'silent' mind and 'hear' what is not manifested in experience, for it might be the implicit that preshapes what will later become manifest in experience and that, generally, significantly defines the ways of the mind.

Our philosophies of mind are selective and discriminatory, written mostly by people whose sole exploratory tool is consciousness. If a philosophy were written from the perspective of the cognitive organism itself, it would display that there are various ways of knowing and that, among other things our brains and bodies do, we also *think*. The proportion of conscious thought in everyday mental activity is rather small. Yet a cognitive organism can behave in a reasoned way even away from awareness, conscious deliberation, and controlled will. That is, however, not to say that thought is ephemeral; it is only to make clear that even as thinkers we do not cease to exist as *embodied* (and largely unconscious) beings. When we design hypotheses, formulate theorems, philosophize, and engage in other forms of creativity, we act as human beings governed by our organismic selves that *know how* to cope with the world in a competent way, understand, appreciate, and enjoy it, without recourse to conscious deliberation.

5. Conclusions

There is a unity of the mental that is primary and pronounced; but for some reason, our theoretical minds remain insensitive to it and instead traditionally favour the mental architecture of Cartesian design, or rather divide. The first and foremost conclusion is that the minds we are investigating do not conform to the categorical strictures theorists have conceptualized. Also, the implication that differences necessarily lead to divisions and oppositions (such as in the case of the conscious and non-conscious) is unfounded, but it has not been realized as such and is the cause of many faulty conceptions.

A further conclusion from the above is that the unconscious does not live isolated in the enclave of the mind but is an integral and vital part of it. It also prompts us to conclude that the conscious 'self' is never alone. It is thus possible to say there is no consciousness without the non-conscious; however, the opposite is true, too – the non-conscious does not appear out of the blue but is powered by once-conscious contents. As such, the non-conscious permanently exercises its impact upon the conscious 'counterpart'. This takes the form of ongoing cross-fertilization; the unconscious shapes and reshapes itself and constantly increases its competence. (If it were merely blind and ignorant, it could never serves us as well as it does.)

The life of mind is a result of the intensive mutuality of both of these spheres, whereby it has to be stressed that most mental activity is by far non-conscious. This allows us to claim further that not only is mind embodied but also body 'enminded'. In order to make the idea of a 'knowing' body or 'bodily wisdom' plausible, we need to dismantle the 'gapped' or counterpointed image of human mentality that still haunts many theoretical minds.

We can then say that we talk with the easiness of when we walk; that we calculate with the ease of when we cook; that our understanding is as simple as grasping; that recalling the past may be as automatic as opening a window; that we predict with the same ease that we throw a ball; that coming to an idea may be as simple as gesturing; that perceiving appears as basic and sponataneous as breathing; that we deal with the fictional with the spontaneity typical of our dealing with reality; that we can think of the 'impossible' in the same way we learn about the existing, etc.

The organismic self is the master of all kinds of acting – from brushing our teeth to using a brush to paint a picture, from using a knife to cut fruit to using a scalpel to perform demanding surgical operations – and is to be merited even for complex cognitive tasks such as decision making, judging, and thinking. Thanks to the capacity of the organismic self to conduct a wide range of behaviour in a 'silent' mode, human agency is empowered to do more as (an unconscious) doer than as (a conscious) decider or thinker. This privilege of not having to process each and every instance of mentality in consciousness enables us to bring our actions forward with the ease of automaticity, during which reasoning benefits most.

An agent governed by the organismic self outgrows the capacity and competences of the conscious 'self' so that he or she can in many cases perform more quickly, more easily, and better in the absence of conscious deliberation (though not always; see, e.g., Frith and Metzinger, 2016; see also Cole, this volume). As unconscious agents, we have the capacity to do more than we realize by means of conscious thought; we can perform better away from awareness; act more promptly without the effort of conscious computing; make more appropriate decisions if the task is delegated to the non-conscious; apply complex skills and intellectual routines in a way that does not require attentive thinking. But this sort of competence does not have miraculous sources; it is powered by once conscious acts.

The feeling Gustav Mahler describes—'I am not composing, I am composed'—or that Bruce Lee alludes to by saying it is not him who is kicking, is also known to all ordinary agents—we feel that what we do outgrows us as conscious doers and that it must be another kind of 'self' in charge, not reducible to deliberate thinking. It is this kind of (mostly unconscious) agency, governed by the (organismic) self that makes us capable of competent coping in the world, so wonderfully variegated and rich, and so beautifully present to experience.

Acknowledgment

I would like to thank IMéRA (Institut d'études avancées, Aix-Marseille Univerisité) for providing extraordinary research facilities and a stimulating environment from which I profited during my stay as a fellow at the Institute. Ideas developed during that period are integrated in the present volume and this chapter.

This chaper is partly the result of the work done within the research project *The Autonomous Mind: Inquiries into Self-generating Unconscious Processes*, funded by the HRZZ (Croatian Endowment for Science).

References

Agassi, J. (1969) Privileged access, *Inquiry*, **12**, pp. 420–426.

Baars, B.J. (1988) *A Cognitive Theory of Consciousness*, Cambridge, MA: Cambridge University Press.

Baars, B.J. (1997) *In the Theater of Consciousness*, New York: Oxford University Press.

Baccarini, M. & Maravita, A. (2013) Beyond the boundaries of the hand: Plasticity of body–space interactions following tool use, in Radman, Z. (ed.) *The Hand, an Organ of the Mind: What the Manual Tells the Mental*, Cambridge, MA: MIT Press.

Bargh, J.A. (1994) The four horsemen of automaticity: Awareness, intention, efficiency, and control in social cognition, in Wyer, R.S. &

Srull, T.K. (eds.) *Handbook of Social Cognition*, vol. 1, Hillsdale, NJ: Erlbaum.

Bargh, J.A. & Chartrand, T.L. (1999) The unbearable automaticity of being, *American Psychologist*, **54** (7), pp. 462–479.

Bargh, J.A. & Morsella, E. (2008) The Unconscious mind, *Perspectives on Psychological Science*, **3** (1), pp. 73–79.

Bronovski, J. (1978) *The Visionary Eye*, Cambridge, MA: MIT Press.

Burkeman, O. (2015) Why can't the world's greatest minds solve the mystery of consciousness?, *The Guardian*, 21 Jan 2015.

Bruner, J. & Kalmar, D.A. (1998) Narrative and metanarrative in the construction of self, in Ferrari, M. & Sternberg, R.J. (eds.) *Self-Awareness: Its Nature and Development*, New York: Guilford Press.

Chalmers, D. (1995) Facing up the problem of consciousness, *Journal of Consciousness Studies*, **2** (3), pp. 200–219.

Chalmers, D. (1996) *The Conscious Mind: In Search of a Fundamental Theory*, New York: Oxford University Press.

Clark, A. (2003) Artificial intelligence and the many faces of reason, in Stich, S. & Warfield, T. (eds.) *The Blackwell Guide to Philosophy of Mind*, Oxford: Blackwell.

Clark, A. (2007) *Supersizing the Mind: Embodiment, Action, and Cognitive Extension*, Oxford: Oxford University Press.

Clark, A. (2013) Gesture as Thought?, in Radman, Z. (ed.) *The Hand, an Organ of the Mind: What the Manual Tells the Mental*, Cambridge, MA: MIT Press.

Clark, A. & Chalmers, D. (1998) The extended mind, *Analysis*, **58** (1), pp. 7–19.

Cole, J. (1999) *About Face*, Cambridge, MA: MIT Press.

Crane, T. (2002) *Elements of Mind*, Oxford: Oxford University Press.

Debner, J.A. & Jacoby, L.L. (1994) Unconscious perception: Attention, awareness, and control, *Journal of Experimental Psychology: Learning, Memory, and Cognition*, **20** (2), pp. 304–317.

Dennett, D.C. (1991) *Consciousness Explained*, Boston, MA: Little Brown & Co.

Dennett, D.C. (2013) *Intuition Pumps and Other Tools of Thinking*, New York: W.W. Norton & Co.

Dijksterhuis, A. (2004) Think different: The merits of unconscious thought in preference development and decision making, *Journal of Personality and Social Psychology*, **87** (5), pp. 586–598.

Dijksterhuis, A. (2016) A case for thinking without consciousness, *Perspectives on Psychological Science*, **11** (1), pp. 117–132.

Dijksterhuis, A. & Nordgren, L.F. (2006) A theory of unconscious thought, *Perspectives on Psychological Science*, **1** (2), pp. 95–109.

Drake, D. (1925) *Mind and Its Place in Nature*, New York: Macmillan.

Dresler, M., A. Sandberg, K. Ohla, C. Bublitz, C. Trenado, A. Mroczko-Wąsowicz, S. Kühn, D. Repantis (2013), Non-pharmacological cognitive enhancement', *Neuropharmacology*, Cognitive Enhancers: molecules, mechanisms and minds, *22nd Neuropharmacology Conference: Cognitive Enhancers*, Elsevier, 64 (January), 529–543.

Dreyfus, H.L. & Dreyfus, S.E. (1986) *Mind over Machine: The Power of Human Intuition and Expertise in the Age of the Computer*, Oxford: Basil Blackwell.

Edelman, G.M. & Tononi, G. (2000) *A Universe of Consciousness*, New York: Basic Books.

Farmer, H. & Tsakiris, M. (2013) Touching hands: A neurocognitive review of intersubjective touch, in Radman, Z. (ed.) *The Hand, an Organ of the Mind: What the Manual Tells the Mental*, Cambridge, MA: MIT Press.

Fechner, G. (1848) *Nanna, oder, über das Seelenleben der Pflanzen*, Leipzig: L. Voss.

Fechner, G. (1906) *Zend-Avista: oder über die Dinge des Jenseits vom Standpunkt der Naturbetrachtung*, 3rd ed., Hamburg: L. Voss.

Feigl, H. (1958) The mental and the physical, in Feigl, H., Scriven, M. & Maxwell, G. (eds.) *Minnesota Studies in the Philosophy of Science II: Concepts, Theories, and the Mind–Body Problem*, Minneapolis, MN: University of Minnesota Press.

Flanagan, O. (1991) *The Science of the Mind*, Cambridge, MA: MIT Press.

Freeman, W.J. (1995) *Societies of Brains: A Study in the Neuroscience of Love and Hate*, Hillsdale, NJ: Lawrence Erlbaum Associates.

Freeman, W.J. (2001) *How Brains Make Up Their Minds*, New York: Columbia University Press.

Frith, C.D. & Metzinger, T. (2016) What's the use of consciousness? How the stab of conscience made us really conscious, in Engel, A.K., Friston, K.J. & Kragic. D. (eds.) *The Pragmatic Turn: Toward Action-Oriented Views in Cognitive Science*, Strüngmann Forum Reports, vol. 18, Lupp, J. (series ed.), Cambridge, MA: MIT Press.

Fuchs, T. (2012) The phenomenology of body memory, in Koch, S.C., Fuchs, T., Summa, M. & Müller, C. (eds.) *Body Memory, Metaphor and Movement*, Advances in Consciousness Research, 84, Amsterdam: John Benjamins.

Gazzaniga, M.S. (1998) *The Mind's Past*, Berkeley, CA: University of California Press.

Gold, J.I. & Shadlen, M.N. (2007) The neural basis of decision making, *Annual Review of Neuroscience*, **30**, pp. 535–574.

Goldin-Meadow, S. (2003) *Hearing Gesture: How Our Hands Help Us Think*, Cambridge, MA: Harvard University Press.

Goldin-Meadow, S., Nussbaum, M., Kelly, S.D. & Wagner, S. (2001) Explaining math: Gesturing lightens the load, *Psychological Science*, **12** (6), pp. 516–522.

Goleman, D. (2005) *Emotional Intelligence: Why It Can Matter More than IQ*, New York: Bantam Books.

Gray, J.A. (2004) *Consciousness: Creeping up on the Hard Problem*, New York: Oxford University Press.

Gretchen, R. (2014) Want to be more creative? Take a walk, *New York Times*, 30 April.

Hameroff, S.R. (2006) The entwined mysteries of anesthesia and consciousness, *Anesthesiology*, **105** (2), pp. 400–412.

Hameroff, S. & Penrose, R. (2014a) Consciousness in the universe: A review of the 'Orch OR' theory, *Physics of Life Reviews*, **11** (1), pp. 39–78.

Hameroff, S. & Penrose, R. (2014b) Reply to criticism of the 'Orch OR qubit' — 'Orchestrated objective reduction' is scientifically justified, *Physics of Life Reviews*, **11** (1), pp. 94–100.

Heidegger, M. (1962) *Being and Time*, Macquairre, J. & Robinson, E. (trans.), New York: Harper & Row.

Holender, D. & Duscherer, K. (2007) Unconscious perception: The need for a paradigm shift, *Perception & Psychophysics*, **66** (5), pp. 872–881.

Iacoboni, M. (2008) *Mirroring People: The New Science of How We Connect with Others*, New York: Farrar, Straux & Giroux.

Jackson, F. (1986) What Mary didn't know, *Journal of Philosophy*, **83**, pp. 291–295.

Jackson, F. (1995) Postscript on 'What Mary didn't know', in Moser, P. & Trout, J. (eds.) *Contemporary Materialism*, London: Routledge.

James, W. (1890/1950) *The Principles of Psychology*, 2 vols., New York: Dover.

Kant, I. (1781–87/1933) *Critique of Pure Reason*, Kemp Smith, N. (trans.), London: Macmillan.

Kirk, R. (1974) Zombies v. materialists, *Proceedings of the Aristotelian Society*, **48**, pp. 135–152.

Koch, C. & Hepp, K. (2006) Quantum mechanics in the brain, *Nature*, **440**, p. 611.

Lakoff, G. & Núñez, R.E. (2000) *Where Mathematics Comes From*, New York: Basic Books.

LeDoux, J. (1998) *The Emotional Brain: The Mysterious Underpinnings of Emotional Life*, New York: Simon and Schuster.

LeDoux, J. (2002) *Synaptic Self: How Our Brains Become Who We Are*, New York: Penguin Books.

LeDoux, J. (2015) *Anxious: Using the Brain to Understand and Treat Fear and Anxiety*, New York: Penguin.

Lewis, C.I. (1929) *Mind and World Order*, New York: Dover.

Lewis, D. (1980) Mad pain and Martian pain, in Block, N. (ed.) *Readings in Philosophy of Psychology, Volume Two*, Cambridge, MA: Harvard University Press.

Libet, B. (1986) Unconscious cerebral initiative and the role of conscious will in voluntary action, *Behavioral and Brain Sciences*, **8**, pp. 529–566.

Libet, B. (2004) *Mind Time: The Temporal Factor in Consciousness*, Cambridge, MA: Harvard University Press.

Litt, A., Eliasmith, C., Frederick, W.K., Weinstein, S. & Thagard, P. (2006) Is brain a quantum computer?, *Cognitive Science*, **30**, pp. 593–603.

Marcel, A.J. (1983) Conscious and unconscious perception: An approach to the relations between phenomenal experience and perceptual processes, *Cognitive Psychology*, **15**, pp. 238–300.

Marr, D. (1982) *Vision: A Computational Investigation into the Human Representation and Processing of Visual Information*, San Francisco, CA: W.H. Freeman.

McGinn, C. (1999) *The Mysterious Flame: Conscious Minds in a Material World*, New York: Basic Books.

McGurk, H. & MacDonald, J. (1976) Hearing lips and seeing voices, *Nature*, **264**, pp. 746–748.

Menary, R. (2013) Cognitive integration, enculturated cognition and the socially extended mind, *Cognitive Systems Research*, **25**, pp. 26–34.

Menary, R. (2015). Mathematical cognition—A case of enculturation, in Metzinger, T. & Windt, J.M. (eds) *Open MIND*, 25 (T), Frankfurt am Main: MIND Group.

Metzinger, T. (2010) *The Ego Tunnel: The Science of the Mind and the Myth of the Self*, New York: Basic Books.

Millikan, R.G. (1990) Truth, rules, hoverflies, and the Kripke-Wittgenstein paradox, *Philosophical Review*, **99** (3), pp. 323–353.

Mlodinow, L. (2012) *Subliminal: How Your Unconscious Mind Rules Your Behavior*, New York: Vintage.

Nagel, T. (1974) What is it like to be a bat?, *Philosophical Review*, **83**, pp. 435–450.

Noë, A. (2005) Against intellectualism, *Analysis*, **65** (4), pp. 278–290.

Noë, A. (2009) *Out of Our Heads*, New York: Hill and Wang.

Morsella, E., Godwin, C.A., Jantz, T.J., Krieger, S.C. & Gazzaley, A. (2015) Homing in on consciousness in the nervous system: An action-based synthesis, *Behavioral and Brain Sciences*, target article.

Oppezzo, M. & Schwartz, D.L. (2014) Give your ideas some legs: The positive effect of walking on creative thinking, *Journal of Experimental Psychology: Learning, Memory, and Cognition*, **40** (4), pp. 1142–1152.

Penfield, W. (2015) *Mystery of the Mind: A Critical Study of Consciousness and the Human Brain*, Princeton, NJ: Princeton University Press.

Penrose, R. (1989) *The Emperor's New Mind: Concerning Computers, Minds and the Laws of Physics*, Oxford: Oxford University Press.

Piech, M.R., Pastorino, M.T. & Zald, D.H. (2010) All I saw was the cake: Hunger effects on attentional capture by visual food cues, *Appetite*, **54**, pp. 579-582.

Pinker, S. (2007) The brain: The mystery of consciousness, *Time*, 29 January 2007.

Polger, T. (2000) Zombies explained, in Ross, D., Brook, A. & Thompson, D. (eds.) *Dennett's Philosophy: A Comprehensive Assessment*, Cambridge, MA: MIT Press.

Popper, K.R. & Eccles, J. (1977) *The Self and its Brain: An Argument for Interactionism*, New York: Routledge.

Preston, J. & Bishop, M. (eds.) (2002) *Views into the Chinese Room: New Essays on Searle and Artificial Intelligence*, Oxford: Oxford University Press.

Prinz, J. (2004) *Gut Reactions: A Perceptual Theory of Emotion*, New York: Oxford University Press.

Prinz, J. & Clark, A. (2004) Putting concepts to work: Some thoughts for the twentyfirst century, *Mind and Language*, **19** (1), pp. 57-69.

Putnam, H. (1973) Meaning and reference, *Journal of Philosophy*, **70**, pp. 699-711.

Putnam, H. (1985) The meaning of 'meaning', *Philosophical Papers, Vol. 2: Mind, Language and Reality*, Cambridge: Cambridge University Press.

Radman, Z. (1995) Rehabilitating the body: Introductory, *Synthesis Philosophica*, **10** (1-2), pp. 3-8.

Radman, Z. (1997) Emotion and cognition, in Radman, Z. (ed.) *Horizons of Humanity: Festschrift in Honour of Ivan Supek*, Frankfurt am Main: Peter Lang.

Radman, Z. (2004) The felt self: Aesthetics and neuroscience in dialog, *Changes in Aesthetics, XVI International Congress of Aesthetics*, Rio de Janeiro, 18-23 July 2004, conference paper.

Radman, Z. (2007) Introductory: Consciousness: modeling the mystery, *Synthesis Philosophica*, **22** (2), pp. 267-271.

Radman, Z. (2012) The background: A tool of potentiality, in Radman, Z. (ed.) *Knowing Without Thinking; Mind, Action, Cognition, and the Phenomenon of the Background*, Basingstoke: Palgrave Macmillan.

Radman, Z. (ed.) (2013a) *The Hand, an Organ of the Mind: What the Manual Tells the Mental*, Cambridge, MA: MIT Press.

Radman, Z. (2013b) Body, brain, and beauty: The place of aesthetics in the world of the mind, *Diogenes*, **59**, pp. 41-51.

Reber, A.S. (1993) *Implicit Learning and Tacit Knowledge: An Essay on the Cognitive Unconscious*, Oxford: Oxford University Press.

Roepstorff, A., Niewohner, J. & Beck, S. (2010) Enculturating brains through patterned practices, *Neural Networks*, **23**, pp. 1051-1059.

Ryle, G. (1949) *The Concept of Mind*, London: Routledge.

Schacter, D.L. (1987) Implicit memory: History and current status, *Journal of Experimental Psychology: Memory, and Cognition*, **13**, pp. 501-518.

Scheper-Hughes, N. & Lock, M.M. (1987) The mindful body: A prolegemenon to future work in medical anthropology, *Medical Anthropology Quaterly*, **1**, pp. 6-41.

Searle, J.R. (1980) Minds, brains, and programs, *Behavioral and Brain Sciences*, **3**.

Searle, J.R. (1983) *Intentionality: An Essay in the Philosophy of Mind*, Cambridge: Cambridge University Press.

Searle, J.R. (1997) *The Mystery of Consciousness*, New York: New York Review of Books.

Searle, J.R. (1998) The mind and education, in Ferrari, M. & Sternberg, R.J. (eds.) *Self-Awareness: Its Nature and Development*, New York: Guilford Press.

Searle, J.R. (2009) Chineese room argument, *Scholarpedia*, **4** (8), 3100.

Seger, C.A. (1994) Implicit learning, *Psychological Bulletin*, **115**, pp. 163-196.

Shear, J. (1997) *Explaining Consciousness: The 'Hard Problem'*, Cambridge, MA: MIT Press.

Singer, W. (2003) *Ein neues Menschenbild?: Gespräche über Hirnforschung*, Frankfurt: Suhrkamp.

Sun, R., Merrill, E. & Peterson, T. (2001) From implicit skills to explicit knowledge: A bottom-up model of skill learning, *Cognitive Science*, **25**, pp. 203-244.

Stanley, J. & Williamson, S. (2001) Knowing how, *Journal of Philosophy*, **98** (8), pp. 411-444.

Strawson, G. (1994) *Mental Reality*, Cambridge, MA: MIT Press.

Strawson, G. (2006) *Consciousness and Its Place in Nature: Does Physicalism Entail Panpsychism?*, Exeter: Imprint Academic.

Tsai, C.-H. (2011) Linguistic know-how: The limits of intellectualism, *Theoria*, **77** (1), pp. 71-86.

Thouless, R.H. (1931) Phenomenal regression to the real object, *The British Journal of Philosophy*, **21** (Part 4), pp. 339-359.

Wartofsky, M. (1972) Pictures, representation, and the understanding, in Rudner, R.S. & Scheffler, I. (eds.) *Logic and Art: Essays in Honor of Nelson Goodman*, Indianapolis, IN: Bobbs-Merrill.

Wittgenstein, L. (1953) *Philosophical Investigations*, Anscombe, G.E.M. (trans.), Oxford: Blackwell.
Zeki, S. (2003) The disunity of consciousness, *Trends in Cognitive Sciences*, **7** (5), pp. 214–218.

Donish Cushing, Reza D. Ghafur
and Ezequiel Morsella

The Interdependence between Conscious and Unconscious Processes

Even the simplest of voluntary actions—the flick of a switch, the pressing of a button, or the uttering of the word 'hello'—require a complex orchestration of conscious and unconscious mechanisms. In this chapter,[1] we review the unconscious components of voluntary action and how these components interact with conscious processes (e.g. percepts, urges, and the sense of agency). Theoretical developments (e.g. Passive Frame Theory) specify which stages of processing, from stimulus input to behavioural output, can occur unconsciously and which cannot transpire without the conscious field. We focus only on the most basic forms of consciousness (e.g. the experience of a smell, visual after-images, tooth pain, or urges to scratch an itch). This form of consciousness has fallen under the rubrics of 'sentience' (Pinker, 1997), 'phenomenal consciousness' (Block, 1995), 'qualia' (Gray, 2004), 'phenomenal states' (Tye, 1999), and 'subjective experience'.

Second, in this chapter, we examine the findings from the Reflexive Imagery Task (RIT). They reveal that, as a function of external control, high-level mental representations can enter consciousness in a reflex-like, automatic, and insuppressible manner. Together, the RIT and Passive Frame Theory shed light on the interdependence between conscious and unconscious processes and on the counter-intuitive, reflex-like nature of the generation of conscious representations. Our reductionistic and mechanistic approach corroborates scientifically what philosophers have long known—that our most basic intuitions regarding the liaison between conscious and unconscious processes are incorrect. These intuitions include the conclusions that (a) during action

---

[1] This chapter is based in part on theorizing that has been presented in Morsella *et al.* (2016), Morsella and Bargh (2011), and Bhangal, Cho, *et al.* (2016).

selection, there is an homuncular-like 'decider' choosing one action plan over another, and that (b) conscious processes, which recent theoretical developments propose to be passive albeit essential, are more 'active' than their unconscious counterparts.

1. Content Generation and the Response to Conscious Contents are Often Mediated Unconsciously

We refer to a thing of which one is conscious (e.g. an after-image) as a *conscious content* (Merker, 2007; Seth, 2007). All the contents of which one is conscious at one time can be construed as composing the *conscious field* (Freeman, 2004; Köhler, 1947; Searle, 2000). The contents of the conscious field change over time (Figure 1). In Passive Frame Theory, the generators of conscious content and the responders to conscious content operate unconsciously, at least most of the time. In most cases, conscious contents 'just happen'. Regarding the unconscious generation of conscious content, there is substantial evidence that low-level perceptual analysis (e.g. motion detection, colour detection, auditory analysis; Zeki and Bartels, 1999) and semantic-conceptual processing (Harley, 1993; Lucas, 2000) occur unconsciously. Unconscious generation of conscious content occurs in feature binding (e.g. the binding of shape to colour; Zeki and Bartels, 1999) and in intersensory binding (Vroomen and de Gelder, 2003). Dramatic demonstrations of the latter are ventriloquism and McGurk effects (McGurk and MacDonald, 1976). (The McGurk effect involves interactions between visual and auditory processes: an observer views a speaker mouthing 'ga' while presented with the sound 'ba'. The observer is unaware of any intersensory interaction, perceiving only 'da'.) (See list of many kinds of intersensory illusions that are consciously impenetrable in Morsella, 2005, Appendix A.)

Action-related urges (e.g. the urge to inhale while underwater), too, are generated unconsciously and, often, simply 'just happen' to the organism. External stimuli often activate conscious urges in a direct, automatic manner. However, overt behaviour is often not influenced in this direct way, for reasons discussed below. Hence, in most scenarios, inclinations triggered by external stimuli are *behaviourally suppressible*, but they are often not *mentally suppressible* (Bargh and Morsella, 2008; Morsella, 2005). Passive Frame Theory proposes that the operating principles within content generators (e.g. for urges) can be consciously impenetrable, as in the unconscious factors that engender addiction-related urges (Baker *et al.*, 2004). Research has shown that people can have inexplicable 'gut feelings' (or 'somatic markers'; *cf.* Damasio *et al.*, 1991) that reflect response tendencies arising from systems whose inner workings are unconscious (LeDoux, 2000; Öhman and Mineka, 2001;

Olsson and Phelps, 2004). Such tendencies can be triggered by subliminal stimuli (see review in Morsella and Bargh, 2011). In short, conscious awareness of the source of a response tendency is distinct from consciousness of the associated action-related urge.

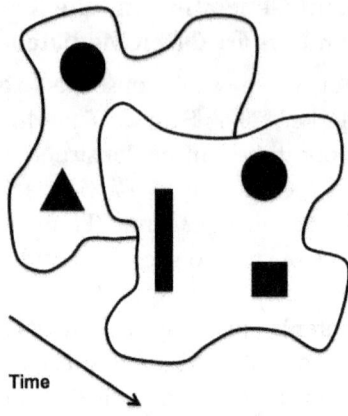

Figure 1. The conscious field, with a different configuration of conscious contents at each moment in time. Each of the two conscious fields, representing two different moments in time, possesses its own mix of conscious contents (the filled shapes). One conscious content (e.g. the circle) could be a smell; another conscious content (e.g. the square) could be an action-related urge.

1.1. Motor programming is mediated unconsciously

Regarding the claim that one is unconscious of the activities of the responders to the conscious contents in the conscious field, there is substantial evidence that one is unconscious of the efference to the muscles that dictates which fibres should be activated at which time (Rosenbaum, 2002). (For evidence regarding the conscious impenetrability of motor programming, see Fecteau *et al.*, 2001; Fourneret and Jeannerod, 1998; Heath *et al.*, 2008; Jeannerod, 2006; Liu, Chua and Enns, 2008; Rossetti, 2001.) For example, in speech production, one is unconscious of the articulatory code that dictates the activities of the vocal apparatus (e.g. the phenomenon of voicing). Moreover, in a simple action such as sniffing, participants believe that the actions are engendered more by the nose than by abdominal sources, even though the latter play an essential role in the constitution of this action (Berger, Bargh and Morsella, 2012; Tortora, 1994). Likewise, in finger and arm flexing, the actor believes that the finger tip or a point on the hand, respectively, is more responsible for the action than more proximal muscular regions (e.g. *m. brachioradialis*), which are responsible for the changes in bodily state (Tortora, 1994).

The sophistication of unconscious motor programming is evident in the phenomena of co-articulation (Levelt, 1989), end-state comfort (Zhang and Rosenbaum, 2008), and *motor equivalence* (Lashley, 1942), in which several different motor acts can lead to the same end state. (With this in mind, it is interesting to ponder that, in speech production, only one of tens of thousands of words is 'selected' for production. Such a selection process could never arise from conscious processes in such a short a span of time [less than a second].) More dramatic demonstrations of the unconscious control of action stem from the neurological literature. First, it is well-documented that several kinds of sophisticated behaviours can occur while subjects are in what appears to be an unconscious state (Laureys, 2005). For example, actions such as automatic ocular pursuit and some reflexes (e.g. pupillary reflex) can occur in some forms of coma and persistent vegetative states (Klein, 1984; Laureys, 2005; Pilon and Sullivan, 1996). In addition, it seems that licking, chewing, swallowing, and other behaviours can occur unconsciously once the incentive stimuli activate the appropriate receptors (Bindra, 1974; Kern et al., 2001). Second, research on the kinds of behaviours (e.g. automatisms) exhibited during epileptic seizures in which the patient appears to be unconscious (or to at least not have any conscious control) has revealed that stereotypic actions such as simple motor acts (Kutlu et al., 2005; Kokkinos et al., 2012), humming (Bartolomei et al., 2002), the removal of clothing (Wortzel et al., 2011), spitting (Carmant et al., 1994), whistling (Raghavendra, Mirsattari and McLachlan, 2010), oroalimentary automatisms (Maestro et al., 2008), and laughing (Enatsu et al., 2011) can occur independently of conscious mediation. More complex acts such as written and spoken (nonsense) utterances (Blanken, Wallesch and Papagno, 1990; Kececi, Degirmenci and Gumus, 2013), singing (Enatsu et al., 2011), sexual behaviours (Spencer et al., 1983), and rolling, pedalling, and jumping (Kaido et al., 2006) have also been found to occur in a reflex-like manner during seizures in which consciousness appears to be absent. Most dramatically, there are cases in which, during seizures, patients sing recognizable songs (Doherty et al., 2002) or express repetitive affectionate kissing automatisms (Mikati, Comair and Shamseddine, 2005). Third, investigations on narcolepsy (Zorick et al., 1979) and somnambulism (Plazzi et al., 2005; Schenk and Mahowald, 1995) reveal that complex behaviours (e.g. successfully negotiating objects) can be mediated unconsciously.

For all motor programs, the component mechanisms are mediated unconsciously, and no memories are formed for these programs. Instead, the programs are computed on the fly, executed online, and then 'scrapped' (Grossberg, 1999; Rosenbaum, 2002). The immediate scrapping of motor programs arises because these plans cannot be reused: the physical spatial relationship between the objects of the world and one's body changes too often (e.g. a glass is sometimes at left or right; Grossberg, 1999). (In addition, bodily effectors, because of changes due to growth, require that old motor programs be replaced by new ones.) Hence, each time an action is performed, new motor programs are generated online in order to deal with the peculiarities of each setting.

During action control, one is unconscious of motor programming, but one is conscious of the perceptual consequences of the actions generated by such programming. This is obvious in our examples above about sniffing and about the flexing of the arms and fingers. In terms of stages of processing, that which characterizes conscious content is, not *efference generation*, but rather that which can be construed as forms of *perceptual afference* (information arising from the world that affects sensory-perceptual systems; Sherrington, 1906), including *corollary discharges* (e.g. when subvocalizing; *cf.* Chambon et al., 2013; Christensen et al., 2007; Jordan, 2009; Obhi et al., 2009; Scott, 2013). According to several frameworks (e.g. Hommel et al., 2001), the perceptual consequences of action are stored as memories (specifically, of 'action-effects'). These memories can, in the future, be activated to generate the motor programs that gave rise to them (as in *ideomotor theory*; see Hommel et al., 2001), which is a form of 'inverse modelling' (Wolpert and Kawato, 1998; cited in Melcher et al., 2013). From this standpoint, to generate intentionally the efference that gives rise to, say, sniffing, one must activate the memory of the perceptual effects of that kind of action. These effects are experienced in the nose. Such perceptual-like activation then leads to the associated efference generation, automatically and unconsciously. Accordingly, in several models concerning perception-and-action (e.g. *cascade* models [McClelland, 1979], *continuous flow* models [Eriksen and Schultz, 1979]), an action-related stimulus will automatically activate the relevant motor programs (Figure 2).

1.2. The conscious field presents both selected and unselected action options

In recent theoretical approaches (e.g. Krisst, Montemayor and Morsella, 2015; Morsella et al., in press; Merker, 2013), the contents composing the conscious field (e.g. urges and action-effects) are construed as 'action options', both those which are selected to influence overt behaviour and those which are not selected. (These recent ideas are based in part on Gibson's, 1979, notion of affordances and Jamesian *functionalism* – see discussion in Morsella, Hoover and Bargh, 2013.) Thus, the conscious field reflects the selected action plan along with the unselected action plans. Because of this, the conscious field, with its often conflicting action-related contents, is very much unlike Morton's (1969) logogen, in which complete summation across diverse inputs yields only one output. In the logogen, only the selected output is represented after summation. Concerning unselected action options, such 'below threshold' activations normally do not result in perceptible, overt actions only because of the simultaneous activation of stronger, competing mental representations, such as those underlying goal-directed actions (Bhangal, Cho et al., 2016).

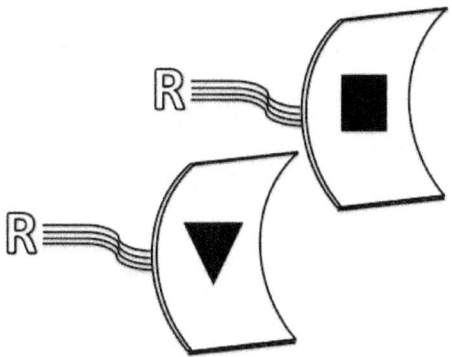

Figure 2. How perceptual-like, conscious contents in the conscious field can activate motor plans in the skeletal muscle output system, as described by ideomotor theory. In this case, each conscious content (filled shapes) in the conscious field can be associated with its own action response. Each conscious content occupies part of the conscious field (Figure 1), as represented by the concave planes on which the contents reside.

With such an architecture, at one moment in time, one could experience multiple action-related urges. Interestingly, one is unaware of the computational products that, should they exist (see Kaufman et al., 2015), determine the general course of observed action. That is, one is unconscious of the mechanism(s) that decide whether to express one behavioural inclination over another. Thus, in the experience of action-related, conscious conflicts, such as *approach-avoidance* conflicts (Miller, 1959) and other forms of intrapsychic conflicts, one is unconscious of the mechanisms by which these conflicts are resolved, that is, of the putative representations that, should they exist, determine the general course of observed action. In stages-of-processing terms, consciousness is associated with stages that, though clearly subsequent to those of sensory processing (Hochberg, 1998; Logothetis and Schall, 1989; Marcel, 1993) and preceding those of motor programming, may precede even that of action selection. (This view is consistent with Jackendoff's, 1990, proposal that consciousness reflects some form of intermediate, action planning stage in between sensory and motor processing.) The conscious field is about presenting the action options. It does not present the mechanisms by which these options are generated nor the mechanisms that determine which option is selected to influence overt behaviour.

In summary, low-level perceptual processes, motor control, and the processes responsible for action selection are usually not presented in the conscious field. Only the options for action selection, and percepts of the external world that serve to constrain that selection process, are represented therein. These options seem to be associated, not with all the effector systems in the body, but, specifically, with the skeletal muscle output system ('motorium', for short). Accordingly, the integrations/conflicts that occupy the conscious field do not concern integrations/conflicts associated with perceptual stages of processing, motor programming, or effector systems other than the motorium (Figure 3).

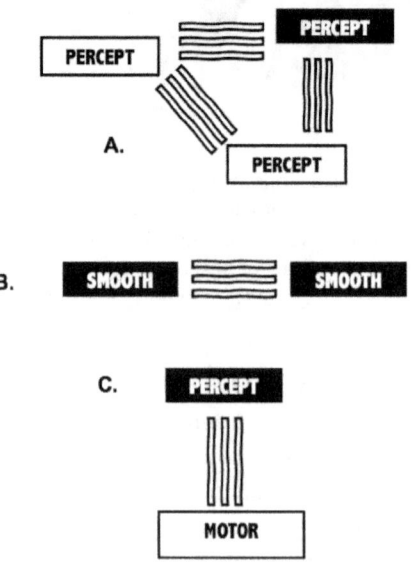

Figure 3. Many forms of integration can occur unconsciously in the nervous system. (A) Unconscious integrations between perceptual systems within a sensory modality (signified by the boxes sharing the same hue) and between different sensory modalities (signified by the boxes bearing distinct hues). Inter-sensory integrations include the ventriloquism effect. (B) Unconscious interactions in the control of smooth muscle effectors, as in the case of the consensual processes in the pupillary reflex. (C) Unconscious interactions between perceptual and motor processes, as in the case of reflexes and motor responses to subliminal stimuli.

1.3. Encapsulation

According to several frameworks (Firestone and Scholl, 2014; in press; Fodor, 1983; Krisst *et al.*, 2015; Pylyshyn, 1984), conscious contents arise from 'encapsulated' processes. These processes are *consciously impenetrable*, and hence cannot, at will, be influenced or introspected about. Perceptual processes giving rise to illusions are often said to be encapsulated: in these illusions, knowledge of the true nature of the perceptual stimuli (e.g. that the two lines of the Müller-Lyer illusion are equal in length) cannot 'turn off' or otherwise modulate the illusion. Similarly, when holding one's breath while underwater, or when running barefoot across the hot desert sand to reach water, the action-related urges arising in such circumstances are encapsulated: one cannot avoid the strong conscious inclinations to inhale or to avoid touching the hot ground, respectively (Morsella, 2005). For most conscious contents, the actor's beliefs or desires cannot modulate the nature of the content or turn off the content. This is the case even when such modulation would be adaptive (Morsella, 2005; Öhman and Mineka, 2001). Thus, we revisit the aforementioned conclusion that, although inclinations triggered by external stimuli are *behaviourally suppressible*, they are often not *mentally suppressible* (Bargh and Morsella, 2008; Morsella, 2005). This is consistent with past theorizing (e.g. James, 1890; N.E. Miller, 1959; Vygotsky, 1962; Wegner, 1989).

It has also been proposed that it is adaptive for content generation to be encapsulated in this manner (Firestone and Scholl, 2014; Merrick *et al.*, 2014; Rolls, Judge and Sanghera, 1977). For example, perceptual representations for instrumental action *should be* unaffected by the organism's beliefs or motivational states (Bindra, 1974; 1978). It would not be adaptive for contents pertaining to incentive/motivational states to be influenced directly by other contents, such as desires and beliefs (Baumeister *et al.*, 2007): simply put, if one's beliefs could lead one to voluntarily 'turn off' pain, guilt, or hunger, then these negative states would lose their adaptive value. In line with this view, Rolls *et al.* (1977) conclude concerning instrumental action, 'It would not be adaptive, for example, to become blind to the sight of food after we have eaten it to satiety' (p. 144). Hence, though beliefs, motivations, and biases may contaminate higher-order processes such as memory, they should have little influence over perceptual contents (Firestone and Scholl, 2014; Pylyshyn, 1984). Violation of this encapsulation rule would cause 'cross-contamination' across contents, thereby compromising the critical influence of incentive/motivational states on behaviour. In addition, for adaptive functioning, each conscious content should not 'know', in a sense, how it is relevant to the other contents composing the field or to ongoing action. Hence, when the conscious field portrays a contrast between a green banana and a yellow banana, for example, the content generators that engender this contrast do not know if the contrast is relevant to other perceptual contents or to urge-related contents (e.g. hunger). Similarly, contrasts provided by the conscious field might be (a) an object appearing to the left of another object, (b) something touching the arm versus touching the leg, and (c) an object differing in some way from an object held in memory.

Because of encapsulation, conscious contents, including percepts and urges, are most often generated in a reflexive manner, triggered by external stimuli or memory-based processes. Conscious contents seldom arise from elaborate chains of thought. Consistent with this view, in one experiment (Godwin, Morsella and Geisler, 2016), subjects performed a concentration exercise (9 min) requiring them to focus only on their breathing. Subjects were instructed to observe spontaneous thoughts and count the number of cognitions/percepts ('links') that they believed led to each spontaneous thought. In three studies using this paradigm, subjects reported less than two links per thought, over 80% of thoughts were attributed to a known cause, and, important for present purposes, roughly half of the thoughts were attributed to external stimuli.

1.4. Collective influence of encapsulated contents on the motorium

If content generation is encapsulated, then how do the processes responsible for different kinds of conscious content (e.g. a colour versus a smell) influence action collectively? It has been proposed that the primary function of the conscious field is to permit for otherwise encapsulated processes to influence skeletomotor action collectively (Morsella et al., 2016). From this standpoint, the collective influence of the medley of contents composing the conscious field is not toward the conscious field itself; instead, it is toward the (unconscious) mechanisms of, specifically, the skeletomotor output system, that is, our 'motorium', which is responsible for what has been regarded as 'voluntary' action. This form of action is a form of *integrated action* (defined below; Morsella and Bargh, 2011). When action is decoupled from the conscious field, the organism yields only *unintegrated actions* (Morsella and Bargh, 2011), such as unconsciously inhaling while underwater or reflexively removing the hand from a hot object. Unintegrated actions, reflecting a lack of integration, appear as if they are not influenced by all the kinds of information by which they should be influenced. In contrast, 'integrated action occurs when two (or more) action plans that could normally influence behavior on their own (when existing at that level of activation) are simultaneously co-activated and trying to influence the same skeletal muscle effector' (Morsella and Bargh, 2011, p. 241). Thus, integrated action occurs when one holds one's breath, suppresses the urge to scratch an itch, refrains from dropping a hot dish, or suppresses a pre-potent response in a laboratory paradigm. Integrated action occurs also when, for example, one makes oneself breathe faster for a reward (Morsella, 2005). Integrated action selection takes into account the 'votes' of the often conflicting component systems, as when one system wants to approach a stimulus and another system wants to avoid the stimulus. These votes can be construed as tendencies based on inborn or learned knowledge.

Through collective influence, complex behaviours can arise. For example, when starved and faced with food that is trapped under a heavy slab of ice, it could be proposed that, in terms of instrumental action, one 'wants' to lift the heavy ice. Such a desire has been referred to as an 'instrumental want' (Morsella, 2005). In contrast, the intuitive interpretation that one simply 'wants food' in such a situation overlooks the fact that an instrumental intention is essential to remove the obstacle and obtain the food. From this standpoint, in contrast to traditional operant conditioning approaches (Skinner, 1953), performing an instrumental act for an incentive simultaneously leads to different kinds of learning (instrumental versus incentive), different kinds of learning which occur in parallel (Bindra, 1974; Öhman and Mineka, 2001; Tolman, 1948).

The present theorizing reveals also that the conscious contents of blue, a smell, or the urge to scratch the skin are the tokens of a mysterious language understood by the unconscious action mechanisms of the motorium. From this standpoint, content generators cannot respond to the contents in the field; instead, they report the contents to a different kind of mechanism: the unconscious response systems of the motorium. This is a form of 'one-way' communication. For adaptive behaviour, the response to one conscious content—and the 'meaning' of that content for ongoing action selection—depends exclusively on the nature of the other conscious contents at that moment in time. This is why, again, it has been proposed (e.g. Morsella *et al.*, 2016) that the primary function of the conscious field is to permit for the response to one content to be made in light of the other contents (see also Tsushima, Sasaki and Watanabe, 2006). For integrated behaviour, action selection should take into account all contents. This is precisely what the conscious field affords (e.g. responding to X in light of Y). With this in mind, it is important to consider that, to this end, the point of the conscious field is to distinguish one token from all items occupying the field *at that moment*. Hence, action selection must always be framed by information about the current environment, as last-minute changes to courses of action often arise in the face of new data about the current environment (Merker, 2013). Hence, the present context is of utmost importance. Context can change the response to a given token. For example, the presence of a bear is scary, unless, however, the bear is behind a pane of glass (e.g. in a zoo environment). Thus, the response to one conscious content varies depending on the other elements composing the field.

Within this functional architecture, things appear the way they do because, in order to benefit action selection, each content must differentiate itself from the other contents of the same modality (e.g. within vision) and as well as from contents of other modalities. Hence, one token (e.g. the colour blue) must be different in nature from tokens of its own kind as well as from tokens of an entirely different kind (e.g. thirst). For the conscious field, the range of possible discriminations, many of which are learned, is set, in a sense, by evolutionary processes. If one is not aware of a certain perceptual contrast, then voluntary behaviour cannot reflect that perceptual discrimination. Hence, the important thing about blue is that, in the field, it leads to behaviours that are different from those toward, say, red objects.

In addition, the properties of a conscious content must reflect, not only contrast between tokens (e.g. a smell versus a colour), but also what is required for that content to be, at some level, comparable to the other contents composing the field: for adaptive action selection, all contents must exist in the same decision space as comparable tokens. Apart from this commonality, tokens need not share all features. For example, whereas spatial coordinates are essential for the adaptive responding to external objects detected by vision, such coordinates are not as critical for the adaptive responses to, for example, urges associated with the viscera. In the conscious field, the representation of an action-related urge, though perceived as different from the self (Brentano, 1874; Merker, 2012; Schopenhauer, 1818/1819), need usually be only of nothing more than the urge itself, with little information provided about, for instance, the spatial reference between the urge and the self. Such information is not that critical for adaptive action selection based on visceral urges. Hence, 'wanting X' can often be, and nothing more than, just 'wanting X'. For example, it is not about how the wanting of X is perceived from one or another vantage point. From a functionalist standpoint, for adaptive action selection, an urge does not usually require any consciously experienced properties beyond the urge itself.

In addition, some contents (e.g. percepts) appear, with respect to the degree to which they spur action, to be more passive than others (e.g. urges). Nevertheless, despite these differences, urges and percepts are functionally similar in that they each constrain action selection. As a conscious content, the urge serves as what the behaviourists referred to as a discriminative stimulus, one that influences action selection. (It is important to differentiate conscious representation of the urge from the actual, action-related mechanisms that influence overt behaviour; Desmurget *et al.*, 2009; Desmurget and Sirigu, 2010.)

In the proposed cognitive architecture, the contents composing the field determine the action selection process *wholly* and *exclusively*. Thus, it is not that conscious contents constrain voluntary action; it is that they determine it fully. Hence, the conscious field functions as a unitary entity in terms of its influence over the motorium. This insight renders moot the debate concerning whether the conscious field should be construed as a unitary or componential entity (*cf.* Searle, 2000): in terms of functional consequences, action-related modules in the motorium (which are unconscious) must treat the mosaic of contents in the field as one thing. From this, one arrives at the conclusion that, if a content is not held in the constellation, by memory or by external stimuli, then that content cannot influence voluntary action in any way. In other words, differentiations not presented in the conscious field cannot yield voluntary actions that reflect such discriminations.

Regarding this proposed architecture, it is important to appreciate that *action selection is always ongoing, as there is no such thing as 'doing nothing'*. ('Doing nothing' is a colloquial and unscientific expression. See similar conclusion in Skinner, 1953.) Hence, for every configuration of conscious contents—whether complete or incomplete, adaptive or maladaptive—there will always be an act of action selection. In addition, when the same configuration of conscious contents is in the conscious field, then the same voluntary action will always be selected (all things being equal regarding unconscious processes). Hence, it could be concluded that the often heard phrase, 'Knowing what I know now, if I could go back in time and do X again, I would certainly do things differently' is true only if new knowledge (e.g. from memory) is permitted to participate in the original conscious field—the field that led to the initial, regretted action. If one actually went back in time into the very exact situation, and experienced the same conscious field as before, then one would have no choice but to repeat the regretted action.

1.5. The nuts and bolts underlying the conscious field

The actor experiencing the conscious field believes that he or she is interacting with an external world, full of colours, smells, and sounds. Because the purpose of conscious contents is to appear as something other than what they actually are (creations from neural activities), the contents do not convey any aspects of their true, underlying neural hardware. One perceives only an apple or a sunset (Brentano, 1874), not the neural activities that give rise to such percepts. Unlike a computer screen, in which, with some effort, one can detect the pixels that form the visual images, the conscious field does not convey information about the grain of the primitive elements composing each content, should such elements exist. Although such information about grain might benefit the scientist in the laboratory, it does not at all benefit action selection in natural environments.

In conclusion, the priority of the conscious content, a true chameleon, is to be completely different from what it is, which renders it mysterious. Thus, how neurons engender the conscious field remains a mystery. How do they (or anything else, for that matter) give rise to any kind of conscious experience? It is important to appreciate that our scientific models of natural phenomena are often limited. For example, a small-scale classroom model of the solar system could not instantiate a critical phenomenon (strong gravitational fields) which is what glues our solar system together. Analogously, our human experience and way of thinking occurs at a timescale far different from that in which consciousness arises (fast neurons interacting with each other). Simply put, there is nothing that we can perceive in everyday life that works as fast, and in such an interactive manner, as the functional nuts and bolts of our brain. What we observe and understand, such as the 'cause and effect' relationships among billiard balls striking each other, occurs at a much slower scale than the workings of neurons. Thus, regarding speed, our conscious experience is far slower than that which gives rise to it. It could be that something arising from the fast speeds of neurons and their networks cannot happen at the slower speed of our everyday existence, the only environment that we evolved to understand. Perhaps our assumptions about how things work (and cannot work) at the 'human scale' of existence may not be applicable to smaller, faster scales.

1.6. Bottlenecks stemming from the requisites of collective influence

Through collective influence, the conscious field permits for a massive many-to-one (or, at least, many-to-few) conversion (Merker, 2013). At one moment in time, the field possesses many action-related contents, though overt behaviour reveals only one 'operant' (e.g. pressing only one of two buttons). Such an outcome may reflect the mechanical limitations of the skeletomotor system, in which only one word can be uttered at a time (Wundt, 1900). For action to be adaptive, the one, resulting operant must be 'integrated'. Such an outcome requires (*A*) that the response systems of the motorium sample all conscious contents at one moment in time (no *access specificity* though there is *processing specificity*; Barrett, 2005), and (*B*) that overt behavior be influenced by the level of activation of each of the response systems that was activated by the contents composing the field. *A* by itself would not necessarily create a bottleneck, for the field could in principle be sampled by systems of the motorium that are working in parallel, with each system sampling at different times. However, the combination of *A* and *B*, with *B* requiring summation of response activation levels within a certain time window, certainly might introduce a bottleneck in which processing can transpire only sequentially. This potential bottleneck, associated with the stage-of-processing of action selection, might have been identified independently by several researchers. For example, Pashler (1993, p. 52) concludes:

> The limitations in carrying out stimulus-response tasks concurrently are not introduced at the level of stimulus perception, nor in production of the motor response. Those mental operations can work in parallel. Rather, the problem is in deciding what the response will be, and this kind of mental operation seems to be carried out in series — that is, one task at a time.

Regarding bottlenecks, it is important to note also that the incompatibility between action plans during intrapsychic conflicts (e.g. approach-approach conflicts) resides, not in the conscious field, in which both action options could be represented simultaneously, but rather in the simultaneous execution of the two action plans. It is a fact of everyday experience that multiple action options are presented in consciousness.

Regarding the perceptual-end of processing in the perception-to-action arc, one bottleneck regarding the latter is the following. It is difficult for the same content generator to produce two different contents at the same time, such as would occur if one attempted to imagine two words being spoken at the same time.

2. The Reflexive Imagery Task

Many of the phenomena discussed above, including the unconscious generation of conscious contents and enapsulation, have been investigated in a new experimental paradigm, the Reflexive Imagery Task (RIT; see review in Bhangal, Cho *et al.*, 2016). The task, based on the notion that action inclinations often are mentally insuppressible and on the experimental approaches of Ach (1905), Wegner (1989), and Gollwitzer (1999), was developed to investigate the unconscious generation of high-level, involuntary conscious contents. In the most basic version of the task (Allen *et al.*, 2013), subjects are instructed to not subvocalize (i.e. say in their head but not aloud) the names of objects (e.g. line drawings from Snodgrass and Vanderwart, 1980; Figure 4) that are presented on a computer screen. For example, in Allen et al. (2013), subjects were presented before each trial with the instruction, 'Don't Think of the Name of the Object'. During the trial, an object appeared for 4 s, during which time subjects indicated by button press if they happened to subvocalize the name of the object. Subjects fail to suppress such subvocalizations on the majority of the trials (86% in Allen et al., 2013; 87% in Cho et al., 2014; and 73% in Merrick et al., 2015). This effect requires the sophisticated process of object naming, in which only one of tens of thousands of phonological representations is selected for production in response to a visual stimulus (e.g. CAT yields /k/, /æ/, and /t/; Levelt, 1989). The effect usually arises only moments (~2 s) after the visual stimulus appears (M = 1,451.27 ms [SD = 611.42] in Allen et al., 2013; M = 2,323.91 ms [SD = 1,183.01] in Cho et al., 2014; M = 1,745.97 ms [SD = 620.86] in Merrick et al., 2015). In more complex versions of the RIT (Merrick et al., 2015), subjects were instructed to (a) not subvocalize the name of the visual stimulus, and (b) not subvocalize the number of letters in the object name. On a significant proportion of trials (.30 [SE = .04]), subjects reported experiencing both kinds of imagery. Each thought arises from distinct, high-level processes (i.e. that of object naming versus object counting).

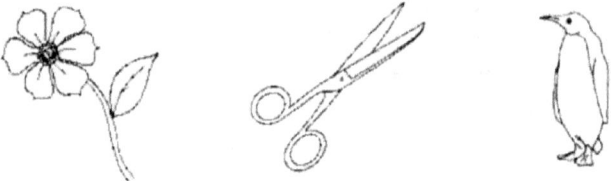

Figure 4. Sample visual stimuli used in the Reflexive Imagery Task. (Not drawn to scale.)

Empirical evidence and theory (including Wegner's, 1994, model of *ironic processing*) suggest that, for subjects, entry of the undesired content 'just happens' and is not simply an artefact of high-level strategic processes. Accordingly, in one version of the RIT, subjects reported on the majority of trials that the involuntary subvocalization felt 'immediate' (Bhangal, Merrick and Morsella, 2015). The view that the effect is automatic is further supported by several observations. First, on some trials, the effect arises too quickly to be caused by strategic processing (Allen et al., 2013; Cho et al., 2014). Second, the effect still arises under conditions of cognitive load, in which it is difficult for subjects to implement strategic processing (Cho *et al.*, 2014). For example, in an experiment by Cho *et al.* (2014), in which the RIT effect still arose even though subjects were instructed to subvocalize a hum through the entirety of the trial, it would have been difficult for participants to carry out strategic processing while experiencing the cognitive load from such a dual-task condition. Third, the effect is unlikely to be due to demand characteristics because the nature of the subvocalizations is influenced systematically by factors such as word frequency (Bhangal *et al.*, 2015). Such an artefact of experimental demand would require for subjects to know how word frequency should influence responses in an experiment. Fourth, the RIT effect habituates (i.e. is less likely to arise) after repeated presentation of the same stimulus object, which suggests that the RIT effect is activated involuntarily and in a reflex-like manner (Bhangal, Allen *et al.*, 2016). Importantly, as in the case of the habituation of a reflex, the habituation effect was stimulus-specific.

Entry in the RIT appears to rely on several component processes, many of which are encapsulated (Allen *et al.*, 2013). First, it seems that the relevant action set must be activated. This activation stems from the instructions provided by the experimenter (*ibid.*). It is unlikely that, without such an activation of set, subjects would experience the phonological representations of the names of the objects that happen to be perceived visually. The action set is held in memory, at least from the beginning of the trial until the onset of the visual object. During this time, the action set held in mind can be regarded as *imageless thought*, because it influences behavioural dispositions without being maintained explicitly in consciousness (*cf.* Woodworth, 1915; see recent, relevant research in Scullin, McDaniel and Einstein, 2010). (Imageless thought was first investigated by theorists of the Würzburg School of Psychology; Schultz and Schultz, 1996.) The final process occurs when the onset of the visual object begins the stages of processing that, somehow, leads to the entry of involuntary imagery (e.g. subvocalization of the object name).

Wegner (1994) proposes that the kind of unintentional, 'ironic' effect in the RIT arises from an interaction between two distinct processes. One process is an *operating* process, which is associated with the conscious intention to maintain a particular mental state. This process actively scans mental contents (e.g. thoughts, sensations) that can help maintain the desired mental state (e.g. to be calm). This process tends to be effortful, capacity-limited, and consciously mediated (Wegner, 1994). The other mechanism is an 'ironic' *monitoring* process that automatically scans activated mental contents to detect contents signalling the failure to establish the desired mental state. When the monitor detects contents that signify failed control of the operating mechanism, it increases the likelihood that the particular content will enter consciousness, so that the operating mechanism then processes the content and changes its own operations accordingly. Pertinent to the RIT, the ironic monitor mechanism is usually unconscious, autonomous, and requires little mental effort. In most cases of cognitive control, the two mechanisms work together harmoniously. However, harmony fails when the goal in mental control is to not activate a particular mental content (e.g. content X), because (a) the operating process can bring only goal-related contents into consciousness and cannot actively exclude contents, and (b) the ironic monitor will reflexively bring into consciousness mental contents (e.g. content X) that are incongruent with the goal. Together, the interaction of the two mechanisms—intentional *operating* process and the ironic *monitoring* process—will lead to the automatic activation of content X in consciousness. (For treatment of the neural correlates of the two mechanisms, see Mitchell *et al.*, 2007; Wyland *et al.*, 2003. For reviews of ironic processing and thought suppression, see Rassin, 2005; Wegner, 1989.)

2.1. Field construction and the RIT effect

Intuitively, it seems that conscious states spanning for a lengthy period of time (e.g. pain) are like a switch that, for a span, has been set in the 'on' position. However, a more accurate depiction of these states is that the relevant content (e.g. pain) is actually generated anew each time the conscious field is refreshed as a whole. For each time the field is constructed by unconscious processes, a decision of sorts must be made regarding whether this content should form part of it. The refresh rate of the conscious field must be slow enough to present the outputs from various modalities and sources (e.g. urge generators) that are important for action selection. These sources may vary with respect to their speed of processing (e.g. haptic processing is faster than visual processing). Hence, field construction must wait, in a sense, for the slowest system. Thus, it is no surprise that, in terms of stages of processing, consciousness occurs late (J.A. Gray, 2004; Libet, 2004). Accordingly, it has been proposed that the well-known lateness of consciousness in processing stems from the fact that the conscious field must integrate information, which is necessary for one system to 'veto' another, from neural sources having different processing speeds (Libet, 2004). The classic psychophysiological research by Libet (1989; 2004) revealed that consciousness occurs relatively late (e.g. over 300 ms) in the processing stream (Soon *et al.*, 2008). Recent research (Pitts and Britz, 2011) corroborates that conscious content arise late (~200–300 ms) in the processing stream. (This neural activity is proposed to involve the inferior occipital-temporal cortex; *ibid.*) At the same time, the construction of the field must be quick enough to influence action selection for fast effector systems, such as those for voluntary eye movements.

With this cognitive architecture in mind, it could be argued that, in the RIT effect, set-related activations have their influence on the conscious field, and thereby furnish their unintentional content (i.e. the subvocalization), during the refreshing of the field. Importantly, the involuntariness of the conscious contents elicited in the RIT reflects, not the exception, but the rule regarding the generation of high-level, conscious contents. (It is obvious that the vast majority of perceptual contents and urges are activated automatically.) Involuntary entry of high-level contents into consciousness is evident in the case of earworms (e.g. a tune playing repeatedly in one's head), in binocular rivalry, and with rivalrous images, for which there are involuntary perceptual 'reversals' (see Montemayor, Allen and Morsella, 2013). (In binocular rivalry [Logothetis and Schall, 1989], an observer is presented with different visual stimuli to each eye—e.g. an image of a house in one eye and of a face in the other. However, an observer experiences seeing only one object at time—a house and then a face—even though both images are presented concurrently; see review in Alais and Blake, 2005.)

2.2. Benefits of using the RIT to investigate the interdependence between conscious and unconscious processes

The RIT has revealed that most people, when presented with certain instructions (inducing a certain action set) and then being presented with a stimulus, cannot suppress the conscious experience of the prospective action-effect (e.g. the phonological form of the word 'triangle'). In the basic effect, the stimulus triggers conscious content that is very different in nature from that of the environmental stimulation that brought the content into existence: the conscious representation is associated, not with vision, but with audition (the auditory-based, phonological representation is a 'late-stage' process in naming; Levelt, 1989). Word reading also reveals how high-level conscious contents can arise from automatic, stimulus-driven processes, which resemble the unconscious inferences in visual perception (Helmholtz, 1856/1961). Thus, the RIT is a paradigm that allows the experimenter to activate, at a specific moment in time and through external stimuli and involuntary mechanisms, entry into consciousness of 'high-level' contents, high-level contents which stem from the interaction between conscious and unconscious processes.

When comparing operations that transpire consciously and intentionally to those that transpire involuntarily (e.g. as in the case of an undesired, automatic process), it is important to consider the amount of time that each kind of operation requires. For example, in a variant of the RIT involving the involuntary counting of the number of letters composing each stimulus word (e.g. CAR = 'three'), Merrick *et al.* (2015) observed that involuntary counting was more likely to arise for short words (e.g. CAR) than for long words (e.g. BALLOON). One difference between the counting of the letters of short words and of long words is that, because the latter usually requires a longer period of time, the involuntary counting of short words by unconscious brain mechanisms is more likely to suffer interference from, and be interrupted by, other ongoing processes that, too, require that mechanism. These interfering, ongoing processes could be top-down (e.g. executive control) or bottom-up (e.g. a salient stimulus).

Additional advantages of using the RIT are that (a) the involuntary nature of the RIT effect renders it similar to a reflex while also diminishing the likelihood of artefacts from strategic processing, idiographic processing, social desirability, and demand characteristics, (b) the effect is reliable and easily replicable, requiring very few experimental subjects and very few trials, (c) one can systematically vary stimulus parameters (e.g. word frequency, number of letters, number of stimulus objects, word valence) and instruction or 'set' parameters, (d) the dependent measure (e.g. object naming) is well studied and has well examined properties (e.g. the phonological form; G. Miller, 1996), (e) one can measure on a trial-by-trial basis several aspects of the response, including latency, neural aspects (e.g. as in neuroimaging), and the sense of agency about the involuntary cognitions (Allen *et al.*, 2013), (f) the RIT effect is predicted by several frameworks (e.g. by Gollwitzer, 1999; N.E. Miller, 1959; Morsella, 2005; Wegner, 1989) and complements previous research (e.g. Gaskell, Wells and Calam, 2001; Wegner *et al.*, 1987; Smári, 2001), and (g) the RIT illuminates contrasts between the capacities of conscious and unconscious processing, as discussed in the next section. Last, (h) it is worth emphasizing that the RIT provides the kind of incremental research that, building on prior research (e.g. Allen *et al.*, 2013; Wegner, 1989) and involving a phenomenon that is robust, multifaceted, and reliable, is important for progress in the field of psychological science (Nosek, Spies and Motyl, 2012).

2.3. Neural correlates of the RIT effect

Investigations on the neural correlates of ironic processing, cognitive control, and phonological processing suggest that, in the basic RIT effect, there may be the involvement of at least three distinct neural systems: those associated with (1) the action set to not subvocalize the name of the object (e.g. the prefrontal cortex; Mitchell *et al.*, 2007), (2) the detection of a discrepancy between desired performance and the ironic effect (e.g. the anterior cingulate cortex; see *ibid.*; Wyland *et al.*, 2003), and (3) the phonological representation of the object name.

Regarding 3, it is important to emphasize that controversy continues to surround the identification of the neural correlates of the phonological representations that are activated by heard, spoken speech (e.g. Hickok, 2009; Schomers *et al.*, 2014). Thus, strong claims cannot yet be made regarding the neural correlates of subvocalized speech (see discussion in Buchsbaum, 2013; Buchsbaum and D'Esposito, 2008). Nevertheless, at this stage of understanding, it seems that the neural correlates of phonological representations involve the left superior temporal cortex (including the superior temporal gyrus and sulcus) and a medley of other regions (supramarginal gyrus, inferior frontal gyrus, and precentral gyrus; DeWitt and Rauschecker, 2012; Eggert and Wernicke, 1874/1977; Gazzaniga, Ivry and Mangun, 2014; Peramunage *et al.*, 2011). Buchsbaum (2013) concludes that subvocalized speech is often associated with activations in both motor-related regions in frontal cortex, such as the inferior frontal gyrus (for phonological planning) and the precentral gyrus (for motor programming), and in perception-related regions that are associated with speech perception (e.g. superior temporal sulcus). Accordingly, Scott (2013) presents evidence that, during subvocalization, corollary discharge provides the conscious sensory content of one's inner speech (Ford *et al.*, 2005). In the research by Ford *et al.* (2005), mismatches involving one's intended speech and what one actually hears oneself say are associated with decreased functional synchrony (a kind of communication) between frontal and temporal lobes.

It remains controversial whether subvocalized speech requires the activation of motor-related regions or whether subvocalized speech and other forms of auditory imagery can arise without these activations (Hickok, 2009; Schomers *et al.*, 2014; see discussion in Buchsbaum and D'Esposito, 2008, and in Mahon and Caramazza, 2008). At present, there is no clear evidence that lesions to motor areas associated with speech production eradicate the capacity for subvocalizing or for other kinds of verbal imagery on the part of the patient (*cf.* Gruber, Gruber and Falkai, 2005; Müller and Knight, 2006; Sato *et al.*, 2004; Vallar, Corno and Basso, 1992). For some evidence of a necessary, causal role of motor areas in speech perception, see Schomers *et al.* (2014).

2.4. The passivity of consciousness

The foregoing theorizing reveals that the role of consciousness in nervous function is more low-level, circumscribed, counter-intuitive, and passive than what has been proposed previously. Consciousness is less purposeful at one moment in time than what intuition suggests: contents do not 'know', in a sense, their relevance to other contents nor to ongoing action. According to Passive Frame Theory, the conscious field contributes only one function (albeit an essential function) to a wide range of processes, much as how the human eye, though involved in various processes, always performs the same function. According to Passive Frame Theory, the conscious field itself has no memory and performs no symbol manipulation; for these high-level mechanisms, it only 'presents', in a sense, the outputs of dedicated memory systems and of executive processes to the motorium. Memorial processes, for example, stem from dedicated systems operating outside the conscious field. Because consciousness contributes to a wide range of processes, it appears to be doing more than it actually does. In short, we confuse what consciousness is for with what we, as humans, happen to use it for in our adult life.

In Morsella et al. (in press), the conscious field has what can be regarded as a fixed architecture with few 'moving parts', even though it is occupied across time by varied configurations of contents. Much of the intelligence of this system is in what can be construed as the building of the field and not in the dynamics that operate between the elements of the field. For example, great intelligence is demonstrated when poisons (e.g. hydrogen peroxide [H_2O_2]) happen to taste *bad* but when water (H_2O), which is chemically similar but not poisonous, happens to taste *good*. An intelligent, behavioural bias toward these chemicals is already 'built into' the system.

2.5. *The conscious field does not require a 'decider'*

In the present theorizing and in ideomotor theory, conscious contents automatically and directly activate the motor plans that enact the actions with which they are associated. At first glance, this theorizing seems to be on the wrong track. In everyday life, it does not seem to be the case that one acts out all the actions that are associated with external stimuli. In response to this criticism, it is important to appreciate that, according to James, the mere thoughts of potential action effects, which are often triggered by external stimuli, produce impulses that, if not curbed or controlled by 'acts of express fiat' (i.e. exercise of veto), result in the performance of the associated actions (James, 1890, pp. 520–4). The 'acts of express fiat' refer, not to an homunculus reining in action, but to the activations of another, competing action plan). For example, according to Lotze, two conditions must be fulfilled in order to carry out a voluntary action:

First, there must be an idea or mental image of what is being willed (*Vorstellung des Gewollten*). Second, all conflicting ideas or images must be absent or removed (*Hinwegräumung aller Hemmungen*). When these two conditions are met, the mental image acquires the power to guide the movements required to realize the intention, thus converting ideas in the mind into facts in the world. (Prinz, Aschersleben and Koch, 2009, p. 38)

From this standpoint, in everyday life one does not enact all of the actions triggered by external stimuli because such actions are prevented, not by an homunculus, but rather by the simultaneous activations of incompatible action plans. Hence, there is no homunculus needed to select one action plan over another: at one moment in time, Action Plan A will compete with Action Plan B; at another moment in time, Action Plan C will compete with Action Plan D, without the need of an omnipresent 'decider'.

In conscious states, there is often the sense of some unitary observer apprehending the field. This sense of the observer arises also in the dream world and during illusions resulting from experimental manipulations (e.g. Ehrsson, 2007). It has been proposed that, for adaptive action selection, the conscious field should appear as if it were viewed from such a single, first-person perspective. Such a perspective allows for the process of action selection to take into account spatial considerations. For example, which of two objects is reached for might depend on the distance between each object and the organism. As well, an object on the left must not be treated the same as an object on the right. Hence, the requirements of adaptive action selection force the field to be designed in a first-person perspective (Merker, 2013). This perspectival requirement for adaptive action also illuminates why representations of external objects in perceptual consciousness (while dreaming or awake) fail to represent all of the three-dimensional information that may be stored about such objects (J. Prinz, 2007). For instance, when foraging, one cannot perceive all sides of a fruit, although one may infer the existence of the hidden parts of the object.

That this sense of self is malleable is evident in the following. When subvocalizing words about the objects of the current environment in one language (e.g. English), one has the strong sense that the 'observer' of the conscious field is an entity that operates in that language (e.g. English). Likewise, when one is subvocalizing about such things in a different language (e.g. Spanish), one then has the strong sense that the observer is a Spanish agent. This reveals that there are many observers, in a sense. The English and Spanish thinkers are only two such observers. As with experimental manipulations, this sense of self is generated online, arising from the peculiarities of the current context (e.g. an English or Spanish task set).

Rather than having a single, 'apprehending homunculus', the conscious field is sampled by the various (unconscious) response systems composing the motorium. Each system has no *access specificity* but does have *processing specificity* (Barrett, 2005). The arm might sample and then react to contents in a manner different from that of, say, the eye. However, for action to be integrated behaviour at one moment in time, the sampling of information by the various systems must be coordinated in some way. As mentioned above, such coordination may introduce functional bottlenecks. From this, one appreciates the passive but essential role of consciousness is not a simple translation function between perception-and-action but a 'many to one' function, in which many conscious contents, from various sources, yield one integrated operant at one moment in time.

3. Conclusion

In this chapter, we reviewed how the unconscious components of voluntary action interact with conscious processes (e.g. percepts and urges). Although essential for adaptive action, which results from the collective influence of conscious contents over the motorium, conscious processes serve a passive and well circumscribed role. Consciousness has few moving parts. The 'active' processes of *content generation* and *responding to contents* arise outside of the conscious field. Together, Passive Frame Theory and the Reflexive Imagery Task shed light on the interdependence between (passive) conscious and (active) unconscious processes. Passive Frame Theory specifies which stages of processing, from stimulus input to behavioural output, can transpire unconsciously and which require involvement of the conscious field. The findings from the RIT reveal that high-level conscious contents can enter consciousness in a reflex-like, automatic, and insuppressible manner. Often, such activations stem from external stimuli triggering encapsulated processes that then, behind the scenes, yield conscious contents. Subsequently, action selection transpires without a 'decider' choosing one action plan over another. This manner of operation counters our intuitions about the generation of conscious thoughts and about the nature of action selection. One intuits that conscious processes, which constitute the totality of our existence, must be more 'active' than their hidden, unconscious counterparts and that, during action selection, there must be an homuncular-like 'decider' choosing one action plan over another.

References

Ach, N. (1905/1951) Determining tendencies: Awareness, in Rapaport, D. (ed.) *Organization and Pathology of Thought*, New York: Columbia University Press.

Alais, D. & Blake, R. (2005) *Binocular Rivalry*, Cambridge, MA: MIT Press.

Allen, A.K., Wilkins, K., Gazzaley, A. & Morsella, E. (2013) Conscious thoughts from reflex-like processes: A new experimental paradigm for consciousness research, *Consciousness and Cognition*, **22**, pp. 1318–1331.

Baker, T.B., Piper, M.E., McCarthy, D.E., Majeskie, M.R. & Fiore, M.C. (2004) Addiction motivation reformulated: An affective processing model of negative reinforcement, *Psychological Review*, **111**, pp. 33–51.

Bargh, J.A. & Morsella, E. (2008) The unconscious mind, *Perspectives on Psychological Science*, **3**, pp. 73–79.

Barrett, H.C. (2005) Enzymatic computation and cognitive modularity, *Mind & Language*, **20**, pp. 259–287.

Bartolomei, F., Wendling, F., Vignal, J.P., Chauvel, P. & Liegeois-Chauvel, C. (2002) Neural networks underlying epileptic humming, *Epilepsia*, **43**, pp. 1001–1012.

Baumeister, R.F., Vohs, K.D., DeWall, N. & Zhang, L. (2007) How emotion shapes behavior: Feedback, anticipation, and reflection, rather than direct causation, *Personality and Social Psychology Review*, **11**, pp. 167–203.

Berger, C.C., Bargh, J.A. & Morsella, E. (2012) The 'what' of doing: Introspection-based evidence for James's ideomotor principle, in Durante, A. & Mammoliti, A. (eds.) *Psychology of Self-Control*, New York: Nova.

Bhangal, S., Merrick, C. & Morsella, E. (2015) Ironic effects as reflexive responses: Evidence from word frequency effects on involuntary subvocalizations, *Acta Psychologica*, **159**, pp. 33–40.

Bhangal, S., Allen, A.K., Geisler, M.W. & Morsella, E. (2016) Conscious contents as reflexive processes: Evidence from the habituation of high-level cognitions, *Consciousness and Cognition*, **41**, pp. 177–188.

Bhangal, S., Cho, H., Geisler, M.W. & Morsella, E. (2016) The prospective nature of voluntary action: Insights from the reflexive imagery task, *Review of General Psychology*, **20**, pp. 101–117.

Bindra, D. (1974) A motivational view of learning, performance, and behavior modification, *Psychological Review*, **81**, pp. 199–213.

Bindra, D. (1978) How adaptive behavior is produced: A perceptual-motivational alternative to response-reinforcement, *Behavioral and Brain Sciences*, **1**, pp. 41–91.

Blanken, G., Wallesch, C.W. & Papagno, C. (1990) Dissociations of language functions in aphasics with speech automatisms (recurring utterances), *Cortex*, **26**, pp. 41–63.

Block, N. (1995) On a confusion about a function of consciousness, *Behavioral and Brain Sciences*, **18**, 227.

Brentano, F. (1874) *Psychology from an Empirical Standpoint*, Oxford: Oxford University Press.

Buchsbaum, B.R. (2013) The role of consciousness in the phonological loop: Hidden in plain sight, *Frontiers in Psychology*, **4**.

Buchsbaum, B.R. & D'Esposito, M. (2008) The search for the phonological store: From loop to convolution, *Journal of Cognitive Neuroscience*, **20**, pp. 762–778.

Carmant, L., Riviello, J.J., Thiele, E.A., Kramer, U., et al. (1994) Compulsory spitting: An usual manifestation of temporal lobe epilepsy, *Journal of Epilepsy*, **7**, pp. 167–170.

Chambon, V., Wenke, D., Fleming, S.M., Prinz, W. & Haggard, P. (2013) An online neural substrate for sense of agency, *Cerebral Cortex*, **23**, pp. 1031–1037.

Cho, H., Godwin, C.A., Geisler, M.W. & Morsella, E. (2014) Internally generated conscious contents: Interactions between sustained mental imagery and involuntary subvocalizations, *Frontiers in Psychology*, **5**, 1445.

Christensen, M.S., Lundbye-Jensen, J., Geertsen, S.S., Petersen, T.H., Paulson, O.B. & Nielsen, J.B. (2007) Premotor cortex modulates somatosensory cortex during voluntary movements without proprioceptive feedback, *Nature Neuroscience*, **10**, pp. 17–419.

Damasio, A.R., Tranel, D. & Damasio, H.C. (1991) Somatic markers and the guidance of behavior: Theory and preliminary testing, in Levin, H.S., et al. (eds.) *Frontal Lobe Function and Dysfunction*, London: Oxford University Press.

Desmurget, M., Reilly, K.T., Richard, N., Szathmari, A., Mottolese, C. & Sirigu, A. (2009) Movement intention after parietal cortex stimulation in humans, *Science*, **324** (5928), pp. 811–813.

Desmurget, M. & Sirigu, A. (2010) A parietal-premotor network for movement intention and motor awareness, *Trends in Cognitive Sciences*, **13**, pp. 411–419.

DeWitt, I. & Rauschecker, J.P. (2012) Phoneme and word recognition in the auditory ventral stream, *Proceedings of the National Academy of Sciences*, **109**, pp. 505–514.

Doherty, M.J., Wilensky, A.J., Holmes, M.D., Lewis, D.H., Rae, J. & Cohn, G.H. (2002) Singing seizures, *Neurology*, **59**, pp. 1435–1438.

Eggert, G.H. & Wernicke, C. (1874/1977) *Wernicke's Works on Aphasia: A Sourcebook and Review*, The Hague: Mouton.

Ehrsson, H.H. (2007) The experimental induction of out-of-body experiences, *Science*, **317** (5841), p. 1048.

Enatsu, R., Hantus, S., Gonzalez-Martinez, J. & So, N. (2011) Ictal singing due to left frontal lobe epilepsy: A case report and review of the literature, *Epilepsy & Behavior*, **22**, pp. 404–406.

Eriksen, C.W. & Schultz, D.W. (1979) Information processing in visual search: A continuous flow conception and experimental results, *Perception and Psychophysics*, **25**, pp. 249–263.

Fecteau, J.H., Chua, R., Franks, I. & Enns, J.T. (2001) Visual awareness and the online modification of action, *Canadian Journal of Experimental Psychology*, **55**, pp. 104–110.

Firestone, C. & Scholl, B.J. (2014) 'Top-down' effects where none should be found: The El Greco fallacy in perception research, *Psychological Science*, **25**, pp. 38–46.

Firestone, C. & Scholl, B.J. (in press) Cognition does not affect perception: Evaluating the evidence for 'top-down' effects, *Behavioral and Brain Sciences*.

Fodor, J.A. (1983) *Modularity of Mind: An Essay on Faculty Psychology*, Cambridge, MA: MIT Press.

Ford, J.M., Gray, M., Faustman, W.O., Heinks, T.H. & Mathalon, D.H. (2005) Reduced gamma-band coherence to distorted feedback during speech when what you say is not what you hear, *International Journal of Psychophysiology*, **57**, pp. 143–150.

Fourneret, P. & Jeannerod, M. (1998) Limited conscious monitoring of motor performance in normal subjects, *Neuropsychologia*, **36**, pp. 1133–1140.

Freeman, W.J. (2004) William James on consciousness, revisited, *Chaos and Complexity Letters*, **1**, pp. 17–42.

Gaskell, S.L., Wells, A. & Calam, R. (2001) An experimental investigation of thought suppression and anxiety in children, *British Journal of Clinical Psychology*, **40**, pp. 45–56.

Gazzaniga, M.S., Ivry, R.B. & Mangun, G.R. (2014) *Cognitive Neuroscience: The Biology of the Mind*, 4th ed., New York: Norton.

Gibson, J.J. (1979) *The Ecological Approach to Visual Perception*, Boston, MA: Houghton-Mifflin.

Godwin, C.A., Morsella, E. & Geisler, M.W. (2016) The origins of a spontaneous thought: EEG correlates and thinkers' source attributions, *AIMS Neuroscience*, **3**, pp. 203–231.

Gollwitzer, P.M. (1999) Implementation intentions: Strong effects of simple plans, *American Psychologist*, **54**, pp. 493–503.

Gray, J.A. (2004) *Consciousness: Creeping up on the Hard Problem*, New York: Oxford University Press.

Grossberg, S. (1999) The link between brain learning, attention, and consciousness, *Consciousness and Cognition*, **8**, pp. 1–44.

Gruber, O., Gruber, E. & Falkai, P. (2005) Neural correlates of working memory deficits in schizophrenic patients: Ways to establish neurocognitive endophenotypes of psychiatric disorders, *Der Radiologe*, **45**, pp. 153–160.

Harley, T.A. (1993) Phonological activation of semantic competitors during lexical access in speech production, *Language and Cognitive Processes*, **8**, pp. 291–309.

Heath, M., Neely, K.A., Yakimishyn, J. & Binsted, G. (2008) Visuomotor memory is independent of conscious awareness of target features, *Experimental Brain Research*, **188**, pp. 517–527.

Helmholtz, H. von (1856/1961) Treatise of physiological optics: Concerning the perceptions in general, in Shippley, T. (ed.) *Classics in Psychology*, pp. 79–127, Oxford: Philosophy Library.

Hickok, G. (2009) Eight problems for the mirror neuron theory of action understanding in monkeys and humans, *Journal of Cognitive Neuroscience*, **21**, pp. 1229–1243.

Hochberg, J. (1998) Gestalt theory and its legacy: Organization in eye and brain, in attention and mental representation, in Hochberg, J. (ed.) *Perception and Cognition at Century's End: Handbook of Perception and Cognition*, 2nd ed., San Diego, CA: Academic Press.

Hommel, B., Müsseler, J., Aschersleben, G. & Prinz, W. (2001) The theory of event coding: A framework for perception and action planning, *Behavioral and Brain Sciences*, **24**, pp. 849–937.

Jackendoff, R.S. (1990) *Consciousness and the Computational Mind*, Cambridge, MA: MIT Press.

James, W. (1890) *The Principles of Psychology*, New York: Dover.

Jeannerod, M. (2006) *Motor Cognition: What Action Tells the Self*, New York: Oxford University Press.

Jordan, J.S. (2009) Forward-looking aspects of perception-action coupling as a basis for embodied communication, *Discourse Processes*, **46**, pp. 127–144.

Kaido, T., Otsuki, T., Nakama, H., Kaneko, Y., Kubota, Y., Sugai, K. & Saito, O. (2006) Complex behavioral automatism arising from insular cortex, *Epilepsy and Behavior*, **8**, pp. 315–319.

Kaufman, M.T., Churchland, M.M., Ryu, S.I. & Shenoy, K.V. (2015) Vacillation, indecision and hesitation in moment-by-moment decoding of monkey cortex, *eLife*, **4**, e04677.

Kececi, H., Degirmenci, Y. & Gumus, H. (2013) Two foreign language automatisms in complex partial seizures, *Epilepsy & Behavior Case Reports*, **1**, pp. 7-9.

Kern, M.K., Jaradeh, S., Arndorfer, R.C. & Shaker, R. (2001) Cerebral cortical representation of reflexive and volitional swallowing in humans, *American Journal of Physiology: Gastrointestinal and Liver Physiology*, **280**, pp. 354-360.

Klein, D.B. (1984) *The Concept of Consciousness: A Survey*, Lincoln, NE: University of Nebraska Press.

Köhler, W. (1947) *Gestalt Psychology: An Introduction to New Concepts in Modern Psychology*, New York: Liveright Publishing Corporation.

Kokkinos, V., Zountsas, B., Kontogiannis, K. & Garganis, K. (2012) Epileptogenic networks in two patients with hypothalamic hamartoma, *Brain Topography*, **25**, pp. 327-331.

Krisst, L.C., Montemayor, C. & Morsella, E. (2015) Deconstructing voluntary action: Unconscious and conscious component processes, in Eitam, B. & Haggard, P. (eds.) *Human Agency: Functions and Mechanisms*, New York: Oxford University Press.

Kutlu, G., Bilir, E., Erdem, A., Gomceli, Y.B., Kurt, G.S. & Serdaroglu, A. (2005) Hush sign: A new clinical sign in temporal lobe epilepsy, *Epilepsy and Behavior*, **6**, pp. 452-455.

Lashley, K.S. (1942) The problem of cerebral organization in vision, in Kluver, H. (ed.) *Visual Mechanisms: Biological Symposia*, **7**, Lancaster, PA: Cattell Press.

Laureys, S. (2005) The neural correlate of (un)awareness: Lessons from the vegetative state, *Trends in Cognitive Sciences*, **12**, pp. 556-559.

LeDoux, J.E. (2000) Emotion circuits in the brain, *Annual Review of Neuroscience*, **23**, pp. 155-184.

Levelt, W.J.M. (1989) *Speaking: From Intention to Articulation*, Cambridge, MA: MIT Press.

Libet, B. (1986) Unconscious cerebral initiative and the role of conscious will in voluntary action, *Behavioral and Brain Sciences*, **8**, pp. 529-566.

Libet, B. (2004) *Mind Time: The Temporal Factor in Consciousness*, Cambridge, MA: Harvard University Press.

Liu, G., Chua, R. & Enns, J.T. (2008) Attention for perception and action: Task interference for action planning, but not for online control, *Experimental Brain Research*, **185**, pp. 709-717.

Logothetis, N.K. & Schall, J.D. (1989) Neuronal correlates of subjective visual perception, *Science*, **245**, pp. 761-762.

Lucas, M. (2000) Semantic priming without association: A meta-analytic review, *Psychonomic Bulletin & Review*, **7**, pp. 618-630.

Maestro, I., Carreno, M., Donaire, A., Rumia, J., Conesa, G., Bargallo, N., Falcon, C., Setoain, X., Pintor, L. & Boget, T. (2008) Oroalimentary automatisms induced by electrical stimulation of the fronto-opercular cortex in a patient without automotor seizures, *Epilepsy and Behavior*, **13**, pp. 410–412.

Mahon, B.Z. & Caramazza, A. (2008) A critical look at the embodied cognition hypothesis and a new proposal for grounding conceptual content, *Journal of Physiology – Paris*, **102**, pp. 59–70.

Marcel, A.J. (1993) Slippage in the unity of consciousness, in Bock, G.R. & Marsh, J. (eds.) *Experimental and Theoretical Studies of Consciousness*, Ciba Foundation Symposium 174, Chichester: Wiley.

McClelland, J.L. (1979) On the time-relations of mental processes: An examination of systems of processes in cascade, *Psychological Review*, **86**, pp. 287–330.

McGurk, H. & MacDonald, J. (1976) Hearing lips and seeing voices, *Nature*, **264**, pp. 746–748.

Melcher, T., Winter, D., Hommel, B., Pfister, R., Dechent, P. & Gruber, O. (2013) The neural substrate of the ideomotor principle revisited: Evidence for asymmetries in action-effect learning, *Neuroscience*, **231**, pp. 13–27.

Merker, B. (2007) Consciousness without a cerebral cortex: A challenge for neuroscience and medicine, *Behavioral and Brain Sciences*, **30**, pp. 63–134.

Merker, B. (2012) From probabilities to percepts: A subcortical 'global best estimate buffer' as locus of phenomenal experience, in Edelman, S., Tomer, F. & Neta, Z. (eds.) *Being in Time: Dynamical Models of Phenomenal Experience*, Amsterdam: John Benjamins.

Merker, B. (2013) The efference cascade, consciousness, and its self: Naturalizing the first person pivot of action control, *Frontiers in Psychology*, **4**, pp. 1–20.

Merrick, C., Godwin, C.A., Geisler, M.W. & Morsella, E. (2014) The olfactory system as the gateway to the neural correlates of consciousness, *Frontiers in Psychology*, **4**, 1011.

Merrick, C., Farnia, M., Jantz, T.K., Gazzaley, A. & Morsella, E. (2015) External control of the stream of consciousness: Stimulus-based effects on involuntary thought sequences, *Consciousness and Cognition*, **33**, pp. 217–225.

Mikati, M.A., Comair, Y.G. & Shamseddine, A.N. (2005) Pattern-induced partial seizures with repetitive affectionate kissing: An unusual manifestation of right temporal lobe epilepsy, *Epilepsy and Behavior*, **6**, pp. 447–451.

Miller, G.A. (1996) *The Science of Words* (Scientific American Library), New York: W.H. Freeman & Co.

Miller, N.E. (1959) Liberalization of basic S-R concepts: Extensions to conflict behavior, motivation, and social learning, in Koch, S. (ed.) *Psychology: A Study of a Science*, vol. 2, New York: McGraw-Hill.

Mitchell, J.P., Heatherton, T.F., Kelley, W.M., Wyland, C.L., Wegner, D.M. & Macrae, C.N. (2007) Separating sustained from transient aspects of cognitive control during thought suppression, *Psychological Science*, **18**, pp. 292–297.

Montemayor, C., Allen, A.K. & Morsella, E. (2013) The seeming stability of the unconscious homunculus, *Sistemi Intelligenti*, **25**, pp. 581–600.

Morsella, E. (2005) The function of phenomenal states: Supramodular interaction theory, *Psychological Review*, **112**, pp. 1000–1021.

Morsella, E. & Bargh, J.A. (2011) Unconscious action tendencies: Sources of 'un-integrated' action, in Cacioppo, J.T. & Decety, J. (eds.) *Handbook of Social Neuroscience*, New York: Oxford University Press.

Morsella, E., Hoover, M.A. & Bargh, J.A. (2013) Functionalism redux: How adaptive action constrains perception, simulation, and evolved intuitions, in Johnson, K.L. & Shiffrar, M. (eds.) *People Watching: Social, Perceptual, and Neurophysiological Studies of Body Perception*, New York: Oxford University Press.

Morsella, E., Godwin, C.A., Jantz, T.J., Krieger, S.C. & Gazzaley, A. (2016) Homing in on consciousness in the nervous system: An action-based synthesis, *Behavioral and Brain Sciences*, **39**, doi: 10.1017/S0140525X15000643.

Morton, J. (1969) Interaction of information in word recognition, *Psychological Review*, **76**, pp. 165–178.

Müller, N.G. & Knight, R.T. (2006) The functional neuroanatomy of working memory: Contributions of human brain lesion studies, *Neuroscience*, **139**, pp. 51–58.

Nosek, B.A., Spies, J.R. & Motyl, M. (2012) Scientific utopia II: Restructuring incentives and practices to promote truth over publishability, *Perspectives on Psychological Science*, **7**, pp. 615–631.

Obhi, S., Planetta, P. & Scantlebury, J. (2009) On the signals underlying conscious awareness of action, *Cognition*, **110**, pp. 65–73.

Öhman, A. & Mineka, S. (2001) Fears, phobias, and preparedness: Toward an evolved module of fear and fear learning, *Psychological Review*, **108**, pp. 483–522.

Olsson, A. & Phelps, E.A. (2004) Learned fear of 'unseen' faces after Pavlovian, observational, and instructed fear, *Psychological Science*, **15**, pp. 822–828.

Pashler, H. (1993) Doing two things at the same time, *American Scientist*, **81**, pp. 48–55.

Peramunage, D., Blumstein, S.E., Myers, E.B., Goldrick, M. & Baese-Berk, M. (2011) Phonological neighborhood effects in spoken word production: An fMRI study, *Journal of Cognitive Neuroscience*, **23**, pp. 593–603.

Pilon, M. & Sullivan, S.J. (1996) Motor profile of patients in minimally responsive and persistent vegetative states, *Brain Injury*, **10**, pp. 421–437.

Pinker, S. (1997) *How the Mind Works*, New York: W.W. Norton & Co.

Pitts, M.A. & Britz, J. (2011) Insights from intermittent binocular rivalry and EEG, *Frontiers in Human Neuroscience*, **5**, 107.

Plazzi, G., Vetrugno, R., Provini, F. & Montagna, P. (2005) Sleepwalking and other ambulatory behaviors during sleep, *Neurological Sciences*, **26**, pp. 193–198.

Prinz, J. (2007) The intermediate level theory of consciousness, in Velmans, M. & Schneider, S. (eds.) *The Blackwell Companion to Consciousness*, pp. 248–260, Oxford: Blackwell.

Prinz, W., Aschersleben, G. & Koch, I. (2009) Cognition and action, in Morsella, E., Bargh, J.A. & Gollwitzer, P.M. (eds.) *Oxford Handbook of Human Action*, pp. 35–71, Oxford: Oxford University Press.

Pylyshyn, Z.W. (1984) *Computation and Cognition: Toward a Foundation for Cognitive Science*, Cambridge, MA: MIT Press.

Raghavendra, S., Mirsattari, S. & McLachlan, R. (2010) Ictal whistling: A rare automatism during temporal lobe seizures, *Epileptic Disorders*, **12**, pp. 133–135.

Rassin, E. (2005) *Thought Suppression*, Amsterdam: Elsevier.

Rolls, E.T., Judge, S.J. & Sanghera, M. (1977) Activity of neurones in the inferotemporal cortex of the alert monkey, *Brain Research*, **130**, pp. 229–238.

Rosenbaum, D.A. (2002) Motor control, in Pashler, H. (series ed.) & Yantis, S. (vol. ed.) *Stevens' Handbook of Experimental Psychology: Vol. 1. Sensation and Perception*, 3rd ed., New York: Wiley.

Rossetti, Y. (2001) Implicit perception in action: Short-lived motor representation of space, in Grossenbacher, P.G. (ed.) *Finding Consciousness in the Brain: A Neurocognitive Approach*, Amsterdam: John Benjamins Publishing.

Sato, M., Baciu, M., Loevenbruck, H., Schwartz, J.-L., Cathiard, M.-A., Segebarth, C. & Abry, C. (2004) Multistable representation of speech forms: A functional MRI study of verbal transformations, *NeuroImage*, **23**, pp. 1143–1151.

Schenk, C.H. & Mahowald, M.W. (1995) A polysomnographically documented case of adult somnambulism with long-distance automobile driving and frequent nocturnal violence: Parasomnia with continuing danger as a noninsane automatism?, *Sleep*, **18**, pp. 765–772.

Schopenhauer, A. (1818/1819) *The World as Will and Representation*, vol. 1, New York: Dover.

Schomers, M.R., Kirilina, E., Weigand, A., Bajbouj, M. & Pulvermüller, F. (2014) Causal influence of articulatory motor cortex on comprehending single spoken words: TMS evidence, *Cerebral Cortex*, pp. 1-9.

Schultz, D.P. & Schultz, S.E. (1996) *A History of Modern Psychology*, 6th ed., San Diego, CA: Harbrace College Publishers.

Scott, M. (2013) Corollary discharge provides the sensory content of inner speech, *Psychological Science*, **24**, pp. 1824-1830.

Scullin, M.K., McDaniel, M.A. & Einstein, G.O. (2010) Control of cost in prospective memory: Evidence for spontaneous retrieval processes, *Journal of Experimental Psychology: Learning, Memory, and Cognition*, **36**, pp. 190-203.

Searle, J.R. (2000) Consciousness, *Annual Review of Neurosciences*, **23**, pp. 557-578.

Seth, A.K. (2007) The functional utility of consciousness depends on content as well as on state, *Behavioral and Brain Sciences*, **30**, 106.

Sherrington, C.S. (1906) *The Integrative Action of the Nervous System*, New Haven, CT: Yale University Press.

Skinner, B.F. (1953) *Science and Human Behavior*, New York: Macmillan.

Smári, J. (2001) Fifteen years of suppression of white bears and other thoughts: What are the lessons for obsessive-compulsive disorder research and treatment?, *Scandinavian Journal of Behavior Therapy*, **30**, pp. 147-160.

Snodgrass, J.G. & Vanderwart, M. (1980) A standardized set of 260 pictures: Norms for name agreement, image agreement, familiarity, and visual complexity, *Journal of Experimental Psychology: Human Learning and Memory*, **6**, pp. 174-215.

Soon, C.S., Brass, M., Heinze, H.-J. & Haynes, J.-D. (2008) Unconscious determinants of free decisions in the human brain, *Nature Neuroscience*, **11**, pp. 543-545.

Spencer, S.S., Spencer, D.D., Williamson, P.D. & Mattson, R.H. (1983) Sexual automatisms in complex partial seizures, *Neurology*, **33**, pp. 527-533.

Tolman, E.C. (1948) Cognitive maps in rats and men, *Psychological Review*, **55**, pp. 189-208.

Tortora, G.J. (1994) *Introduction to the Human Body: The Essentials of Anatomy and Physiology*, 3rd ed., New York: Harper Collins.

Tsushima, Y., Sasaki, Y. & Watanabe, T. (2006) Greater disruption due to failure of inhibitory control on an ambiguous distractor, *Science*, **314**, pp. 1786-1788.

Tye, M. (1999) Phenomenal consciousness: The explanatory gap as cognitive illusion, *Mind*, **108**, pp. 705-725.

Vallar, G., Corno, M. & Basso, A. (1992) Auditory and visual verbal short-term memory in aphasia, *Cortex*, **28**, pp. 383–389.

Vroomen, J. & de Gelder, B. (2003) Visual motion influences the contingent auditory motion aftereffect, *Psychological Science*, **14**, pp. 357–361.

Vygotsky, L.S. (1962) *Thought and Language*, Cambridge, MA: MIT Press.

Wegner, D.M. (1989) *White Bears and Other Unwanted Thoughts*, New York: Viking/Penguin.

Wegner, D.M. (1994) Ironic processes of thought control, *Psychological Review*, **101**, pp. 34–52.

Wegner, D.M., Schneider, D.J., Carter, S.R. & White, T.L. (1987) Paradoxical effects of thought suppression, *Journal of Personality and Social Psychology*, **53**, pp. 5–13.

Wolpert, D.M. & Kawato, M. (1998) Multiple paired forward and inverse models for motor control, *Neural Networks*, **11**, pp. 1317–1329.

Woodworth, R.S. (1915) A revision of imageless thought, *Psychological Review*, **22**, pp. 1–27.

Wortzel, H., Strom, L., Anderson, A., Maa, E. & Spitz, M. (2011) Disrobing associated with epileptic seizures and forensic implications, *Journal of Forensic Science*, **57**, pp. 550–552.

Wundt, W. (1900) *Die Sprache*, Leipzig: Engelmann.

Wyland, C.L., Kelley, W.M., Macrae, C.N., Gordon, H.L. & Heatherton, T.F. (2003) Neural correlates of thought suppression, *Neuropsychologia*, **41**, pp. 1863–1867.

Zhang, W. & Rosenbaum, D.A. (2008) Planning for manual positioning: The end-state comfort effect for abduction-adduction of the hand, *Experimental Brain Research*, **184**, pp. 383–389.

Zeki, S. & Bartels, A. (1999) Toward a theory of visual consciousness, *Consciousness and Cognition*, **8**, pp. 225–259.

Zorick, F.J., Salis, P.J., Roth, T. & Kramer, M. (1979) Narcolepsy and automatic behavior, *Journal of Clinical Psychiatry*, **40**, pp. 194–197.

Axel Cleeremans

As Above, So Below
Tangled Loops between Consciousness and the Unconscious

1. Introduction

Understanding what comes 'before consciousness' is instrumental in understanding consciousness itself, inasmuch as defining it necessitates understanding the differences between what happens with consciousness and without it. 'What comes before consciousness' is a fascinating way to think about it because it compels us to think about the dynamics of the transition between unconscious information processing and conscious information processing. Further, this dynamical perspective can be cast at different timescales: over evolution, over development, over a specific learning episode, or even within a single trial in a cognitive psychology experiment. Let us turn to each in turn.

Albeit the panpsychist views that have recently gained new popularity (Skrbina, 2005) would lead us to think otherwise, many organisms are presumably unconscious. However, people's intuitions differ strongly in this respect. Are dogs conscious? Most would say 'yes', though I found myself taken aback when I recently asked this question from an audience of about forty and only one person raised his hand! Are mice conscious? Again, if I were to speculate, I would tend to say 'yes', essentially because I can imagine that there is something it is like it to be a mouse. I should stress that my intuition here is entirely pre-theoretical, as there are little grounds beyond the gross anatomy of the mouse brain to argue one way or the other. Are bees conscious? Some (Klein and Barron, 2016) argue that they minimally have the required brain anatomy—though we do not know precisely which anatomical features of brains cause consciousness, or even if consciousness depends on particular anatomical features. What about invertebrates such as earthworms and slugs? Or unicellular organisms such as bacteria? My intuitions stop there. But they are mere intuitions, for there are no clear criteria to ascribe consciousness to non-human animals. The same challenge that presents itself in a comparative project is at play when evolutionary considerations are taken into account—we do not know *when* conscious experience emerged over evolution, and we have few arguments to build on (but see Jaynes, 1976). Perhaps the most interesting question to ask here is not *when* conscious organisms appeared but *why*. In other words, which functional features does consciousness enable that were important enough to be amplified through natural selection? In either case, thinking about consciousness in terms of evolution leads us to espouse a perspective that is rooted in the dynamics characteristic of such long timescales, and this focus on change is a productive perspective to hold.

More or less the same questions play out over development. Are infants conscious? Most would agree that they are as matter of course, though one can find a few dissenting voices (e.g. Macphail, 1998). Beyond intuitions, infant consciousness has recently been the focus of intense scrutiny. The work of Kouider and colleagues (e.g. Kouider *et al.*, 2013), for instance, has demonstrated that the electrophysiological response to briefly presented face pictures already exhibits the temporal dynamics characteristic of perceptual awareness in infants aged 5 months to 15 months, albeit those dynamics are very much slowed with respect to those exhibited by adults. Likewise, recent work by the same authors (Goupil and Kouider, 2016) highlights the fact that infants aged 18 months already exhibit behaviour that is indicative of metacognition, that is, of their ability to accurately judge whether or not they are correct when making a decision. For instance, infants will spend more time ('persistence time') searching a box when they believe a reward is hidden in it than when they believe the box is empty. To explore the dynamics of confidence in this situation, Goupil and Kouider manipulated the temporal interval that lapses between the moment an object is placed in one of two boxes and the moment where infants are asked to point to the correct box and offered the possibility to search it. Importantly, they showed that the accuracy of first-order decisions and confidence (as measured through persistence time) are not influenced in the same way by the delay, which rules out the alternative hypothesis that both simply depend on the quality of the memory trace and instead suggests that persistence time truly reflects a second-order, metacognitive process.

In another direction, we showed (Legrain, Destrebecqz and Cleeremans, 2010) that self-awareness, short of being the monolithic ability that we think it is, instead follows a continuous and graded developmental course that one can characterize, following Rochat (2003), as involving different levels, amongst which three levels of 'explicit' self-awareness—the identified self, the permanent self, and the external self. In an attempt to go beyond the mirror task (Gallup, 1970), we designed tasks specifically aimed at distinguishing between these different levels: mirror self-recognition for the first level, picture self-recognition for the second, and picture self-recognition while wearing a mask for the third. Our results showed a clear age-dependent rate of success on each of these three tasks with toddlers aged 22–32 months old, thus confirming the idea that self-awareness takes time to develop, and that different aspects of it emerge at different points during development.

These and other findings are suggestive that different aspects of adult consciousness develop over time in specific patterns that are now being documented and characterized all the way down to the neural level. Exploring the developmental tipping points that mark the transition between 'before' and 'after' a particular feature emerges will undoubtedly prove to be a fruitful strategy through which to piece it all together.

Turning now to learning, it is striking to note that practice and expertise can create as well as eliminate contents from phenomenal experience. Thus, tasting wine for the first time is a wholly different experience than that of an oenologist (Smith, 2006), whose phenomenology has been enriched through expertise. But expertise can also eliminate phenomenal contents from awareness, as in the 'find the F's' illusion, wherein observers are asked to count the number of instances of the letter 'F' in a sentence. Observers often fail to reach the correct answer because reading expertise has eliminated function words from awareness. There are many other examples of such 'predictive attenuation' mechanisms: tickling oneself is far less effective than being tickled (Blakemore, Frith and Wolpert, 1999), for when we tickle ourselves (but not when we are tickled) our brain can predict the consequences of our actions. Cognitive development also highlights how some changes go unheeded (i.e. the fact that our action and perceptual systems remain adapted despite our limbs growing spectacularly during the first few years) whereas other changes have profound phenomenal consequences (i.e. learning to read). Thus, learning shapes conscious experience and conscious experiences shapes learning. Consciousness, in this light, is a profoundly dynamical process through which our experience of the world is constantly shaped both by incoming sensory inputs and by our learned expectations about what is coming next ('priors').

In cognitive psychology, the possibility of learning without awareness now has a long and controversial history, and no satisfactory resolution yet. It may be the case, as Shanks and St John (1994) have argued, that 'human learning is almost systematically accompanied by conscious awareness' (p. 394), and yet, this perspective (see also Newell and Shanks, 2014) is hard to reconcile with the fact that learning is clearly possible in the absence of awareness: even simple organisms such as *Aplysia* are susceptible to conditioning despite presumably lacking conscious experience, and there are innumerable artificial systems, such as connectionist networks, capable of learning. In humans, different recent experiments also support the idea that learning is possible without awareness (e.g. Atas *et al.*, 2014; Pessiglione *et al.*, 2008; Alamia *et al.*, 2016; amongst others). The field of implicit learning (see Cleeremans *et al.*, 1998, for a review) has long been a battleground of sorts where different theoretical and methodological perspectives have confronted each other, and it would be much too long to offer an overview here. The key point I wish to highlight here, however, is again one about the dynamics of learning in such situations: solid evidence that learning is possible in the absence of awareness would consist of establishing that, over the course of training, behavioural sensitivity to some state of affairs appears before ability to consciously report knowledge of the relevant contingencies. I discuss one possible such demonstration (Alamia *et al.*, 2016) below.

Finally, the dynamics of what happens over the course of a single trial is also informative with respect to the processes associated with conscious awareness. Dehaene (e.g. Dehaene *et al.*, 2003; Dehaene, 2014) for instance, with the global neuronal workspace hypothesis, has proposed that the main 'signature' of conscious access consists of a characteristic 'double peak' in electrophysiological signals, whereby feedforward processing of a (e.g. visual) stimulus over the first 300 ms is then followed by a late and sustained amplification ('ignition') of the corresponding sensory areas.

These different examples vividly illustrate why focusing on the dynamics of consciousness, that is, on the differences between what happens 'before' and 'after' consciousness at different timescales, is such a useful experimental strategy. But what about theory? Here, I will attempt to delineate the deep connections between a focus on dynamics and a novel theory of consciousness in which prediction-driven learning plays a central role.

2. Consciousness as a Prediction-Driven Redescription Process

Prediction is a ubiquitous computational principle in the brain (Bar, 2009; Clark, 2013). As Clark puts it: 'Brains, it has recently been argued, are essentially prediction machines.' Clark fleshes out this claim by highlighting the lineage between the early ideas of Helmoltz (1860/1962) and the much more recent ideas associated with contemporary connectionist models (McClelland and Rumelhart, 1986; Rumelhart, Hinton and Williams, 1986), in particular the so-called generative models described by Hinton (Dayan, Hinton and Neal, 1995; Hinton, 2007) and by Friston and colleagues (Friston, 2006; 2010). There is considerable evidence for predictive mechanisms in the human brain (Bar, 2009). This idea, in fact, forms the core of the Bayesian perspective on information processing and is at the heart of Friston's free energy principle (Friston, 2006), according to which the brain continuously attempts to minimize 'surprise' or conflict by anticipating its own future activity based on learned priors.

Helmholz first proposed the idea that perception involves a form of prediction-driven inference through which the mind attempts to reconstruct the sensory causes of bodily effects. Perception, in this view, is thus an active process of elaborating the best possible representations of the input based on both the sensory evidence and relevant prior knowledge (the 'priors') rather than a mere bottom-up process. The work of Hinton, Friston, and colleagues elaborated on this view by showing how it is possible to conceive learning rules able to shape top-down connection weights to as to minimize 'prediction error', that is, the difference between expected and observed inputs. In such models, the top-down flow of information thus attempts to 'explain the sensory input away', leaving only information about the residual errors (the 'prediction error') to flow upwards in a hierarchy of interconnected layers. This is the core principle of 'predictive coding' (Rao and Ballard, 1999), through which hierarchical systems can simultaneously learn about their inputs and about the best internal models of these same inputs. As Clark (2013) puts it, this dampens the distinction between perception and belief to the point that they appear almost identical to each other:

> To perceive the world just is to use what you know to explain away the sensory signal across multiple spatial and temporal scales. The process of perception is thus inseparable from rational (broadly Bayesian) processes of belief fixation, and context (top down) effects are felt at every intermediate level of processing. As thought, sensing, and movement here unfold, we discover no stable or well-specified interface or interfaces between cognition and perception. Believing and perceiving, although conceptually distinct, emerge as deeply mechanically intertwined. (p. 29)

These ideas have now gained wide interest, both in the philosophy of mind (e.g. Hohwy, 2013) and in the cognitive sciences (e.g. Seth, Suzuki and Critchley, 2011).

A central aspect of the entire hierarchical predictive coding approach, though this is not readily apparent in the corresponding literature, is the emphasis it puts on learning mechanisms. In other works (Cleeremans, 2008; 2011), I have defended the idea that consciousness is itself the result of learning. From this perspective, agents become conscious in virtue of learning to redescribe their own activity to themselves. Taking the proposal that consciousness is inherently dynamical seriously opens up the mesmerizing possibility that conscious awareness is itself a product of plasticity-driven dynamics. In other words, from this perspective, we learn to be conscious. To dispel possible misunderstandings of this proposal right away, I am not suggesting that consciousness is something that one learns like one would learn about the Hundred Years War, that is, as an academic endeavour, but rather that consciousness is the result (vs. the starting point) of continuous and extended interaction with the world, with ourselves, and with others. The brain, from this perspective, continuously (and unconsciously) learns to anticipate the consequences of its own activity on itself, on the environment, and on other brains, and it is from the practical knowledge that accrues in such interactions that conscious experience is rooted. This perspective, in short, endorses the enactive approach introduced by O'Regan and Noë (2001), but extends it both inwards (the brain learning about itself) and further outwards (the brain learning about other brains), so connecting with the central ideas put forward by the predictive coding approach to cognition. In this light, the conscious mind is the brain's (implicit, enacted) theory about itself, expressed in a language that other minds can understand.

The theory rests on several assumptions and is articulated over three core ideas. A first assumption is that information processing as carried out by neurons is intrinsically unconscious. There is nothing in the activity of individual neurons that make it so that their activity should produce conscious experience. Important consequences of this assumption are (1) that conscious and unconscious processing must be rooted in the same set of representational systems and neural processes, and (2) that tasks in general will always involve both conscious and unconscious influences, for awareness cannot be 'turned off' in normal participants.

A second assumption is that information processing as carried out by the brain is graded and cascades (McClelland, 1979) in a continuous flow (Eriksen and Schultz, 1979) over the multiple levels of a heterarchy (Fuster, 2008) extending from posterior to anterior cortex as evidence accumulates during an information processing episode. An implication of this assumption is that consciousness takes time.

The third assumption is that plasticity is mandatory: the brain learns all the time, whether we intend to or not. Each experience leaves a trace in the brain (Kreiman, Fried and Koch, 2002). With these assumptions in place, the theory is articulated around three core ideas that I will now briefly expose.

3. Quality of Representation

The first core idea is that consciousness depends on quality of representation (see Figure 1). 'Quality of representation' (QoR), here, designates graded properties of neural representations, specifically their Strength, their Stability in time, and their Distinctiveness. QoR depends both on bottom-up factors such as stimulus properties and on top-down factors such as attention. QoR determines the extent to which a representation is available to (1) influence behaviour, (2) form the contents of awareness, (3) be the object of cognitive control and other high-level processes. Crucially, QoR changes as a function of learning and plasticity over different timescales (processing within a single trial, learning, and development), as depicted in Figure 1. The first region of the figure, labelled 'Implicit Cognition', corresponds to the point at which processing starts in the context of a single trial, or to some early stage of development or skill acquisition. This stage is characterized by weak, poor-quality representations. Implicit representations are capable of influencing behaviour, but only weakly so (e.g. through priming).

The second region corresponds to the emergence of higher-quality explicit representations, here defined as representations over which one can exert control. Such representations are good candidates for redescription and can thus be recoded in different ways, e.g. as linguistic propositions (supporting verbal report).

The third region involves what I call automatic representations, that is, representations that have become so strong that their influence on behaviour can no longer be inhibited (e.g. as in the Stroop situation). Such representations exert a mandatory influence on processing. Importantly, however, and unlike the weak representations characteristic of implicit cognition, one is (at least potentially) aware of possessing such strong representations and of their influence on processing. Thus, both the weak representations characteristic of implicit cognition and the very strong representations characteristic of automaticity cannot be controlled, but for very different reasons. This leaves intermediate-quality (explicit) representations, that is, representations that are strong enough that their influence on behaviour needs to be monitored yet not sufficiently adapted that they can be 'trusted', as those representations that require the most cognitive control. Crucially, this also predicts that intermediate-quality representations are the most susceptible to be influenced by other sources of knowledge, as they are the most flexible. One would thus expect non-monolithic effects as expertise develops, in different paradigms ranging from perception to motor skill learning.

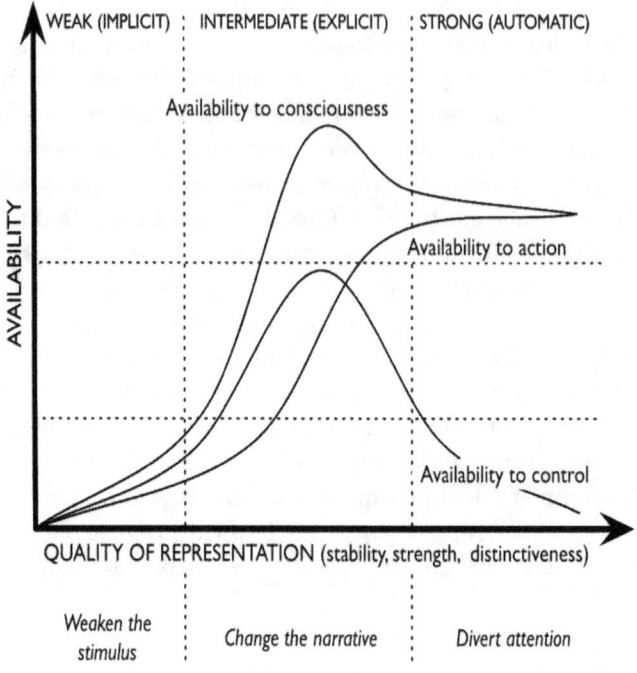

Figure 1. The 'QoR' framework.

4. Metarepresentation

The second core idea is that consciousness depends on the involvement of metarepresentations. Indeed, quality of representation cannot be the only factor that shapes availability to different aspects of consciousness, however. Even strong stimuli can fail to enter conscious awareness — this is what happens in change blindness (Simons and Levin, 1997), in the attentional blink (Shapiro, Arnell and Raymond, 1997), or in inattentional blindness (Mack and Rock, 1998). States of altered consciousness like hypnosis, and pathological states such as blindsight (Weiskrantz, 1986) or hemineglect, likewise suggest that high-quality percepts can fail to be represented in awareness while remaining causally efficacious. This suggests that quality of representation, while necessary for conscious awareness, is not sufficient.

One way of understanding what is missing is to appeal to the central hypothesis of the higher-order thought (HOT) theory of consciousness (Rosenthal, 1997; see also Lau and Rosenthal, 2011), namely that a representation is a conscious representation when one knows that one is conscious of the representation. This roots conscious awareness in a system's capacity to redescribe its own states to itself, a process ('representational redescription') also viewed as central during cognitive development (Karmiloff-Smith, 1992) and metacognition in general (Nelson and Narens, 1990). A system's ability to redescribe its own knowledge to itself depends (1) on the existence of recurrent structures that enable the system to access its own states, and on (2) the existence of predictive models (metarepresentations) that make it possible for the system to characterize and anticipate the occurrence of first-order states (Bar, 2009; Friston, 2006; Wolpert, Doya and Kawato, 2004). Such redescription is also uniquely facilitated, in humans, by language, viewed here as the metarepresentational tool *par excellence*. A natural spot for such metarepresentations to play their functions is the prefrontal cortex (i.e. Crick and Koch's 'the front is looking at the back' principle — Crick and Koch, 2003). Importantly however, here, such metarepresentational models (1) may be local and hence occur anywhere in the brain, (2) can be subpersonal, are (3) are subject, just like first-order representations, to plasticity and hence can themselves become automatic. Metacognition, just like cognition, can thus involve implicit, explicit, or automatic metarepresentations.

The theory thus proposes a novel conception of skill acquisition that links automaticity with the observation that conscious awareness seems to proceed from the top down (i.e. Crick and Koch's 'the high levels first' principle, see Crick and Koch, 2003): we become aware of the higher-level aspects (the gist) of a scene before becoming aware of its lower-level features. I suggest that this stems from the fact that, from a computational point of view, metarepresentations implement what one could call cortical reflexes or shortcuts: a system that has learned to redescribe the activity of an entire feedforward pathway can now also anticipate the consequences of early activity in such a chain on its output faster than the pathway itself can compute the output. As a result, adapted metarepresentations (and only adapted metarepresentations) make it possible to bypass the first-order pathway altogether. I surmise that this accounts not only for the fact that the time course of (expert) perception seems to follow a reverse hierarchy (Ahissar and Hochstein, 2004), but also for the fact that automaticity entails loss of access to the contents computed along the first-order pathway. By the same token, this also opens up the possibility for postdictive effects in conscious experience, as metarepresentations are shaped by first-order processing. This top-down view of automaticity contrasts with extant theories (Chein and Schneider, 2012).

With these ideas in place, we can now ask: 'When is knowledge unconscious?' Figure 2 shows a simple network organized in two different pathways: a (horizontal) first-order pathway comprising five layers of units, and a (vertical) second-order pathway simplified here to show only a single layer of 'metarepresentational units' (see Pasquali *et al.*, 2007; 2010, for implemented instantiations of such networks). The figure is aimed to distinguish four stages that such networks traverse through as they learn to carry out a particular task. These four stages correspond to four different ways in which knowledge may remain unconscious.

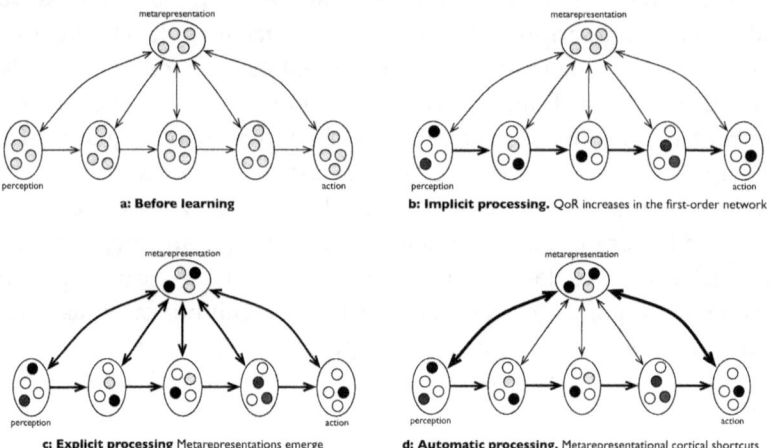

Figure 2. Implicit, explicit, and automatic processing.

First (Figure 2a), knowledge embedded in synapses is assumed not to be accessible at all, for such knowledge fails to be instantiated in the form of active patterns of neural activity (Koch, 2004), a necessary condition for their contents to be available to awareness. The provocative idea here is that the brain does not know, for example, that SMA activity consistently precedes M1 activity. To represent this causal link to itself, it therefore has to learn to redescribe its own activity so that the causal link is now represented explicitly as a metarepresentation. Second, weak representations (Figure 2b), while they can influence behaviour, remain unconscious for they fail to be sufficiently strong to be the target of metarepresentations. Third, when sufficiently strong, first-order representations can begin to be redescribed into metarepresentations (Figure 2c), yet other conditions (e.g. lack of attention induced by distraction, failure to properly redescribe first-order contents) may make such redescription impossible or difficult. Fourth, the very strong representations characteristic of automaticity (Figure 2d) are not necessary anymore to drive behaviour since the learned meta-representations now implement a faster 'shortcut' pathway from input to output. This also accounts for the fact that metacognitive accuracy often lags first-order performance initially, but precedes first-order performance with expertise (i.e. I know that I know the answer to a query before I can actually answer the query).

The distinctions introduced here overlap partially with the distinctions introduced by existing theories of consciousness: Dehaene's conscious–preconscious–unconscious taxonomy (Dehaene et al., 2006), Lamme's Stages 1/2/3/4 framework (Lamme, 2006), and Kouider's partial awareness hypothesis (Kouider et al., 2010), but uniquely frames the transitions dynamically as resulting from leaning.

3. Theory of Mind and Self-Awareness

The third core idea is that consciousness depends on theory of mind (Schilbach et al., 2013; Timmermans et al., 2012). The emergence of an agent's ability to redescribe its own representations to itself in the way sketched above, I argue, critically depends on the agent being embedded in interaction with other agents. From this perspective, as Frege pointed out, conscious experience cannot be understood independently from the agent who experiences these experiences. Yet, as obvious as this may seem, neuroscientists have approached the question as though the differences between conscious and unconscious representations could be understood independently of the subject, from a purely 'objective', third-person point of view. The entire 'search for the neural correlates of consciousness' is, in this sense at least, misguided. As Donald (2001) put it, 'the human mind is unlike any other on this planet, not because of its biology, which is not qualitatively unique, but because of its ability to generate and assimilate culture' (p. xiii). Thus, I build a model of myself not only by developing a non-conceptual understanding of how my goals are eventually expressed in action, but also by understanding how agents similar to me react to actions directed towards them. It is thus essential that we strive to understand how interactions with other agents shape our own conscious experiences.

Putting the three core ideas together, we end up with the radical plasticity thesis (Cleeremans, 2008; 2011), that is, with the idea that consciousness emerges in cognitive systems that are capable of learning to redescribe their own activity to themselves. In other words, one 'learns to be conscious'.

4. Experimental Strategies

The framework outlined above suggests different experimental strategies through which to manipulate the extent to which processing is conscious vs. unconscious. To develop this, consider that higher-order thought theories conjure an image of consciousness as a monitoring system—a radar of sorts, that continuously sweeps and redescribes our own representations, so carrying out what Lau (2008) dubbed 'signal detection on our mind'.

If one accepts this admittedly very rough metaphor, it is interesting to consider what it takes to fool the radar, that is, to escape detection. There are essentially three strategies, each corresponding to one of three regions (implicit, explicit, automatic) depicted in Figure 1. One is to be small enough to simply fail to register. While an aeroplane will register on an airspace radar, for instance, this would not be the case for a bird or a small drone. In cognitive psychology, this strategy amounts to *weakening the stimulus*, that is, reducing its energy or its temporal duration, or making it less distinguishable from noise, in such a way as to render it phenomenologically unconscious. This is the strategy that most experimental designs based on masking or crowding leverage, as in subliminal priming for instance.

The second strategy one could use to fool a radar is to *divert its attention*, or rather, the attention of the radar operator. Here, there is something that has sufficient energy to be detected, yet it fails to be noticed because the radar operator is not looking at it or because his attentional resources are otherwise engaged. This is precisely what happens in paradigms such as inattentional blindness, change blindness, or the attentional blink, but also in any paradigm that is designed to overload working memory capacity. While *weakening the stimulus* is a bottom-up strategy, *diverting attention* is clearly a top-down strategy. In the first case, one manipulates the quality of incoming stimuli in such a way that they are not available to form the contents of consciousness even when attention is directed towards them, whereas in the other case, a stimulus that has the necessary quality to be available to form the contents of awareness fails to become conscious because insufficient attention is dedicated to it.

There is a second top-down strategy that is somewhat more unusual: it consists of *changing the narrative*. To return to our radar analogy, this would consist of making the radar operator believe, not that there is *nothing* out there, but rather that what is there is *not what he thinks it is*. For instance, one aeroplane may pass itself off as another; or a flock of birds might organize itself so as look like an aeroplane (though why birds would do that is open to question). This is connected to the well-known issue of misrepresentation in the philosophy literature (e.g. Dretske, 1986). In cognitive psychology, perhaps the best-known contemporary example is the phenomenon of choice blindness (Johansson *et al.*, 2005), whereby participants are made to believe that a photograph of a person they have just *not chosen* is in fact the one they chose. This trick makes a number of them confabulate reasons for their choice, just as in Nisbett and Wilson's (1977) famous 'nylon stockings' experiment. But this kind of paradigm is just one instance of a wide variety of such manipulations of people's conscious beliefs. One also immediately thinks, for instance, of hypnosis, and of the placebo effect. The latter is particularly striking, since it demonstrates that the mere belief that a pill contains an active ingredient is sufficient to profoundly modify people's subjective appraisal of their symptoms. Likewise, in hypnosis, people can be convinced that their actions are not their own; that their arm is lifting up 'on its own'. This reduced sense of agency under hypnosis is, according to some (i.e. Dienes and Perner, 2007; Lush, Naish and Dienes, 2016), the defining feature of the hypnotic phenomenon. Such high-level manipulations of what people believe to be the case can penetrate deeply in the cognitive hierarchy. Raz and collaborators, for instance, have shown that a hypnotic suggestion that 'words would appear as mere gibberish' is sufficient to reduce Stroop interference in a colour-naming task. We have replicated this finding ourselves using a placebo-aided nonhypnotic suggestion that colour perception would either be enhanced or deteriorated by a fake apparatus to which people were connected (Magalhães de Saladanha da Gama *et al.*, 2013).

Each of these three strategies to fool consciousness comes with its own challenges, which are often thorny and extremely difficult to properly address. Subliminal perception, and specifically the claim that a stimulus can be weak enough to fail to enter consciousness yet strong enough that it remains causally efficacious, such as in its ability to prime subsequent decisions, continues to elicit vivid debate today. This should come as no surprise, because it turns out it is exceedingly challenging to come up with experimental designs in which a stimulus can, precisely, be weak enough to 'fly under the radar' yet strong enough to be causally efficacious. This is the quandary that faces all researchers interested in demonstrating unconscious cognition. I call it the *strength-efficacy dilemma*, and it is a dilemma that comes up in any paradigm leveraging the strategy of weakening the stimulus. This gets complicated by the further challenges involved in properly assessing awareness itself. Should we ask participants to rate visibility on each trial, or ask them to take part in a visibility test after the main (e.g. priming) task is completed? In either case, should we use binary (seen/unseen) measures or graded reports such as the Perceptual Awareness Scale (PAS, see Ramsøy and Overgaard, 2004) or a fully continuous scale (Sergent and Dehaene, 2004)? Crucially, these methodological decisions, which at first sight appear to be secondary, are demonstrably crucial in that they change our conclusions about the extent to which processing was indeed unconscious.

To wit, consider the fact that a systematic comparison between different ways of collecting subjective judgments about visibility can sometimes yield strikingly different results for each method. Thus, Sandberg et al. presented participants with briefly displayed masked shapes (a square, a circle, a triangle, and a lozenge) in a psychophysical paradigm on each trial of which they had to (1) identify the shape by pressing on one of four buttons and (2) express a subjective judgment of visibility or confidence, using one of three different scales all involving four points: PAS, a confidence judgment, or post-decision wagering (Persaud, McLeod and Cowey, 2007). Briefly put, the study revealed that PAS was the most exhaustive scale: when using PAS, people reported experiencing something ('a brief glimpse') — at shorter stimulus durations than they did with the other two scales. This finding has obvious implications for our interpretation of dissociation results (e.g. between priming and visibility), for they suggest that many such dissociation findings may be a mere consequence of lack of sensitivity.

Further, the study also demonstrated that while people tend to use post-decision wagering in a binary manner, betting either high or low and mostly refraining from using the intermediate scale points, this was not the case with PAS, which showed a much more distributed use of the different scale points. This suggests that intermediate or graded states of awareness are possible and that people are quite willing to report them when offered the possibility.

In recent research, Windey, Gevers and Cleeremans (2013) focused specifically on this question, that is, whether consciousness should be taken to be an all-or-none or a graded phenomenon. Using a simple task consisting of judging either the magnitude or the colour of briefly presented, masked arabic numerals, we found that the shape of the psychophysical functions relating stimulus duration with subjective visibility assessed by PAS was influenced by level of processing: such functions were more linear when judging colour than when judging magnitude, which is suggestive that both stimulus features and level of processing modulate the extent to which perceptual awareness appears graded or dichotomous. This turns out to be important when comparing different studies, for different authors typically use different stimuli. Lamme and colleagues (see e.g. Fahrenfort et al., 2008), for instance, who defend the idea that phenomenal consciousness has a graded character, have often used low-level stimuli such as gratings and gabor patches, whereas researchers who defend the idea that consciousness involves a sharply non-linear transition (see e.g. Del Cul *et al.*, 2007) tend to use high-level stimuli such as numbers or words. Further studies in this vein confirmed and extended our results (see Anzulewicz *et al.*, 2015; Windey *et al.*, 2014).

One may think that such measurement issues are only problematic when subjective measures such as visibility or confidence are collected. However, even objective measures such as d' can exhibit substantial variability. Thus, Vermeiren and Cleeremans (2012) showed that d', as measured in a prime visibility test administered after a priming task involving metacontrast masked arrows, was influenced by different factors such as (1) the delay that lapsed between stimulus presentation and response, (2) whether the target (visible) stimulus was neutral or oriented, and (3) whether attention was divided between the prime and the target. Observed differences in d' magnitude could be as large as 1.0 in some conditions, thus casting doubt on the use of d' as a neutral and objective measure of visibility, and again bearing strong implications for the interpretation of dissociation findings.

Thus, the general take-home message is that the measure we use to assess awareness matters very much, for different measures reveal different dynamics and can lead to very different conclusions about the extent to which processing was unconscious or not.

The second strategy, namely *diverting attention*, has challenges of its own. The most important challenge presents itself in the form of *observer paradox*, that is, the fact that, in cognitive systems, observing a process actually changes that very process. Thus, if my goal is to draw people's attention away from a stimulus, I simply cannot ask participants 'whether they noticed something', for doing so attracts their attention precisely to what I wish them to remain unaware of. This limits possible designs to single-trial experiments, as in Mack and Rock's well-known inattentional blindness experiments, or forces us to collect subjective reports after the entire experiment is over, a method that is problematic in and of itself, as is also the case for subliminal perception experiments.

This particular challenge is one of several listed by Newell and Shanks (2014) in a recent review of unconscious decision making. According to the authors, any measure of awareness should fulfil the following four criteria. First, the measure should be *reliable*, that is, independent from experimental demands or social desirability. Second, it should be *relevant*, that is, probe participants about the very same knowledge that is involved in subtending performance. Third, it should *immediate*, so as to avoid forgetting and interference. Ideally, thus, any measure of awareness (e.g. a visibility or a confidence judgment) should be administered on a trial-by-trial basis rather than after the entire main task is completed. Fourth, the measure should be *sensitive*, which is to say based on the same material used to elicit behaviour and fine-grained enough that it can be as exclusive and exhaustive as possible (Reingold and Merikle, 1988). On the face of it, very few paradigms can claim to fulfil all four criteria.

The third strategy — *changing the narrative* — presents the challenge of being able to develop a convincing cover story to describe the situation to participants. It is also prone to the 'retrospective assessment' problem (Shanks and St John, 1998), as it is impossible to satisfy the immediacy criterion: one cannot test people's beliefs online without undermining such beliefs, the solidity of which is crucial for the effects to obtain. Nevertheless, in recent work, we used this strategy in the hope of demonstrating implicit learning. Implicit learning (see Cleeremans *et al.*, 1998, for a review) is a notoriously challenging domain in so far as demonstrating the involvement of unconscious knowledge is concerned, for it presents most of the challenges we have examined so far.

Thus, in a recent series of experiments, Alamia *et al.* (2016) asked participants to detect the direction most of the dots contained in a random-dot kinematogram were moving. On some trial blocks, the entire patch of dots was coloured either green, red, or blue. Crucially, two of the colours were predictive of the response and the third was neutral. There was thus a very simple way for participants to improve their performance in this otherwise very difficult task (difficulty was staircased individually). Crucially, however, participants were not told about the predictive value of the colours. Instead, they were told that the colours were introduced to make the task more difficult, and that they would have to report the colour on some of the trials.

We found that people quickly learned to use the colours to improve their performance, albeit most were (1) unable to report noticing the association between colour and motion direction, (2) unable to carry out generation tasks asking them to report the colour associated with a particular motion direction, and (3) unable to recognize whether a colour/motion association was familiar or not. Thus, albeit all of these tests of awareness had to be administered after the main experiment was over so as to avoid the observer paradox, we are confident that these results demonstrate that people were able to unconsciously use available predictive information to improve their performance, thus demonstrating implicit learning. The central feature that makes this design successful is probably the use of a convincing story to characterize the function of the colours. By *changing the narrative* in this way, we made sure people actually pay attention to the relevant stimuli (which is necessary for learning to take place), while at the same time ensuring that they form inaccurate metarepresentations about the role that such stimuli play in the task.

5. Conclusion

From the perspective presented here, the brain is continuously and unconsciously learning to anticipate the consequences of action or activity on itself, on the world, and on other people. Thus, we have three closely interwoven loops (Figure 3) all driven by the very same prediction-based mechanisms. A first internal or 'inner loop' involves the brain redescribing its own representations to itself as a result of its continuous unconscious attempts at predicting how activity in one region influences activity in other regions. In this light, consciousness amounts to the brain performing signal detection on its own representations (Lau, 2008), so continuously striving to achieve a coherent (prediction-based) understanding of itself. It is important to keep in mind that this inner loop in fact involves multiple layers of recurrent connectivity, at different scales throughout the brain. A second 'perception-action loop' results from the agent as a whole predicting the consequences of its actions on the world. The third loop is the 'self-other loop', and links the agent with other agents, again using the exact same set of mechanisms as involved in the other two loops ('As above, so below'). The existence of this third loop is constitutive of conscious experience, I argue, for it is in virtue of the fact that as an agent I am constantly attempting to model other minds that I am able to develop an understanding of myself.

In the absence of such a 'mind loop', the system can never bootstrap itself into developing the implicit, embodied, transparent (Metzinger, 2003) model of itself that forms the basis, through higher-order thought theory, of conscious experience. The processing carried out by the inner loop is thus causally dependent on the existence of both the perception-action loop and the self-other loop, with the entire system thus forming a 'tangled hierarchy' (e.g. Hofstadter's concept of 'a strange loop', see Hofstadter, 2007) of predictive internal models (Pacherie, 2008; Wolpert et al., 2004).

Figure 3. Tangled loops.

To conclude, I would thus like to defend the following claim (see also Cleeremans, 2014): conscious experience occurs if and only if an information processing system has learned about its own representations of the world. Consciousness, in this light, is thus the brain's implicit, embodied theory about itself, gained through experience interacting with itself, with the world, and with other people. It is subtended by continuously operating prediction-driven learning mechanisms applied to all levels of a representational hierarchy that make it possible for cognitive agents to know themselves — something that first-order systems are simply incapable of achieving. What comes 'before' consciousness thus involves the very same information processing and representations than what comes 'after' — with the crucial difference brought about by the involvement of specific kinds of learned meta-representations geared towards redescribing first-order knowledge in increasingly informative ways.

Acknowledgments

A.C. is a Research Director with the Fund for Scientific Research (F.R.S.-FNRS, Belgium). This work was supported by ERC Advanced Grant 'RADICAL' to A.C. and by IAUP Grant P3/77 from the Belgian Science Policy Office (Grant P3/77). The author is grateful for the patience of all involved. This work borrows substantially, with permission, from Cleeremans, A. (2014) Prediction as a computational correlate of consciousness, *International Journal of Anticipatory Computing Systems*, **29**, pp. 3–13.

References

Ahissar, M. & Hochstein, S. (2004) The reverse hierarchy theory of visual perceptual learning, *Trends in Cognitive Sciences*, **8** (10), pp. 457–464.

Alamia, A., de Xivry, J.-J., San Anton, E., Olivier, E., Cleeremans, A. & Zenon, A. (2016) Unconscious associative learning with conscious cues, *Neuroscience of Consciousness*, **2016** (1), pp. 1–10.

Atas, A., Faivre, N., Timmermans, B., Cleeremans, A. & Kouider, S. (2013) Nonconscious learning from crowded sequences, *Psychological Science*, **25** (1), pp. 113–119.

Anzulewicz, A., Asanowicz, D., Windey, B., Paulewicz, B, Wierzchon, M. & Cleeremans, A. (2015) Does level of processing affect the transition from unconscious to conscious perception?, *Consciousness and Cognition*, **36**, pp. 1–11.

Bar, M. (2009) Predictions: A universal principle in the operation of the human brain, *Philosophical Transactions of the Royal Society B*, **364**, pp. 1181–1182.

Blakemore, S.J., Frith, C.D. & Wolpert, D.M. (1999) Spatiotemporal prediction modulates the perception of self-produced stimuli, *Journal of Cognitive Neuroscience*, **11** (5), pp. 551–559.

Chein, J.M. & Schneider, W. (2012) The brain's learning and control architecture, *Current Directions in Psychological Science*, **21** (2), pp. 78–84.

Clark, A. (2013) Whatever next? Predictive brains, situated agents, and the future of cognitive science, *Behavioral and Brain Sciences*, **36** (3), pp. 181–204.

Cleeremans, A. (2008) Consciousness: The radical plasticity thesis, *Progress in Brain Research*, **168**, pp. 19–33.

Cleeremans, A. (2011) The radical plasticity thesis: How the brain learns to be conscious, *Frontiers in Psychology*, **2**, pp. 1–12.

Cleeremans, A. (2014) Connecting conscious and unconscious cognition, *Cognitive Science*, **38** (6), pp. 1286–1315.

Crick, F.H.C. & Koch, C. (2003) A framework for consciousness, *Nature Neuroscience*, **6** (2), pp. 119–126.

Dayan, P., Hinton, G.E. & Neal, R.M. (1995) The Helmholz machine, *Neural Computation*, **7**, pp. 889–904.

Dehaene, S. (2014) *Consciousness and the Brain: Deciphering How the Brain Codes Our Thoughts*, New York: Viking.

Dehaene S., Sergent, C. & Changeux, J.-P. (2003) A neuronal network model linking subjective reports and objective physiological data during conscious perception, *Proceedings of the National Academy of Sciences USA*, **100**, pp. 8520–8525.

Dehaene, S., Changeux, J.-P., Naccache, L., Sackur, J. & Sergent, C. (2006) Conscious, preconscious, and subliminal processing: A testable taxonomy, *Trends in Cognitive Sciences*, **10** (5), pp. 204–211.

Del Cul, A., Baillet, S. & Dehaene, S. (2007) Brain dynamics underlying the nonlinear threshold for access to consciousness, *PLoS Biology*, **5** (10), 417e260.

Dienes, Z. & Perner, J. (2007) The cold control theory of hypnosis, in Jamieson, G. (ed.) *Hypnosis and Conscious States: The Cognitive Neuroscience Perspective*, pp. 293–314, Oxford: Oxford University Press.

Donald, M. (2001) *A Mind So Rare*, New York: W.W. Norton.

Dretske, F. (1986) Misrepresentation, in Bodgan, R. (ed.) *Belief: Form, Content and Function*, pp. 17–36, Oxford: Oxford University Press.

Eriksen, C.W. & Schultz, D.W. (1979) Information processing in visual search: A continuous flow conception and experimental results, *Attention, Perception & Psychophysics*, **25** (4), pp. 249–263.

Fahrenfort, J.J., Scholte, H.S. & Lamme, V.A.F. (2008) The spatiotemporal profile of cortical processing leading up to visual perception, *Journal of Vision*, **8** (1), pp. 42712–42712.

Friston, K. (2006) A free energy principle for the brain, *Journal of Physiology (Paris)*, **100**, pp. 70–87.

Friston, K. (2010) The free energy principle: A unified brain theory?, *Nature Reviews Neuroscience*, **11** (2), pp. 127–138.

Fuster, J.M. (2008) *The Prefrontal Cortex*, 4th ed., London: Academic Press.

Gallup, G.G. (1970) Chimpanzees: Self-recognition, *Science*, **167** (1970), pp. 86–87.

Helmholz, H. (1860/1962) *Handbuch der physiologgischen optic*, Southall, J.P.C. (ed.), English translation ed., vol. 3, New York: Dover.

Hinton, G.E. (2007) Learning multiple layers of representation, *Trends in Cognitive Sciences*, **11**, pp. 428–434.

Hofstadter, D.R. (2007) *I Am a Strange Loop*, New York: Basic Books.

Karmiloff-Smith, A. (1992) *Beyond Modularity: A Developmental Perspective on Cognitive Science*, Cambridge, MA: MIT Press.

Hohwy, J. (2013) *The Predictive Mind*, Oxford: Oxford University Press.

Jaynes, J. (1976) *The Origins of Consciousness and the Breakdown of the Bicameral Mind*, New York: Houghton Mifflin.

Johansson, P., Hall, L., Sikström, S. & Olsson, A. (2005) Failure to detect mismatches between intention and outcome in a simple decision task, *Science*, **310** (5745), pp. 116–119.

Klein, C. & Barron, A.B. (2016) Insects have the capacity for conscious experience, *Animal Sentience*, p. 100.

Koch, C. (2004) *The Quest for Consciousness: A Neurobiological Approach*, Englewood, CO: Roberts & Company Publishers.

Kouider, S., de Gardelle, V., Sackur, J. & Dupoux, E. (2010) How rich is consciousness: The partial awareness hypothesis, *Trends in Cognitive Sciences*, **14** (7), pp. 301-307.

Kreiman, G., Fried, I. & Koch, C. (2002) Single-neuron correlates of subjective vision in the human medial temporal lobe, *Proceedings of the National Academy of Sciences of the USA*, **99**, pp. 8378-8383.

Lamme, V.A.F. (2006) Toward a true neural stance on consciousness, *Trends in Cognitive Sciences*, **10** (11), pp. 494-501.

Lau, H. (2008) A higher-order Bayesian decision theory of consciousness, in Banerjee, R. & Chakrabarti, B.K. (eds.) Models of brain and mind. Physical, computational and psychological approaches, *Progress in Brain Research*, **168**, pp. 35-48, Amsterdam: Elsevier.

Lau, H. & Rosenthal, D. (2011) Empirical support for higher-order thought theories of consciousess, *Trends in Cognitive Sciences*, **15** (8), pp. 365-373.

Lush, P., Naish, P. & Dienes, Z. (2016) Metacognition of intentions in mindfulness and hypnosis, *Neuroscience of Consciousness*, pp. 1-10.

Mack, A. & Rock, I. (1998) *Inattentional Blindness*, Cambridge, MA: MIT Press.

Macphail, E. (1998) *The Evolution of Consciousness*, Oxford: Oxford University Press.

McClelland, J.L. (1979) On the time-relations of mental processes: An examination of systems in cascade, *Psychological Review*, **86**, pp. 287-330.

Magalhães de Saladanha da Gama, P., Slama, H., Caspar, E., Gevers, W. & Cleeremans, A. (2013) Placebo-suggestion modulates conflict resolution in the Stroop task, *PLoS One*, **8** (10), e75701.

McClelland, J.L. & Rumelhart, D.E. (1986) *Parallel Distributed Processing: Explorations in the Microstructure of Cognition*, vol. 2: Psychological and Biological Models, Cambridge, MA: MIT Press.

Metzinger, T. (2003) *Being No One: The Self-Model Theory of Subjectivity*, Cambridge, MA: Bradford Books, MIT Press.

Nelson, T.O. & Narens, L. (1990) Metamemory: A theoretical framework and new findings, *The Psychology of Learning and Motivation*, **26**, pp. 125-173.

Newell, B.R. & Shanks, D.R. (2014) Unconscious influences on decision-making: A critical review, *Behavioural and Brain Sciences*, **37** (1), pp. 1-19.

O'Regan, J.K. & Noë, A. (2001) A sensorimotor account of vision and visual consciousness, *Behavioral and Brain Sciences*, **24** (5), pp. 883-917.

Pacherie, E. (2008) The phenomenology of action: A conceptual framework, *Cognition*, **107**, pp. 179–217.

Pasquali, A., Timmermans, B. & Cleeremans, A. (2010) Know thyself: Metacognitive networks and measures of consciousness, *Cognition*, **117**, pp. 182–190.

Persaud, N., McLeod, P. & Cowey, A. (2007) Post-decision wagering objectively measures awareness, *Nature Neuroscience*, **10** (2), pp. 257–261.

Pessiglione, M., Petrovic, P., Daunizeau, J., Palminteri, S., Dolan, R.J. & Frith, C.D. (2008) Subliminal instrumental conditioning demonstrated in the human brain, *Neuron*, **59**, pp. 561–567.

Ramsøy, T.Z. & Overgaard, M. (2004) Introspection and subliminal perception, *Phenomenology and the Cognitive Sciences*, **3** (1), pp. 1–23.

Rao, R. & Ballard, D. (1999) Predictive coding in the visual cortex: A functional interpretation of some extra-classical receptive field effects, *Nature Neuroscience*, **2** (1), p. 79.

Reingold, E.M. & Merikle, P.M. (1988) Using direct and indirect measures to study perception without awareness, *Perception and Psychophysics*, **44**, pp. 563–575.

Rosenthal, D. (1997) A theory of consciousness, in Block, N., Flanagan, O. & Güzeldere, G. (eds.) *The Nature of Consciousness: Philosophical Debates*, Cambridge, MA: MIT Press.

Rochat, P. (2003) Five levels of self-awareness as they unfold early in life, *Consciousness and Cognition*, **12**, pp. 717–731.

Rumelhart, D.E., Hinton, G.E. & Williams, R. (1986) Learning representations by back-propagating errors, *Nature*, **323**, pp. 533–536.

Sandberg, K., Bibby, B.M., Timmermans, B., Cleeremans, A. & Overgaard, M. (2011) Measuring consciousness: Task accuracy and awareness as sigmoid functions of stimulus duration, *Consciousness & Cognition*, **20**, pp. 1659–1675.

Sandberg, C., Timmermans, B., Overgaard, M. & Cleeremans, A. (2010) Measuring consciousness: Is one measure better than the other?, *Consciousness & Cognition*, **19**, pp. 1069–1078.

Schilbach, L., Timmermans, B., Reddy, V., Costall, A., Bente, G., Schlicht, T. & Vogeley, K. (2013) Toward a second-person neuroscience, *Behavioral and Brain Sciences*, **36** (4), pp. 393–414.

Sergent, C. & Dehaene, S. (2004) Is consciousness a gradual phenomenon? Evidence for an all-or-none bifurcation during the attentional blink, *Psychological Science*, **15** (11), pp. 720–728.

Seth, A.K., Suzuki, K. & Critchley, H.D. (2011) An interoceptive predictive coding model of conscious presence, *Frontiers in Psychology*, **2**, art. 395.

Shanks, D.R. & St John, M. (1994) Characteristics of dissociable human systems, *Behavioral and Brain Sciences*, **17**, pp. 367–447.

Shapiro, K.L., Arnell, K.M. & Raymond, J.E. (1997) The attentional blink, *Trends in Cognitive Sciences*, **1**, pp. 291–295.

Simons, D.J. & Levin, D.T. (1997) Change blindness, *Trends in Cognitive Sciences*, **1**, pp. 261–267.

Skrbina, D.F. (2005) *Panpsychism in the West*, Cambridge, MA: MIT Press.

Smith, B.C. (2006) *Questions of Taste: The Philosophy of Wine*, New York: Oxford University Press.

Timmermans, B., Schilbach, L., Pasquali, A. & Cleeremans, A. (2012) Higher order thoughts in action: Consciousness as an unconscious re-description process, *Philosophical Transactions of the Royal Society B*, **367**, pp. 1412–1423.

Weiskrantz, L. (1986) *Blindsight: A Case Study and Implications*, Oxford: Oxford University Press.

Vermeiren, A. & Cleeremans, A. (2012) The validity of d' measures, *PLoS One*, **7** (2), e31595.

Windey, B., Vermeiren, A., Atas, A. & Cleeremans, A. (2014) The graded and dichotomous nature of visual awareness, *Philosophical Transactions of the Royal Society B: Biological Sciences*, **369** (1641).

Windey, B., Gevers, W. & Cleeremans, A. (2013) Subjective visibility depends on level of processing, *Cognition*, **44** (2), pp. 404–409.

Wolpert, D.M., Doya, K. & Kawato, M. (2004) A unifying computational framework for motor control and social interaction, in Frith, C.D. & Wolpert, D.M. (eds.) *The Neuroscience of Social Interaction*, pp. 305–322, Oxford: Oxford University Press.

David Rosenthal

Higher-Order Awareness, Misrepresentation, and Function

1. Introduction

Theories of the consciousness of mental states fall into two broad types. One type consists of the so-called higher-order theories, and the other type comprises what have come, in contrast, to be called first-order theories. Higher-order theories all explain what it is for states to be conscious by appeal to an awareness of that state; because it is an awareness of another state, we can call it a higher-order awareness (HOA). No state of which one is not in any way aware is a conscious state. First-order theories, in contrast, deny that a state's being conscious involves any such HOA.

Being aware of a state resembles in some ways the awareness that occurs in metacognitive functioning. So it will be important to see in what ways the HOAs that higher-order theories posit resemble and differ from metacognition, strictly so-called.

Because the HOA that such theories invoke is something distinct from other mental properties of conscious states, it seems possible that a HOA could misrepresent what mental states one is in. And first-order theorists have pressed this as an apparently unintuitive consequence of higher-order theories, undermining their credibility.

Another issue sometimes raised against higher-order theories is whether they allow for a convincing explanation of the function of consciousness, that is, of the utility that conscious states have specifically in respect of being conscious. This is a second challenge to higher-order theories that is important to evaluate.

In §2, I outline the way higher-order theories seek to explain the consciousness of mental states, and the basic arguments in favour of that kind of theory, and in §3, I consider whether the HOA such theories posit is relevantly similar to metacognitive functioning. In §4, then, I argue that the possibility of misrepresentation by such HOAs is not after all a disadvantage of such theories. Indeed, it is likely that consciousness does actually often misrepresent our thoughts, desires, and experiences. And in §5, I argue that a state's being conscious adds little, if any, utility to that which results simply from being in such states when they are not conscious. My argument for this somewhat surprising conclusion does not rely on adopting a higher-order theory of consciousness, but the conclusion fits well with those theories.

2. Higher-Order Theories of Conscious Awareness

A theory of consciousness may serve various explanatory purposes. It may tell us which neural processes subserve a mental state's being conscious. Independent of that, such a theory might tell what it is for a mental state to be conscious, why any states are conscious, and perhaps even why particular neural processes do subserve the conscious states they do. However, whatever other explanations a theory of consciousness may provide, it must at a minimum tell us how mental states that are conscious differ from those that are not. Unless it does that, it must remain unclear that the theory is about consciousness at all (Rosenthal, 2005; Lau, 2008).

2.1. Higher-order theories

There is a natural way of understanding how conscious states differ from mental states that are not conscious. No mental state is conscious if the individual that is in that state is in no way aware of it. If somebody thinks, desires, or feels something but is wholly unaware of doing so, then that thought, desire, or feeling is not a conscious state.

Experimental work on non-conscious perception typically exploits this common-sense observation. Participants sometimes deny seeing a stimulus even when there is evidence, say from priming, that the relevant visual information has affected psychological processing. The effect on subsequent psychological processing, moreover, typically reflects the perceptual discriminations that are characteristic of conscious visual states, e.g. among colours and shapes. We commonly conclude that the visual state occurred but without being conscious. In such cases, a participant's denial of seeing the stimulus reflects not a failure to see, but simply a lack of awareness of seeing. Things are the same outside experimental work. If a person denies wanting something but acts as people typically do when they want that thing, then we see the person as having that desire, though a desire that is not conscious. Novelists and dramatists have described such situations for centuries.

Higher-order theories take this common-sense observation as basic to understanding how conscious states differ from mental states that are not conscious. Because no mental state of which one is wholly unaware is conscious, conscious states are mental states we are in some suitable way aware of. Higher-order theories differ among themselves about just what kind of awareness is required for a mental state to be conscious, but they are agreed that a state's being conscious involves some form of HOA.

When somebody perceives something subliminally, so that the perception is not conscious, there is nonetheless a kind of awareness of the perceived stimulus. It may sound awkward to speak of a non-conscious state that nonetheless makes one aware of something, but we can distinguish the conscious and non-conscious cases in a completely natural way. When one subliminally perceives something, one is aware of that thing but not consciously aware of it; when one consciously perceives the stimulus, one is consciously aware of it.

Traditional theorists (Locke, 1700/1975, p. 115; Kant, 1787/1998, p. A22/B37) have typically held that the required HOA is perceptual or quasi-perceptual, a view also sometimes championed today (Armstrong, 1980; Lycan, 1996). Some have argued, however, that the HOA is more likely a thought that one is in the relevant state (Rosenthal, 2005; Weisberg, 2010; 2011). Still others have urged that the HOA is internal to the state one is aware of (Brentano, 1874/1973; Kriegel, 2009), though most see the HOA as distinct from the state it makes one aware of. But all these versions of higher-order theory hold that conscious states differ from mental states that are not conscious in virtue of some HOA (Rosenthal, 2004).

2.2. First-order theories

Theories that deny that a state's being conscious consists of one's being aware of it in some suitable way are typically called first-order theories, though these theories often have little in common beyond that denial. Such theories all arguably face difficulty in explaining how conscious states differ from mental states that are not conscious.

Consider the first-order theory owing to Dretske (1993), in which a state is conscious if being in the state results in one's being aware of something else; no awareness of the state is needed. A perception is conscious, then, because it makes one aware of the stimulus. But, Dretske (2006) recognizes that this will not do as it stands; subliminally perceiving something makes one aware of it, just not consciously aware of it. To meet this difficulty, Dretske stipulates that a perception is conscious only if the individual can cite the content of the perception as a justifying reason for doing something (*ibid.*). However, one cannot cite something one is unaware of; so citing the content of a perception requires one to be aware of that perception. Dretske has ended up invoking higher-order considerations to explain how conscious states differ from mental states that are not conscious.

Other first-order theories seek to explain that difference without any higher-order factors. According to Block's (1995) notion of access consciousness, a state is conscious if its content is available for use in reasoning and the rational control of action and speech. In a similar spirit, Baars' (1998) global workspace theory and Dehaene et al.'s (Deahaene and Naccache, 2001; Dehaene, Sergent and Changeux, 2003) neuronal global workspace theory provide that a state is conscious if it is globally available for psychological processing. Mental states that are not conscious, on these views, lack the relevant availability.

But problems face these explanations. For one thing, it is unclear what such global availability has to do with a state's being conscious. The two seem, at first sight, independent, and they probably are. Visual states towards the periphery of the conscious visual field are presumably not globally available, nor are many conscious but stray passing thoughts. And it is likely that many beliefs and desires that are not conscious nonetheless have widespread, substantial effects on action and psychological processing. So, it is far from clear that global availability can explain how conscious states differ from mental states that are not conscious.

Block (1995) also sees conscious qualitative states, which he refers to as phenomenal consciousness, in first-order terms. No HOA figures or is needed on his view for such states to be conscious. Block holds that such states are conscious in virtue of particular sorts of cortical activation; representations in visual cortex, for example, constitute phenomenal consciousness in virtue of suitable cortical activity and connections in visual areas (Block, 2009; 2007).

Because neural activity subserves all mental functioning, including the consciousness of mental states, there will be some difference in neural activation for every difference in mental functioning. So knowing that neural activation differs when there is a difference in mental functioning will help us understand such differences in mental functioning.

2.3. Assessing first- and higher-order theories

But such appeal to neural activation cannot explain what a difference in mental functioning consists of. For that, we need to describe the mental functioning in distinctively psychological terms. The appeal to neural activation can help only when we already have an accurate psychological description of the mental functioning. For conscious qualitative states, we need a grasp in psychological terms of what it is for such states to be conscious. And Block (1978, §1.3) concedes that, on his first-order view, there may be little or nothing informative to say about that.

These considerations aside, there is in any case reason to doubt that activation in visual areas is all that matters to visual states' being conscious. Lau and Passingham (2006) have elegantly isolated cases in which visual performance is matched despite the presence of conscious awareness in some cases but not in others. The only difference functional MRI revealed between cases with and without conscious awareness was activation in mid-dorsolateral prefrontal cortex (PFC). This suggests that the consciousness of perceptual states is not solely due to activation in visual areas, but also to PFC activity. There is substantial additional evidence in support of that conclusion (Lau and Rosenthal, 2011).

The hypothesis that PFC activation is needed for conscious awareness fits well with the higher-order theories of consciousness. Activation in visual areas provides both visual content and visual qualitative character, but by itself does not result in conscious visual awareness. Only when there is also relevant PFC activation do those visual states become conscious. Evidently, PFC activation subserves the relevant awareness of the visual states, which on their own occur without being conscious. Discoveries about neural activation do not by themselves explain how conscious states differ from mental states that are not conscious, but they sustain a higher-order explanation cast in distinctively psychological terms.

When visual content is transmitted from visual areas to PFC, the relevant visual states are conscious (Dehaene and Naccache, 2001; Dehaene, Sergent and Changeux, 2003). But that finding cannot by itself decide between the neuronal global workspace theory and a higher-order theory of consciousness. Neural activation in PFC might be relevant because it makes the representational content of the visual states globally available, but it might instead subserve an awareness of the relevant perceptual states.

PFC subserves many distinct psychological processes. So PFC activation when states are conscious but not when they fail to be conscious cannot by itself decide between a higher-order theory and the neuronal global workspace theory. But further investigation is needed to determine whether the PFC activation that accompanies states' being conscious serves to make their content globally available or instead subserves HOAs of those states. Making a state's content globally available is a distinct psychological process from producing an awareness of that state; the neuronal global workspace theory posits that the first type of process is responsible for a state's being conscious, whereas higher-order theories explain that by appeal to the second type of process.

In any case, we have seen that there is reason, independent of any neural findings, to prefer the higher-order theory. Many conscious states are doubtless globally accessible to various cortical systems, and many non-conscious states are not thus accessible, as work by Dehaene and co-workers (Dehaene and Naccache, 2001; Dehaene, Sergent and Changeux, 2003) has made evident. But because counter-examples exist to both generalizations, the appeal to global availability cannot explain why some states are conscious and others not. Nor, as already noted, is that surprising; even if most conscious states are globally available and most non-conscious states are not, it is unclear what a state's being globally accessible has specifically to do with its being conscious. So, even though PFC activation by itself lends support both to higher-order and to global workspace theories, only the appeal to HOAs underwrites a satisfactory explanation of why some states are conscious and others not.

In our everyday dealings, we have little, if any, concern with mental states that are not conscious. Circumstances must become somewhat special before we ask whether somebody has a thought, desire, perception, or sensation that is not conscious. Such circumstances apart, when somebody denies being in such a state, we take that denial at face value. We may sometimes have reason to override such denials, perhaps more often with desires and emotions than with bodily sensations, but we rarely have reason with any kind of mental state to doubt others' sincere denials.

This everyday practice of crediting what others say about what mental states they are in can encourage seeing conscious states as the norm, and non-conscious states, such as subliminal perceptions, as at best degenerate cases of such states. Crediting others' sincere reports about their mental states leads to our discounting the possibility that they are in mental states they are unaware of, and so that any of their mental states fail to be conscious. We rely on their first-person awareness of mental states to determine what mental states they are in, *tout court*, thereby collapsing the distinction between mental states and conscious mental states.

All this may seem to support the adoption of a first-order theory of consciousness. If, as Block (2011a) suggests, the default for mental states is that they are conscious, then perhaps we should not seek an explanation cast in distinctively psychological terms of how conscious states differ from mental states that are not conscious. And then there may seem no need to invoke awareness of some mental states to distinguish the conscious cases from those that are not conscious.

This explanatory retreat would make it appealing to find some purely neuronal explanation of what it is for a state to be conscious, or an explanation that appeals to global availability. We would no longer be concerned to explain how conscious states differ from mental states that are not conscious, but only to give some neuronal condition to occur whenever conscious states occur. In addition, if we took being conscious to be the default for mental states, it might seem tempting to regard any HOA that does figure in mental states' being conscious as somehow intrinsic or internal to mental states (Kriegel, 2009), a theory Block has called same order (2007; 2011b).

But we should not permit our everyday practice of taking people's sincere views about what mental states they are in to influence theorizing about consciousness. We have ample evidence of individuals being in mental states they are unaware of, and their views and remarks about what states they are in can at best reflect only their awareness of what states they are in. Verbal reports indicate reliably what states an individual is aware of, but not what states occur without the individual's being aware of them (Weiskrantz, 1997; 1998). We can understand the difference between conscious and non-conscious mental states only by appeal to whether there is a suitable HOA.

Nor is there good reason to see such HOAs as intrinsic or internal to the states they make one aware of. For one thing, findings by Libet *et al.* (1983) and Haggard (1999) show that the neural occurrence that leads to a particular action occurs in advance of any awareness of a volition to perform that action. We can best regard the antecedent neural event as the volition to perform an action, initially occurring without being conscious and only subsequently becoming conscious (Rosenthal, 2002). It is difficult to square these findings with the hypothesis that the HOA of a volition or other mental state is intrinsic to that state.

More generally, a theory on which HOAs are intrinsic to the states they make one aware of must provide an independent reason to individual mental states that yields this result. The default assumption would be that a HOA is distinct from the first-order state because they are about different things. We can therefore best see the hypothesis that HOAs are intrinsic to the states they are about as an attempt to split the difference between first- and higher-order theories. The hypothesis concedes that no state is conscious without some HOA, but nonetheless joins first-order theories in positing nothing beyond the conscious state itself. But there is no theoretical advantage to this attempted marriage of first- with higher-order theory; the explanatory work is all done by the HOA.

3. Metacognition and Higher-Order Awareness

The term 'metacognition' covers a broad range of phenomena in which individuals have some knowledge or sense of their own cognitive functioning. That suggests that the awareness we have of mental states when those states are conscious may be a type of metacognition. Many first-order mental states, perhaps all of them, represent the world as being one way or another, and are in that way cognitive. So being aware of oneself as being in some such state looks at first sight like a form of metacognition. The states posited by higher-order theorists are higher-order in virtue of their higher-order intentional content—that is, content that is about another mental state. And the judgments that figure in metacognition also have higher-order content, that is, content about first-order cognitive states.

3.1. Differences between metacognition and higher-order awareness

However, the phenomena typically classified as metacognitive differ in crucial ways from these HOAs. Indeed, there is reason to believe that standard types of metacognition occur without any conscious awareness at all (Koriat, 2007).

Standard types of metacognition have to do with whether something currently being learned or having previously been learned will be readily recalled in the future. Nelson and Narens (1994) describe judgments about information that cannot currently be recalled, but likely will be in the future, as feeling-of-knowing (FOK) judgments; they describe judgments that information currently available will continue to be as judgments of learning (JOL). Others distinguish the two in slightly different ways. In the tip-of-the-tongue (TOT) phenomenon, one may feel one knows, e.g. what George Eliot's real name was, though one cannot currently recall it; some see TOT as a type of FOK judgment, and others not (Nelson, 1992).

Each of these types of metacognitive judgment concerns conscious availability of some first-order informational content. But those judgments differ strikingly from the HOAs that result in one's being aware of a mental state. In TOT and FOK, the first-order informational state is not currently conscious, and the metacognitive judgments concern only future conscious availability of the relevant information. In JOL, though the first-order informational state is currently conscious, the metacognitive judgment again pertains only to future conscious availability. And although all these judgments are about first-order informational states, they are in each case predictions about future recall of the relevant information. Like HOAs, metacognitive judgments have higher-order content because they are about other mental states, but unlike HOAs, they do not operate to make one aware that one is in the state they are about.

These metacognitive judgments differ in another crucial way from the HOAs posited in higher-order theories. We are seldom aware of any such HOAs. Many mental states are conscious, and when they are, we are aware of those states. However, we are rarely also aware of any higher-order states directed upon them. Higher-order theories predict this; no HOA would itself be conscious unless, in addition to the HOA, there were a third-order awareness that made one aware of that second-order awareness. We can expect that such third-order awarenesses are rare. When one introspects some conscious state that one is in, one is then aware of focusing attentively on that state; so one is aware of one's awareness of the state. A third-order state occurs in introspective awareness of a conscious state, but not otherwise. It is a disadvantage of the view that HOAs are intrinsic to the states they make one aware of that they make the wrong prediction about this, holding that we are aware of all HOAs (Kriegel, 2009).

The metacognitive judgments that occur in FOK, TOT, and JOL, in contrast, are all conscious judgments. In experimental work on metacognition, participants are plainly aware of having those judgments (Nelson, 1992; Metcalfe and Shimamura, 1994). But these judgments are not HOAs because their content does not describe one in terms of current mental states. Rather, they are judgments about likely future recall of information, that is, about what states one may come to be in. So although the metacognitive judgments are conscious, it does not involve third-order, introspective awareness of a current state.

The contrast just drawn between HOAs and metacognitive judgments may strike one as paradoxical. HOAs are seldom themselves conscious states, yet they result in the relevant first-order states' being conscious; metacognitive judgments are often if not always conscious, and yet typically do not result in first-order states' being conscious. How can that be? Moreover, how can HOAs that are not themselves conscious make the first-order states they are about conscious?

On higher-order theories, first-order states do not inherit the property of being conscious from higher-order states. On such theories, the property of a state's being conscious consists of one's being aware of oneself as being in that state, and the higher-order states constitute those awarenesses. The HOA does not pass along the property of being conscious to the first-order state; it simply serves to make one aware of that state in the right way, and that is what the state's being conscious consists of.

In contrast, many metacognitive judgments do not make one aware of oneself as being in some first-order state. Rather, the metacognition states in JOL and FOK are judgments about what knowledge one has acquired and what knowledge one will retain and be able to produce at will. Although metacognitive judgments, such as HOAs, are about first-order states, many metacognitive judgments do not have the content that one is now in a particular state.

3.2. Tip-of-the-tongue phenomena

Some care is needed in understanding how consciousness figures in TOT. Although one is unable in such cases to recall the specified information, e.g. George Eliot's real name, one has a sense, often quite compelling, that one will at some time readily recall it. So, one regards oneself as having the information despite one's current inability to recall it.

Because one is aware of oneself as having the information, one regards oneself as having the belief, say, that George Eliot's real name is Mary Anne Evans even though one cannot just now get at the content of that belief. Indeed, the sense that one has such a belief is typically quite compelling. So, one is aware of oneself as being in the relevant mental state. Is this a counter-example to higher-order theories, on which being suitably aware of being in a mental state suffices for that state to be conscious? If not, why is the state that carries the information that George Eliot's real name is Mary Anne Evans not conscious in the TOT situation?

In TOT, one is aware of a mental state that, as it happens, carries that information; one is aware of being in a state whose informational content would say what George Eliot's real name is. But one is not aware of that state in respect of that specific information. One is aware of the state only in respect of one aspect of its intentional content, namely that it has intentional content that would answer the question of what George Eliot's real name is. It is the awareness of the state in respect of only an aspect of its intentional content that is responsible for the notorious subjective oddness of the TOT situation.

Does that awareness result in the state's being conscious? It may not seem obvious what to say. The awareness we have of our conscious states does not always reveal all their representational character; indeed, HOAs seldom capture every aspect of the first-order state. Still, it seems misleading, at best, to say that the TOT feeling results in the informational state's being conscious. That is because the HOA leaves out too much of the state's content; one would regard the state as conscious only if one were aware of the state in respect of that aspect of the content that is of current interest (Rosenthal, 2000).

Metacognition is not confined to TOT, FOK, and JOL. Change blindness consists of a failure to detect many visual changes in a scene, often changes that are relatively salient. However, people are also typically unaware of such failures to detect such changes. Levin *et al.* (2000) regard this as a metacognitive failure, which they refer to as change blindness blindness. It is a metacognitive failure in reflecting a judgment that people generally detect any significant change that occurs in a visual scene. And many authors think of metacognition in similarly broad terms, as even encompassing HOAs of one's conscious states (Overgaard and Sandberg, 2012). The present discussion will confine itself, however, to types of metacognition along the lines of JOL, FOK, and TOT.

3.3. Metacognition and conscious awareness

Not every way of being aware of a mental state results in that state's being conscious. As many have noted (Locke, 1700/1975, p. 335; Descartes, 1641/1984, pp. 77, 171), the awareness must be subjectively unmediated; when one's awareness of being in a state seems subjectively mediated, the state is not conscious.

Suppose I behave happily or angrily or as though I would like to have a particular thing, but I sincerely deny that I am happy or angry or that I desire that thing. We have reason to conclude that I am in the relevant state, though the state is not conscious. Now you tell me that I am behaving in a way that suggests that I am actually in the state and, having confidence in your judgments about such things, I believe you. I become aware of being in the state, but as long as I am thus aware only because I believe what you tell me, that awareness does not result in the state's being conscious.

But it could happen instead that your telling me triggers an independent awareness of being in the relevant state. Although I would not have been aware of the state without your remark, I am now aware of the state independently of taking your word for it; I would be aware of it even if I came to see your judgment as in some way ill-founded. My awareness is subjectively unmediated, and the state is conscious. We can explain this by saying that the awareness does not seem subjectively to depend on any inference or self-observation; if asked, I would not cite such factors as a basis for my awareness that I am in the state. One does not, contrary to Block's (2011c) suggestion, also need an additional HOA to the effect that there is no inference.

Traditional theorists (Locke, 1700/1975, pp. 335, 592; Brentano, 1874/1973; Descartes, 1641/1984, p. 113) maintained that one's awareness of a state must actually be direct, but that overshoots; the awareness need only seem to be direct. So the awareness might actually depend on some inference or self-observation so long as it does not subjectively seem that way to the individual.

Applying this condition requires care. Suppose a therapist tells one that one is in a deeply repressed state, and one believes that. And suppose that one then forgets who told one, and hence how one learned of the repressed state, but one still believes one is in it. Block (2011c, p. 446) argues that this should result in its seeming to one that one's awareness is independent of any inference or self-observation. But simply forgetting the source would by itself not result in one's awareness of the state seeming subjectively not to depend on any inference or self-observation. Unless something else occurred, it would still seem that one's awareness rested on some such mediating factors, despite one's inability to recall which. Such a case would, in that way, resemble TOT; one would have a compelling sense that one had learned from some source that one was in the state, but could not recall what the source was. Only if one became aware of the state in a way that did not seem subjectively to rely on some such source or other would the state come to be conscious.

Like the HOAs that result in mental states being conscious, metacognitive judgments seem unmediated or direct; it does not seem subjectively that they depend on any inference or self-observation. The crucial difference between HOAs and metacognitive judgments lies in their content. The metacognitive judgments considered earlier are judgments about what information one is likely to recall in the future; HOAs all have the content that one is currently in the state in question.

Although metacognitive judgments are typically conscious (Fernandez-Duque, Baird and Posner (2000a), Kentridge and Heywood (2000) have noted that metacognitive processing can occur without conscious awareness at all. G.Y., an individual with blindsight, was able to adapt his visual processing in response to a visual cue presented within the blind field. As Kentridge and Heywood note, such adaptation is typically taken to reflect monitoring of one's psychological processing, and hence is seen as metacognitive in nature. Although striking, this finding should occasion no surprise, because metacognitive judgments are distinct from the type of awareness that results in mental states' being conscious.

Some who deny that a state's being conscious consists of being aware of that state in a suitable way have stigmatized any such awareness as metacognitive awareness, irrelevant to the state's being conscious. Thus, Seth (2008) writes that 'sensory content need not be overlain by metacognitive content in order to be conscious' (p. 981). And Block (1995) sees HOAs as occurring only when a state is reflectively or introspectively conscious, not when it is conscious in a non-introspective way. This evidently fails to take into account that the awareness that higher-order theories posit is seldom itself conscious. When a

HOA is conscious, the state is introspectively conscious, and the awareness of it does seem something like standard types of conscious metacognitive judgment. A state of which one is wholly unaware is not in any intuitive way a conscious state, but the needed awareness typically falls well short of metacognitive processing as that is usually understood.

3.4. Reports, confidence measures, and post-decision wagering

Reliance on an individual's report of being in a state or not being in it to determine whether the state is conscious (Weiskrantz, 1997; 1998; Marcel, 1983) is known as a subjective measure (Dienes, 2008). Such measures reflect a higher-order theory of what it is for mental states to be conscious, because a report that one is in a state expresses an awareness of being in that state, and similarly for a report that one is not.

An alternative measure, also subjective in nature, is to have participants rate the degree of confidence they have about a particular stimulus. Instead of relying on explicit reports about whether participants are in particular states, one can ask them to rate how confident they are that a perceptual decision about a stimulus is accurate. High confidence suggests that participants have conscious awareness of the stimulus. Low confidence, in contrast, suggests that participants take themselves just to be guessing; a highly accurate perceptual decision with low confidence suggests that a participant is perceptually detecting the stimulus, but not consciously. Such confidence ratings may therefore reveal whether a HOA is present, and they differ from cases of metacognition, such as JOL and FOK, because such ratings are described in terms of accuracy of a current perceptual judgment, not in terms of whether one has or will have command over some particular information.

Persaud, McLeod and Cowey (2007) have developed a test for mental states' being conscious that arguably improves on such confidence ratings. Instead of explicitly rating their degree of confidence, participants are asked to place a wager on a perceptual decision they have made, say, about whether a stimulus is present or a string of letters exhibits a particular pattern. Wagering presumably reflects degree of confidence, but motivates subjects to act on that degree of confidence and so may avoid methodological issues about confidence ratings (Koch and Preuschoff, 2007).

Schurger and Sher (2008) and Dienes and Seth (2010) independently pointed to the difficulty that loss aversion in some subjects (Mellor, 1971) creates for the wagering test, and have suggested ways to adjust for that. But Dienes and Seth have shown that even when wagering is adjusted for loss aversion, it is no more sensitive than traditional confidence ratings as an indicator of conscious awareness. See also Overgaard and Sandberg (2012) for a useful review of the wagering method for determining confidence ratings.

There is, moreover, a question about whether the non-conscious information that influences conscious perceptual decisions might also affect conscious wagering behaviour. Presumably, it does not typically do so, because post-decision wagering results conform well to traditional subjective reports (Persaud, McLeod and Cowey, 2007; Dienes and Seth, 2010), and people may typically base wagers only on information they are consciously aware of. But unconsciously seen stimuli do affect inhibitory mechanisms (van Gaal *et al.*, 2008), and so might also affect wagering. That would influence traditional confidence ratings as well as post-decision wagering, leaving subjective reports as the standard against which other measures must be calibrated.

4. Higher-Order Misrepresentation

Metacognitive judgments are by no means always accurate. JOL may misgauge how much one will recall (Schnitzpahn *et al.*, 2011), and feelings of knowing can misrepresent how much information one has command over (Paynter, Reder and Kieffaber, 2009). Indeed, various techniques have been proposed for enhancing the accuracy of metacognitive judgments (Pieschl, 2009). And change blindness illustrates a routine, persistent error in the metacognitive expectation that we readily detect visual changes (Scholl, Simons and Levin, 2004; Levin *et al.*, 2000).

Metacognitive error is hardly surprising. Although metacognitive judgments are often sensitive to actual cognitive abilities and achievements, such judgments are distinct from the cognitive states they are about. So, there will always be room for metacognitive judgments to be wrong about those cognitive states.

4.1. Misrepresentation by conscious awareness

This suggests a parallel question about the consciousness of mental states. On higher-order theories, a state is conscious in virtue of one's being aware of that state, and that awareness is something distinct from the mental properties it makes one aware of. This is clear on standard higher-order theories, which hold that the HOA and the state it makes one aware of are distinct states. But even on higher-order theories that maintain that the HOA is intrinsic to the target state (Brentano, 1874/1973; Kriegel, 2009), the HOA is distinct from those mental properties it makes one aware of and in respect of which we taxonomize the conscious state. Because the HOA is distinct from the state one is aware of, or at least distinct from the mental properties in respect of which one is aware of it, such theories allow for possible misrepresentation by consciousness. In allowing for such misrepresentation, consciousness resembles metacognition.

Consciousness is the way our mental lives appear to us; it is mental appearance. And because appearances are not always accurate, there should be no surprise that consciousness can misrepresent the mental states that actually occur in us. Consciousness may make it appear subjectively that one is in a mental state somewhat different from the state one is actually in.

But there is a traditional tendency to think that, unlike other appearances, the appearances of consciousness cannot diverge from reality, that in respect of the appearances of consciousness the mind is transparent to itself (Locke, 1700/1975, pp. 364, 592; Kant, 1787/1998, p. B132; Brentano, 1874/1973; Descartes, 1641/1984, pp. 77, 171). If so, consciousness could never misrepresent the mental reality it makes us aware of; when it comes to consciousness, there is no distinction between appearance and reality (Nagel, 1974). Consciousness would not merely be the way our mental lives appear to us; it would actually constitute that mental reality. Anything we think we know about the mind would then have to be tested against the deliverances of consciousness. If appearance and reality coincide for consciousness and mind, then misrepresentation by consciousness is plainly impossible. Because higher-order theories accommodate that possibility, such theories could not then be correct.

It is worth stressing that the apparent difficulty for higher-order theories is not about whether consciousness does sometimes misrepresent what mental states individuals are in, though there is reason to think it actually does. But the alleged difficulty concerns whether such misrepresentation is even possible. A number of theorists have pursued this criticism, urging that the mere possibility of misrepresentation leads to incoherent or absurd consequences (Block, 2011b,c; Levine, 2001; Neander, 1998; Wilberg, 2010). Any theory that allows consciousness to misrepresent mental reality must, they maintain, be mistaken about the very nature of consciousness and mind.

4.2. Kinds of higher-order misrepresentation

The idea that misrepresentation is absurd or incoherent recalls the traditional idea that first-person access to mental states is infallible. Only if first-person access is infallible would misrepresentation by consciousness be impossible. But it is plain that first-person access is not infallible. Mental states sometimes occur without one's being in any way aware of them, as in subliminal perception and other cases.

Consciousness in these cases erroneously rules that no such states occur. There are mental states we can detect even when consciousness leads subjects to deny their occurrence.

One can always seek to defend a theory by redescribing the data. So one might deny that subliminal states are mental at all, perhaps describing them as subpersonal states, mere 'events of content fixation', as Dennett (1991) urges. Or one could urge, with Block (1995; 2007; 2009), that even some perceptual states that one is wholly unaware of and sincerely denies the occurrence of are conscious in a special way that does not require any such awareness. But states that occur in subliminal perception function in many or even most of the ways that conscious perceptual states function (Ro et al., 2009). So denying that subliminal states are qualitative or even mental at all is simply redescribing common-sense phenomena to save a theoretical preconception. And if first-person access is not infallible, there can be nothing incoherent or absurd about the possibility of misrepresentation by consciousness.

A finding by Breitmeyer, Ro and Singhal (2004) (see also Breitmeyer et al., 2007) may seem to provide a substantive reason to question whether genuinely qualitative states do occur in subliminal perception. Breitmeyer et al. found that in metacontrast masking of unsaturated blue, green, and white stimuli, the priming effects of the white stimulus resemble those of the green stimulus more than the blue. Because the white stimulus had greater contributions of green wavelengths than of blue, Breitmeyer, Ro and Singhal concluded that priming effects in these masked cases reflect earlier visual processing in area V1 that is mainly responsive to wavelength properties. This effect was not found when stimuli were unmasked and consciously visible. Green stimuli in those cases did not prime like white stimuli, presumably reflecting cortical activity in higher visual areas specialized for colour. Breitmeyer et al. described the masked effect as wavelength-dependent, in contrast with what they described as the percept-dependent behaviour of the unmasked primes.

One might take this to show that the mental qualities characteristic of conscious vision do not occur in subliminal processing. But the findings do not support that conclusion. Processing in the higher visual areas specialized for colour may typically be conscious, but these findings do not show that it always is. Nor is it obvious that such processing never fails to be conscious. So if the masked cases do not exhibit mental qualities, that need not be because those cases fail to be conscious; it may instead be because they do not involve processing beyond area V1. It may even be that mental qualities do figure in the masked cases, but that the lack of higher visual processing results in those qualities being taxonomized in the more coarse-grained ways reflected in the findings by Breitmeyer, Ro and Singhal. More would be needed to show that mental qualities do not occur at all in subliminal vision.

Consciousness may misrepresent mental reality, then, by failing to reveal non-conscious mental states of which one is altogether unaware. But how about states that are conscious? Can consciousness fail to represent those states accurately? Not only is that possible, it occurs routinely. When we consciously take in a visual scene, we see the colours that things have, but we are seldom aware of those colours in respect of their exact shades. We consciously see things as light blue or dark red or bright green, but we are not consciously aware of the exact shades we would discern if we attended to the objects and their colours.

One might question whether being aware of specific shades of visual sensations in merely generic terms is properly described as a case of misrepresentation; we do not generally describe cases of incomplete representation as misrepresentation. But in the cases under consideration, first-order states simply have no generic mental properties of the sort that HOAs represent those first-order states as having; the mental properties of first-order states are all specific. So HOAs do sometimes represent the first-order states inaccurately.

There are, in any case, more dramatic cases that we would plainly describe as misrepresentation. Grimes (1996) used eye trackers to switch displays during participants' saccades, when no retinal signal reaches visual cortex. Participants often were not consciously aware of changes in salient visual features, e.g. in 18 per cent of cases, a dramatic change between red and green central to a display. Because retinal input to the visual cortex resumes after saccades, first-order states in the visual cortex presumably did change in ways that reflected the change in display, despite participants' reporting no awareness of such change. Participants' HOAs of their first-order visual states here plainly misrepresent those states.

But the visual states themselves, independent of conscious awareness, must reflect the more finely differentiated shades, because those shades are readily available whenever we attend to them. The same is true of the mental qualities we are aware of inattentively in other perceptual modalities. Mental qualities occur in more finely differentiated ways than consciousness typically reveals. This is one way in which consciousness misrepresents mental reality, though the misrepresentation is innocuous and is corrected at will by attending. Consciousness misrepresents in connection with other types of mental state as well; Desmurget and co-workers (Desmurget and Sirigu, 2009;

Desmurget *et al.*, 2009) have found evidence of awareness of intentions without intentions, and conversely; here, consciousness represents volitional states inaccurately.

It is worth noting a possible difference between these cases and those reported by Wegner (2002), in which participants are aware of themselves as doing something and thereby causing something to happen even though those effects are due to independent factors. These participants are aware of themselves as being in relevant volitional states and also aware of themselves as causing the target effects. But those two cases of awareness are distinct and can occur independently. Participants might have the volitions they are aware of having even though those volitions do not figure in causing the target effects. Additional evidence would be needed to show that they also lack those volitions.

4.3. Attention, conscious awareness and misrepresentation

Attending cannot by itself always yield a more finely differentiated awareness of mental qualities. We are consciously aware of shades of colour in more finely differentiated ways when we have two or more very close shades to compare at one time than when those shades are presented in succession (Pérez-Carpinell *et al.*, 1998; Halsey and Chapanis, 1951). This also holds for conscious discrimination of auditory pitch (Seashore, 1967). We have no reason to hold that the visual and auditory cortex represent shades and pitches in less fine-grained ways when they occur successively rather than together. So when stimuli do occur in succession, consciousness fails to reveal fine-grained differences that actually occur among the relevant mental qualities. And this effect is altogether independent of attention.

Attention enhances the way we are aware of our perceptual experiences. But one might urge that some attention, perhaps quite low, is needed for a HOA to form in the first place. This is unlikely. For one thing, there is now ample evidence of a double dissociation between attention and mental states' being conscious (van Bostel, Tsuchiya and Koch, 2010; van Gaal and Fahrenfort, 2008; Schurger *et al.*, 2008; Koch and Tsuchiya, 2007; Kentridge, Heywood and Weiskrantz, 2004); attention occurs in connection with states that are not conscious, and is absent with many states that are conscious. And it is in any case subjectively implausible that attention is needed for states to be conscious; many peripheral visual states are conscious but seemingly unattended. One could urge that there is always some attention, however slight; when a state is conscious, but without independent evidence of attention, it is unclear what that claim amounts to.

If attention is not responsible for the generation of HOAs, what is? I argue in §5 that HOAs contribute little or no utility over and above the utility of the first-order states themselves; so we cannot appeal to evolutionary selection pressures to explain why HOAs occur. This issue would go beyond the scope of the present study, though I have addressed it elsewhere (Rosenthal, 2005, chapter 7, §6, chapter 10, §5].

4.4. Other doubts about higher-order misrepresentation

Block (2011c) has urged in reply to me (Rosenthal, 2011) that one cannot be consciously aware of a stimulus in respect of a less finely differentiated shade than is represented in the visual cortex. But he misconstrues the claim, arguing just that there cannot 'be an experience of red but not of any shade of red' (Block, 2011c, p. 445). The claim is not that a conscious experience of red might not represent the shade as red at all, but that it might represent it in a way that is indeterminate with respect to a range of specific shades of red. One is aware of the experience as an experience of red, but not, e.g. as more like maroon than crimson. This kind of conscious awareness is not only possible; it is routine.

It is important to note a possible ambiguity of the term 'experience'. By 'experience', one might mean a sensation or perception of red, independent of whether and if so how that state is conscious. Any such state will exhibit a mental quality of a specific shade of red. But things are different if by 'experience' one means instead a conscious sensation or perception. Even though the sensation or perception itself, independent of how it is conscious, will exhibit a mental quality of a specific shade, the way we are aware of the state will often fail to reflect that specific shade. So if by 'experience' we mean the qualitative state as one is subjectively aware of it, an experience of red may well simply be of a generic shade.

Failing to capture specific qualitative character is not the only way that consciousness misrepresents our mental states. People sometimes confabulate the thoughts or desires as having led to choices, despite evidence that those thoughts and desires could not have been operative. They are, in these cases, aware of themselves as having thoughts and desires that they do not actually have (Nisbett and Wilson, 1977; Johansson *et al.*, 2005; Frith, 2007). Moreover, expectations distort the way we are subjectively aware of our qualitative experiences, as when one drinks apple juice expecting the taste of iced tea. Expectations can affect even the subjective experience of pain (Koyama *et al.*, 2005). In such cases, one's subjective awareness misrepresents the qualitative state one is actually in.

Why, in the face of such commonplace examples as iced tea and apple juice, would anybody doubt that consciousness sometimes misrepresents actual mental occurrences? Consciousness is the way those mental occurrences subjectively appear to one; why think it cannot get things wrong?

Perhaps, if the subjective awareness were intrinsic not simply to the state but to the represented mental properties themselves, that would prevent consciousness from misrepresenting. There is a compelling sense that nothing mediates between the painfulness of a conscious pain and one's subjective awareness of it; the painfulness and the awareness seem indistinguishable. And if they are not only indistinguishable but the same, perhaps that awareness cannot misrepresent the state in respect of such painfulness.

Its seeming that mental qualities are indistinguishable subjectively from awareness of those qualities appears to point toward a first-order theory of consciousness, on which qualitative states are conscious independent of any awareness of them. And when mental states are conscious, the mental properties in virtue of which they are conscious can seem subjectively inseparable from their being conscious.

But such apparent inseparability does not show that consciousness is built into those other mental properties, independent of one's awareness of those properties. When a state is conscious, one is aware of the state in respect of various individuating mental properties, such as its qualitative character. But one is seldom also aware of being aware of those other mental properties. So the state's being conscious typically seems subjectively inseparable from the other mental properties in virtue of which one is aware of it. And the occurrence of states that have those mental properties without being conscious, which pose a challenge for first-order theories, underscores the independence of a state's being conscious from its other mental properties.

There is independent reason to reject a first-order theory on which a state's being conscious is inseparable from the state's other mental properties. Whatever our subjective impressions about conscious states, we can give full accounts of those other mental properties without appeal to consciousness. This is widely recognized for cognitive and volitional states. Despite many disagreements about how best to explain intentionality, virtually every theory with any level of broad support seeks to explain intentionality independent of the relevant states' being conscious.

Things may seem less clear in the case of qualitative states, because many theorists see mental qualities as inexplicable apart from the way those qualities present themselves to conscious awareness. But as these theorists often acknowledge, this approach permits virtually nothing informative to be said about mental qualities (Block, 1978; Strawson, 2008). So there is reason not to rely on consciousness in giving an account of mental qualities (Rosenthal, 2010).

And there is an inviting alternative. Instead of understanding mental qualities in terms of the way they present themselves to consciousness, we can understand them by appeal to the roles they play in perception. We discriminate among various ranges of perceptible properties of stimuli, and we are able to do this because of the mental qualities special to each perceptual modality. For every perceptual discrimination between two perceptible properties, there must be a difference between corresponding mental qualities. So when we construct the quality space of perceptible properties discriminable by a particular modality, we have an account of the mental qualities that enable those discriminations (Rosenthal, 2005; 2010). Because perceptual discrimination can occur subliminally, this account again makes no appeal to qualitative states' being conscious.

When mental states are conscious, we individuate them not by appeal to their being conscious, because all conscious states have that in common, but rather by appeal to their other mental properties. The mental properties in terms of which we individuate conscious states are the intentional and qualitative properties we are subjectively aware of those states as having. The availability of satisfactory theoretical accounts of those properties independent of consciousness trumps whatever subjective appearance there is that being conscious is intrinsic to those properties. And that leaves us with no reason to doubt that consciousness does sometimes misrepresent the states it makes us aware of.

4.5. Neuronal global workspace theory

As noted in §2, global workspace theory is unusual among first-order theories in providing a distinction between mental states that are conscious and those that are not. So it should not be surprising that, unlike other first-order theories, it does not preclude misrepresentation by consciousness. (I am grateful to Hakwan Lau for raising this point.) A state is conscious on global workspace theory if its content comes to be present in the global workspace and thereby available to many cortical systems. So consciousness in effect represents the state as having whatever content is available to those systems. But the content that reaches the global workspace might sometimes differ from the content of the original state. And consciousness would then, on the theory, misrepresent the state that is conscious.

One might question why, given various similarities between higher-order theories and the neuronal global workspace theory, we should not see them as theories of the same type. As noted in §2, both theories receive support from PFC activation when states are conscious and relative absence of such activation when states are not. Both theories allow, moreover, for a type of misrepresentation by consciousness.

But there are pivotal differences. Most importantly, the neuronal global workspace theory sees a state's being conscious as consisting of the global availability of its content, whereas higher-order theories see a state's being conscious as consisting of one's being aware of oneself as being in that state. These are strikingly different ways to understand the property of a state's being conscious.

The two models of what a state's being conscious consists, moreover, result in distinct ways in which consciousness can misrepresent. On higher-order theories, one is aware of oneself as being in a state in some ways distinct from the state one is actually in. On the neuronal global workspace theory, in contrast, the misrepresentation is due rather to a difference between the content that the global workspace makes available and the content of the original state itself.

The neuronal global workspace is posited to make content available to many systems, and these may well include whatever system subserves HOA and the consequent ability to report one's own conscious states. So there is that much overlap between higher-order theories and the neuronal global workspace theory. But it is the HOA that matters for a state's being conscious, according to higher-order theories, not availability to other cortical systems. And experimental testing for most theories must rely on subjects' verbal reports or the equivalent to determine whether a state is conscious.

4.6. Misrepresentation and what it is like

On a higher-order theory, a state is conscious in virtue of one's being aware of oneself as being in that state. On a standard construal of the phrase, 'what it is like' (Nagel, 1974), what it is like for one to be in a state is a matter of how it is for one to be in that state, that is, how one is subjectively aware of oneself as being in the state. So the HOA one has of being in a state will determine what it is like for one to be in it.

On some higher-order theories, my own included, that HOA consists of a thought to the effect that one is in the relevant state. And one might question whether such thoughts could be responsible for there being something it's like for one (Block, 2011c, p. 444). But there is reason to think that it can (Rosenthal, 2005). Learning new words for one's qualitative experiences can sometimes by itself result in coming to be aware of one's experiences in more fine-grained ways, as with learning terms for tastes of wines or for musical instruments one could not previously distinguish.

Learning new words involves learning how to apply the concepts that the new words express. And learning a new concept enables one to have a new range of thoughts that employ that concept. The new words and concepts in these cases pertain to one's qualitative experiences; so learning the new words enables one to have new thoughts about one's own qualitative experiences, thoughts that draw more fine-grained distinctions among them.

There is no other effect that learning new words might have that would enable one to be subjectively aware of those more fine-grained qualitative distinctions. So the best explanation for this effect of learning new words that draw more fine-grained qualitative distinctions is that the new concepts enable one to have more fine-grained higher-order thoughts about the relevant qualitative states. This demonstrates that differences in the intentional content of higher-order thoughts makes a corresponding difference to the way we are subjectively aware of qualitative states. It therefore lends credibility to the claim that higher-order thoughts are indeed responsible for there being something it's like for one to be in conscious qualitative states in the first place.

This effect is seldom evident in adulthood, because it is rare past young childhood that we learn new terms for our qualitative states, and hence new ways to be aware of those states. So it is useful to focus on such unusual cases as coming to have more fine-grained conscious experiences of wines or musical instruments.

It may seem that higher-order-thought theory makes mistaken predictions about what conscious experiences are possible. As Block (2011c, p. 445) urges, because nothing rules out having a thought that one has an experience of red and green fully occupying the same region at once, perhaps higher-order thought theory predicts that such an experience is possible.

Surprisingly, something like that conscious experience seems actually to occur; in a striking study, Billock, Gleason and Tsou (2001) found that when multi-coloured, equiluminant images are stabilized on the retina, some subjects report seeing reddish greens or bluish yellows, presumably owing to filling-in mechanisms. But higher-order theories are in any case not committed to such predictions. Higher-order thoughts must have an assertoric mental attitude, because doubting and wondering do not by themselves result in one's being aware of anything. And one cannot have any assertoric thought one chooses. It is doubtful that one can assertorically think that 2 + 3 = 57, or even that one can have an assertoric thought that it is raining as one gazes out on a clear, cloudless sky. So it is likely that what mental states we can be in constrain in various ways what assertoric higher-order thoughts we can have.

5. Function, Content, and Psychological Ascent

Metacognitive judgments have often considerable utility; it is plainly useful for a person or other creature to have some sense of how much has been learned (JOL) and how much will be able to be recalled (FOK). Moreover, metacognitive regulation of various sorts is often (Fernandez-Duque, Baird and Posner, 2000a,b), if not always (Kentridge and Heywood, 2000), conscious. That suggests a parallel question about mental states' being conscious independent of distinctively metacognitive processes. Is there any utility to a mental state's being conscious over and above that mental state's occurring without being conscious? And if so, what does that utility consist of?

Metacognition and consciousness are distinct phenomena; so metacognitive utility will not automatically carry over to the consciousness of mental states. But similarities between the two suggest that because metacognition is highly useful, consciousness may have some utility as well.

Metacognitive judgments are judgments about representational states, and mental states' being conscious arguably always involves one's being aware of those states. On a higher-order theory, mental states' being conscious consists of such awareness, and even on first-order theories a state's being conscious may typically facilitate such awareness. So it seems inviting to speculate that there will be some utility to mental states' being conscious beyond the utility of mental states that occur without being conscious.

And there are considerations apart from the analogy with metacognition that may lead us to expect that consciousness plays some useful role. We often seek to understand features of the psychology of people as well as other animals by appeal to some utility those features have. So it is natural to look for a positive role played by mental states' being conscious. The search for such utility may result from a focus on evolutionary selection pressures, but it need not; it is often productive, independent of issues about evolutionary origin, to look for some utility that aspects of psychological functioning exhibit.

But the theoretical hunch that we will find some utility must not mislead us. We often understand aspects of psychological functioning by appeal not to their utility, but simply to the factors that cause the psychological functioning under investigation. So we must hold open the possibility that mental states' being conscious adds little or no utility to what those states would have, even without being conscious. In that case, we would have to explain why many states are conscious by appeal to those factors that cause those states to be conscious.

There are two broad areas that theorists have examined as promising candidates for such utility, both suggested by the analogy with metacognition. One is internal to the individual, having to do with rationality generally, and more specifically with planning, inference, and executive function. The other pertains to social interactions, including informing others what mental states one is in and other potential ways of facilitating social cohesion.

5.1. Utility owing to rationality

Rationality is a matter of how one's thoughts, desires, goals, and other cognitive and volitional states cohere with one another. Similarly, for the efficacious laying of plans and the drawing of inferences, and executive function; these are all a matter of how one's cognitive and volitional states interact. And the relevant interactions hinge on the intentional content of those states. Rationality and inference depend entirely on rational connections among the contents of cognitive and volitional states, as does the efficacy of our plans. And executive function, which involves inhibitory control and the fine-tuning and adjusting of the volitional states that lead to behaviour, again hinges on the content of those volitional states.

The exclusive dependence of these processes on intentional content is reflected in the widespread recognition, despite other theoretical disagreements, that such content must involve causal connections among states with suitably related content. Otherwise, the interactions among such states on which rationality, planning, and executive control must rely would remain mysterious. Content goes hand in hand with causal potential.

These processes do often occur consciously; we are frequently aware of cognitive and volitional states that constitute these processes. But cognitive and volitional states also often occur without being conscious. And because rationality and efficacy of planning and executive control all hinge on the intentional content of the relevant states, it is unclear why the states' being conscious would in any way enhance those processes.

Even when the processes do involve conscious states, the efficacy of the processes depends on the content of the relevant states, not on the independent property of the states' being conscious (Rosenthal, 2008). Because volitional states occur without being conscious before they come to be conscious (Libet et al., 1983; Haggard, 1999), the same may happen with cognitive states. If so, consciousness will not have any significant utility in either case. This fits well with findings by Dijksterhuis et al. (2006), controversial (Waroqiuer et al., 2010) but recently replicated by Usher et al. (2011), on which decisions about complex matters often turn out better when they result from thought processes that are largely not conscious. Indeed, conscious monitoring of one's thought processes is often awkward and inefficient.

Adopting a traditional inner-sense picture of consciousness, on which awareness of our conscious states is perceptual, makes it tempting to hold that consciousness does play a beneficial role in these rationality and related processes. Perceiving things enables us to negotiate our way among them, pick them up and move them about in ways that suit our purposes. So a perceptual model of the way in which we are aware of our conscious states encourages the idea that being thus aware of those states should enable us to rearrange them in similarly advantageous ways.

A perceptual model of mental states' being conscious is unconvincing, independent of these considerations. Perceiving involves mental qualities that occur in response to perceptible properties and enable us to discriminate among them. But there are no higher-order mental qualities that figure in the way we are aware of our mental states; such awareness is not perceptual (Rosenthal, 2004).

Independent of any such perceptual model, however, mental states are simply not the sorts of things we can rearrange to suit our purposes as we do with the physical objects we perceive. We do revise and adjust our thinking and our volitional states, and executive function facilitates fine-tuning of such states. But such revising and adjusting typically or always hinges on interactions among the relevant states solely as a result of their content. We revise poor plans and irrational thinking because of the conflicts in content they lead to and the causal potential states have exclusively in virtue of their content. Awareness of the states is seldom, if ever, needed (Rosenthal, 2008).

These considerations apply not only to cognitive and volitional states, but also to qualitative states, such as perceptual sensations. Mental qualities, on the account sketched in §4, have distinctive perceptual roles. And the utility a perceptual state has for the organism will be due mainly or even solely to that perceptual role. Because that perceptual role is independent of qualitative states' being conscious, we have no reason to think that their being conscious would add any utility beyond the utility they already have due to perceptual role.

As noted in §4, learning new words that draw more fine-grained distinctions among one's qualitative states sometimes results in one's being aware of those states in more fine-grained ways; one comes to be able to draw more distinctions among one's qualitative states solely in virtue of one's first-person access to those states. But there is again no reason to think that qualitative states' being conscious in respect of these more fine-grained differences would add significant utility.

Differences among qualitative states already enable fine-grained perceptual discriminations independent of the states' being conscious in respect of those fine-grained differences.

So it is unclear what utility is added by the states' coming to be conscious in respect of those differences apart from a measure of aesthetic pleasure. And that pleasure may well occur without being conscious; people sometimes exhibit an aesthetic preference without conscious awareness of what it is in virtue of which they have that preference. So the added utility would consist, at most, of the pleasure's coming to be conscious, something arguably marginal in the context of one's overall psychological functioning.

Because many mental processes occur without conscious awareness, it is natural to look for some utility of conscious awareness by seeing which mental processes are invariably accompanied by such awareness (Frith, 2010). But that is not sufficient. Greater signal strength may independently result in a state's being conscious and in some other effect, much as variations in atmospheric pressure cause changes both in weather and in barometer readings, effects unconnected except for having a common cause. Even when some useful psychological function always occurs consciously, we need evidence that conscious awareness is required for that function and is not simply an irrelevant by-product, an epiphenomenon, as that term is used in medical contexts.

Weiskrantz (1997, chapter 7) notes that flexible thinking seems not to recruit the perceptual contents that fail to occur consciously in disorders such as blindsight, prosopagnosia, and amnesia. And he suggests that a benefit of those perceptual contents' being conscious may be their availability for such flexible thinking. But such disorders involve deficits in mental processing in addition to the failure of some perceptual states to occur consciously; so the unavailability of such states for flexible thinking may not be due to those states' not being conscious. In addition, individuals with these disorders are habituated to recruiting conscious perceptions in their flexible thinking, perhaps only because so much conscious perceiving is available. And with individuals for whom no relevant perceptions are conscious, those that are not conscious can figure in flexible behaviour (de Gelder et al., 2008).

It is important to note a potential ambiguity in the term 'function'. The question of concern here is whether mental states' being conscious has any utility for the organism beyond the occurrence of those states without their being conscious. But we also speak of function as simply the observable or discernible effect something has, as Cohen and Dennett (2011) evidently do. Their argument that consciousness cannot be separated from function hinges on noting that conscious states can be reported (Weiskrantz, 1997; 1998; Marcel, 1983) and that they typically have other behavioural effects. And as they argue, conscious states that cannot be detected also cannot be studied. But that by itself does not help show that mental states' being conscious adds any utility for the organism beyond the utility of those states' occurring without being conscious.

5.2. Utility owing to social interaction

The ability to report conscious states raises the second type of utility that theorists have explored for mental states' being conscious: the utility of enhancing social interactions. If I can tell you what states I am in, that will facilitate interactions between us. And if a mental state is not conscious, I will be unable to tell you about it (Frith, 2010, p. 539).

It is often useful to know what mental states others are in. But being told is not the only way to learn that. Suppose I think it is raining. One way you can learn that I think that is by my explicitly saying, 'I think it is raining'. And I cannot tell you that unless the thought I report is conscious; indeed, my saying 'I think that it is raining' expresses my awareness of that thought, the HOA in virtue of which the thought is conscious.

But you could also learn that I think that it is raining not by my explicitly saying that I think it is, but by my simply saying instead, 'It is raining'. I need not tell you that I have that thought; I can just express my thought verbally. And if I do just say 'It is raining', I do not thereby express my awareness of that thought; I convey the thought, but I do not explicitly mention it. So in this case my thought's being conscious plays no role in my letting you know what I think. Consciousness is not needed for the social utility of letting others know about our thoughts.

The difference between expressing and reporting seems small in the case of an assertoric thought, such as the thought that it is raining, small enough that it is sometimes overlooked (Wittgenstein, 2009). But the difference is vivid in other cases. If one wonders whether it is raining, one expresses one's wondering simply by asking, 'Is it raining?', whereas one reports the state by saying assertorically, 'I wonder whether it is raining'. The difference stands out even more with emotions; typically, we learn of somebody's having some emotion from that person's expressions of emotion, verbal and nonverbal, rather than from the person's explicitly saying, 'I am feeling angry [sad, happy, and so forth]' (Rosenthal, 2005, chapter 11).

The most straightforward way to convey that one thinks something is to say whatever it is that one thinks. Similarly, with doubting, wondering, and so forth, one can always simply express these states verbally.

But one can also move a step up, and explicitly say that one has the thought. We can describe this move from verbally expressing to explicitly reporting as psychological ascent, by analogy with the move from asserting a sentence to saying that the sentence is true, which Quine (1960) calls 'semantic ascent'. And just as describing a sentence as true has no cognitive advantage over simply asserting the sentence, so explicitly saying that one has a particular thought has no utility beyond simply expressing the thought verbally.

Typically, when one verbally expresses a thought, that thought is conscious. But its being conscious is not what enables the thought to be expressed. Verbally expressed thoughts are typically conscious because our ability to describe our own thoughts induces a propensity to be aware of any thoughts one verbally expresses (Rosenthal, 2005, chapter 10). The thought's being conscious plays no role in enabling one to express the thought verbally.

Frith (2010, p. 539) notes that there can be a distinctively social aspect to subjects' reporting conscious awareness of a stimulus. Reports are communicative acts, and expectations that experimenters or others have may sometimes influence those reports. Still, the reporting is not a joint endeavour; a report expresses an individual's awareness or lack of awareness of a mental state, and so reflects whether particular states occur consciously.

Frith (*ibid.*) has also developed a challenging view on which conscious awareness has a distinctively social utility that derives from the conscious awareness of one's own and others' agency. Conscious awareness of our own agency and that of others, he urges, enhances social cooperation. Distinguishing voluntary from non-voluntary behaviour in others is plainly important for cooperative social interactions, and we often distinguish those things consciously.

But young infants also draw that distinction (Tomasello, 2000; Behne *et al.*, 2005), casting doubt on whether conscious awareness is required to do so; indeed, even newborn chicks seem to draw a distinction between animate and inanimate movement (Mascalzonia, Regolina and Vallortigara (2010), suggesting that the discerning of voluntary behaviour may be too primitive to rely on conscious perception. So it is unclear what additional benefit there might be to doing so consciously. Perhaps Frith (2010, p. 535) sees added utility because he holds that conscious awareness is required in deciding for oneself what to do. But it is also unclear that such decisions cannot occur without being conscious (Desmurget and Sirigu, 2009; Desmurget *et al.*, 2009); so conscious awareness may well add little, if any, utility even there.

It is sometimes held that our ability to tell what mental states others are in has some connection with the ability to be aware of our own mental states in the way we are when those states are conscious (Carruthers, 2000). But the two abilities have little, if anything, in common beyond their both making use of relevant concepts of the relevant states. The mind-reading ability that enables us to tell what states others are in relies on nonverbal behaviour, whereas the awareness of one's own conscious states is independent of self-observation. And although metacognitive judgments may enhance mind-reading abilities (Koriat and Ackerman, 2010), both are independent of conscious awareness.

It is worth stressing that the considerations that undermine the widespread conviction that mental states' being conscious adds significant utility do not rest on one's having adopted a higher-order theory of what it is for mental states to be conscious. A number of the considerations that point to an absence of added utility also support higher-order theories. But those considerations lend support to higher-order theories independent of casting doubt on there being significant added utility to a state's being conscious.

It may strike many as surprising, for reasons noted at the outset of the section, that there should be no significant utility due exclusively to mental states' being conscious, utility that does not occur when those states occur without being conscious and is not an independent by-product of the processes that lead to their being conscious. But theory and empirical investigation seem so far not to have uncovered any such benefit.

References

Armstrong, D.M. (1980) What is consciousness?, in Armstrong, D.M., *The Nature of Mind*, St. Lucia, Queensland: University of Queensland Press.

Baars, B. (1998) *A Cognitive Theory of Consciousness*, Cambridge: Cambridge University Press.

Behne, T., Carpenter, M., Call, J. & Tomasello, M. (2005) Unwilling versus unable: Infants' understanding of intentional action, *Developmental Psychology*, **41**, pp. 328–337.

Billock, V.A., Gleason, G.A. & Tsou, B.H. (2001) Perception of forbidden colors in retinally stabilized equiluminant images: An indication of softwired cortical color opponency?, *Journal of the Optical Society of America A*, **18**, pp. 2398–2403.

Block, N. (1978) Troubles with functionalism, in Savage, C.W. (ed.) *Minnesota Studies in the Philosophy of Science*, vol. IX, Minneapolis, MN: University of Minnesota Press.

Block, N. (1995) On a confusion about a function of consciousness, *Behavioral & Brain Sciences*, **18**, pp. 227–247.

Block, N. (2007) Consciousness, accessibility, and the mesh between psychology and neuroscience, *Behavioral & Brain Sciences*, **30**, pp. 481–548.

Block, N. (2009) Comparing the major theories of consciousness, in Gazzaniga, M. (ed.) *The Cognitive Neurosciences IV*, pp. 1111–1122, Cambridge, MA: MIT Press.

Block, N. (2011a) The Anna Karenina theory of the unconscious, *Neuropsychoanalysis*, **13**, pp. 34–37.

Block, N. (2011b) The higher order approach to consciousness is defunct, *Analysis*, **71**, pp. 419–431.

Block, N. (2011c) Response to Rosenthal and Weisberg, *Analysis*, **71**, pp. 443–448.

Breitmeyer, B.G., Ro, T. & Singhal, N.S. (2004) Unconscious color priming occurs at stimulus—not percept—dependent levels of processing, *Psychological Science*, **15**, pp. 198–202.

Breitmeyer, B.G., Ro, T., Öğmen, H. & Todd, S. (2007) Unconscious, stimulus-dependent priming and conscious, percept-dependent priming with chromatic stimuli, *Perception & Psychophysics*, **69**, pp. 550–557.

Brentano, F. (1874/1973) *Psychology from an Empirical Standpoint*, Rancurello, A.C., Terrell D.B. & McAlister, L.L. (eds.), New York: Routledge & Kegan Paul.

Carruthers, P. (2000) *Phenomenal Consciousness: A Naturalistic Theory*, Cambridge: Cambridge University Press.

Cohen, M.A. & Dennett, D.C. (2011) Consciousness cannot be separated from function, *Trends in Cognitive Sciences*, **15**, pp. 358–364.

de Gelder, B., Tamietto, M., van Boxtel, G., Rainer Goebel, R., Sahraie, A., van den Stock, J., Stienen, B.M.C., Weiskrantz, L. & Pegna, A. (2008) Intact navigation skills after bilateral loss of striate cortex, *Current Biology*, **18**, pp. R1128–R1129.

Dehaene, S. & Naccache, L. (2001) Towards a cognitive neuroscience of consciousness: Basic evidence and a workspace framework, *Cognition*, **79**, pp. 1–37.

Dehaene, S., Sergent, C. & Changeux, J.P. (2003) A neuronal network model linking subjective reports and objective physiological data during conscious perception, *Proceedings of the National Academy of Sciences USA*, **100**, pp. 8520–8525.

Dennett, D.C. (1991) *Consciousness Explained*, Boston, MA: Little, Brown.

Descartes, R. (1641/1984) *The Philosophical Writings of Descartes*, vol. 2, Cottingham, J., Stoothoff, R. & Murdoch, D. (eds.), Cambridge: Cambridge University Press.

Desmurget, M., Reilly, K.T., Richard, N., Szathmari, A., Mottolese, C. & Sirigu, A. (2009) Movement intention after parietal cortex stimulation in humans, *Science*, **324**, pp. 811-813.

Desmurget, M. & Sirigu, A. (2009) A parietal-premotor network for movement intention and motor awareness, *Trends in Cognitive Sciences*, **13**, pp. 411-419.

Dienes, Z. (2008) Subjective measures of unconscious knowledge, *Progress in Brain Research*, **168**, pp. 49-64.

Dienes, Z. & Seth, A. (2010) Gambling on the unconscious: A comparison of wagering and confidence ratings as measures of awareness in an artificial grammar task, *Consciousness & Cognition*, **19**, pp. 674-681.

Dijksterhuis, A., Bos, M.W., Nordgren, L.F. & van Baaren, R.B. (2006) On making the right choice: The deliberation-without-attention effect, *Science*, **311**, pp. 1005-1007.

Dretske, F.I. (1993) Conscious experience, *Mind*, **102**, pp. 263-283.

Dretske, F.I. (2006) Perception without awareness, in Gendler, T.S. & Hawthorne, J. (eds.) *Perceptual Experience*, Oxford: Oxford University Press.

Fernandez-Duque, D., Baird, J.A. & Posner, M.I. (2000a) Awareness and metacognition, *Consciousness & Cognition*, **9**, pp. 324-326.

Fernandez-Duque, D., Baird, J.A. & Posner, M.I. (2000b) Executive attention and metacognitive regulation, *Consciousness & Cognition*, **9**, pp. 288-307.

Frith, C. (2007) *Making Up the Mind: How the Brain Creates our Mental World*, Malden, MA: Blackwell Publishing.

Frith, C. (2010) What is consciousness for?, *Pragmatics & Cognition*, **18**, pp. 497-551.

Grimes, J. (1996) On the failure to detect changes in scenes across saccades, in Akins, K. (ed.) *Perception*, New York: Oxford University Press.

Haggard, P. (1999) Perceived timing of self-initiated actions, in Aschersleben, G., Bachmann, T. & Müsseler, J. (eds.) *Cognitive Contributions to the Perception of Spatial and Temporal Events*, Amsterdam: Elsevier.

Halsey, R.M. & Chapanis, A. (1951) On the number of absolutely identifiable spectral hues, *Journal of the Optical Society of America*, **41**, pp. 1057-1058.

Johansson, P., Hall, L., Sikström, S. & Olsson, A. (2005) Failure to detect mismatches between intention and outcome in a simple decision task, *Science*, **310**, pp. 116-119.

Kant, I. (1787/1998) *Critique of Pure Reason*, Guyer, P. & Wood, A.W. (eds.) Cambridge: Cambridge University Press.

Kentridge, R.W. & Heywood, C.A. (2000) Metacognition and awareness, *Consciousness & Cognition*, **9**, pp. 308–312.

Kentridge, R.W., Heywood, C.A. & Weiskrantz, L. (2004) Spatial attention speeds discrimination without awareness in blindsight, *Neuropsychologia*, **42**, pp. 831–835.

Koch, C. & Preuschoff, K. (2007) Betting the house on consciousness, *Nature Neuroscience*, **10**, pp. 140–141.

Koch, C. & Tsuchiya, N. (2007) Attention and consciousness: Two distinct brain processes, *Trends in Cognitive Sciences*, **11**, pp. 16–22.

Koriat, A. (2007) Metacognition and consciousness, in Zelazo, P.D., Moscovitch, M. & Thompson, E. (eds.) *The Cambridge Handbook of Consciousness*, Cambridge: Cambridge University Press.

Koriat, A. & Ackerman, R. (2010) Metacognition and mindreading: Judgments of learning for self and other during self-paced study, *Consciousness & Cognition*, **19**, pp. 251–264.

Koyama, T., McHaffie, J.G., Laurienti, P.J. & Coghill, R.C. (2005) The subjective experience of pain: Where expectations become reality, *Proceedings of the National Academy of Sciences USA*, **102**, pp. 12950–12955.

Kriegel, U. (2009) *Subjective Consciousness: A Self-Representational Theory*, Oxford: Oxford University Press.

Lau, H. (2008) Are we studying consciousness yet?, in Weiskrantz, L. & Davies, M. (eds.) *Frontiers of Consciousness: Chichele Lecture*, Oxford: Oxford University Press.

Lau, H. & Passingham, R.E. (2006) Relative blindsight in normal observers and the neural correlate of visual consciousness, *Proceedings of the National Academy of Sciences USA*, **103**, pp. 18763–18768.

Lau, H. & Rosenthal, D. (2011) Empirical support for higher-order theories of conscious awareness, *Trends in Cognitive Sciences*, **15**, pp. 365–373.

Levin, D.T., Momen, N., Drivdahl, S.B. & Simons, D.J. (2000) Change blindness blindness: The metacognitive error of overestimating change-detection ability, *Visual Cognition*, **7**, pp. 397–412.

Levine, J. (2001) *Purple Haze: The Puzzle of Consciousness*, New York: Oxford University Press.

Libet, B., Gleason, C.A., Wright Jr, E.W. & Pearl, D.K. (1983) Time of conscious intention to act in relation to onset of cerebral activity (readiness potential), *Brain*, **106**, pp. 623–642.

Locke, J. (1700/1975) *An Essay Concerning Human Understanding*, Nidditch, P.H. (ed.), Oxford: Oxford University Press.

Lycan, W.G. (1996) *Consciousness and Experience*, Boston, MA: MIT Press.

Marcel, A.J. (1983) Conscious and unconscious perception: An approach to the relations between phenomenal experience and perceptual processes, *Cognitive Psychology*, **15**, pp. 238-300.
Mascalzonia, E., Regolina, L. & Vallortigara, G. (2010) Innate sensitivity for self-propelled causal agency in newly hatched chicks, *Proceedings of the National Academy of Sciences USA*, **107**, pp. 4483-4485.
Mellor, D.H. (1971) *The Matter of Chance*, Cambridge: Cambridge University Press.
Metcalfe, J. & Shimamura, A.P. (eds.) (1994) *Metacognition: Knowing About Knowing*, Cambridge, MA: MIT Press.
Nagel, T. (1974) What is it like to be a bat?, *Philosophical Review*, **83**, pp. 435-450.
Neander, K. (1998) The division of phenomenal labor: A problem for representational theories of consciousness, *Philosophical Perspectives*, **12**, pp. 411-434.
Nelson, T. (ed.) (1992) *Metacognition: Core Readings*, Needham Heights, MA: Allyn and Bacon.
Nelson, T. & Narens, L. (1994) Why investigate metacognition?, in Metcalfe, J. & Shimamura, A.P. (eds.) *Metacognition: Knowing About Knowing*, Cambridge, MA: MIT Press.
Nisbett, R.E. & Wilson, T.D. (1977) Telling more than we can know: Verbal reports on mental processes, *Psychological Review*, **84**, pp. 231-259.
Overgaard, M. & Sandberg, K. (2012) Kinds of access: Different methods for report reveal different kinds of metacognitive access, *Philosophical Transactions of the Royal Society B*, **367**, pp. 1287-1296.
Paynter, C.A., Reder, L.M. & Kieffaber, P.D. (2009) Knowing we know before we know: ERP correlates of initial feeling-of-knowing, *Neuropsychologia*, **47**, pp. 796-803.
Pérez-Carpinell, J., Baldoví, R., de Fez, M.D. & Castro, J. (1998) Color memory matching: Time effect and other factors, *Color Research & Application*, **23**, pp. 234-247.
Persaud, N., McLeod, P. & Cowey, A. (2007) Postdecision wagering objectively measures awareness, *Nature Neuroscience*, **10**, pp. 257-261.
Pieschl, S. (2009) Metacognitive calibration: An extended conceptualization and potential applications, *Metacognition & Learning*, **4**, pp. 3-31.
Quine, W.V. (1960) *Word and Object*, Cambridge, MA: MIT Press.
Ro, T., Singhal, N.S., Breitmeyer, B.G. & Garcia, J.O. (2009) Unconscious processing of color and form in metacontrast masking, *Attention, Perception & Psychophysics*, **71**, pp. 95-103.
Rosenthal, D. (2000) Consciousness, content, and metacognitive judgments, *Consciousness & Cognition*, **9**, pp. 203-214.

Rosenthal, D. (2002) The timing of conscious states, *Consciousness & Cognition*, **11**, pp. 215-220.
Rosenthal, D. (2004) Varieties of higher-order theory, in Gennaro, R.J. (ed.) *Higher-Order Theories of Consciousness*, Amsterdam: John Benjamins.
Rosenthal, D. (2005) *Consciousness and Mind*, Oxford: Clarendon Press.
Rosenthal, D. (2008) Consciousness and its function, *Neuropsychologia*, **46**, pp. 829-840.
Rosenthal, D. (2010) How to think about mental qualities, *Philosophical Issues*, **20**, pp. 368-393.
Rosenthal, D. (2011) Exaggerated reports: Reply to Block, *Analysis*, **71**, pp. 431-437.
Schnitzspahn, K.M., Zeintl, M., Jäger, T. & Kliegel, M. (2011) Metacognition in prospective memory: Are performance predictions accurate?, *Canadian Journal of Experimental Psychology*, **65**, pp. 19-26.
Scholl, B.J., Simons, D.J. & Levin, D.T. (2004) Change blindness blindness: An implicit measure of a metacognitive error, in Levin, D.T. (ed.) *Thinking and Seeing: Visual Metacognition in Adults and Children*, Cambridge, MA: MIT Press.
Schurger, A., Cowey, A., Cohen, J., Treisman, A. & Tallon-Baudry, C. (2008) Distinct and independent correlates of awareness and attention in a hemianopic patient, *Neuropsychologia*, **46**, pp. 2189-2197.
Schurger, A. & Sher, S. (2008) Awareness, loss aversion, and post-decision wagering, *Trends in Cognitive Sciences*, **12**, pp. 209-210.
Seashore, C.E. (1967) *The Psychology of Music*, New York: Dover Publications.
Seth, A.K. (2008) Post-decision wagering measures metacognitive content, not sensory consciousness, *Consciousness & Cognition*, **17**, pp. 981-983.
Strawson, G. (2006) Realistic monism: Why physicalism entails panpsychism, *Journal of Consciousness Studies*, **13**, pp. 3-31.
Tomasello, M. (2000) Culture and cognitive development, *Current Directions in Psychological Science*, **9**, pp. 37-40.
Usher, M., Russo, Z., Weyers, M., Brauner, R. & Zakay, D. (2011) The impact of the mode of thought in complex decisions: Intuitive decisions are better, *Frontiers in Psychology*, **2**, aert. 37.
van Boxtel, J.J.A., Tsuchiya, N. & Koch, C. (2010) Opposing effects of attention and consciousness on afterimages, *Proceedings of the National Academy of Sciences USA*, **107**, pp. 8883-8888.
van Gaal, S. & Fahrenfort, J.J. (2008) The relationship between visual awareness, attention, and report, *Journal of Neuroscience*, **28**, pp. 5401-5402.

van Gaal, S., Ridderinkhof, K.R., Fahrenfort, J.J., Scholte, H.S. & Lamme, V.A.F. (2008) Frontal cortex mediates unconsciously triggered inhibitory control, *Journal of Neuroscience*, **28**, pp. 8053–8062.
Waroquier, L., Marchiori, D., Klein, O. & Cleeremans, A. (2010) Is it better to think unconsciously or to trust your first impression? A reassessment of unconscious thought theory, *Social Psychological & Personality Science*, **1**, pp. 111–118.
Wegner, D.M. (2002) *The Illusion of Conscious Will*, Cambridge, MA: MIT Press.
Weisberg, J. (2010) Misrepresenting consciousness, *Philosophical Studies*, **154**, pp. 409–433.
Weisberg, J. (2011) Abusing the notion of what-it's-like-ness: A response to Block, *Analysis*, **71**, pp. 438–443.
Weiskrantz, L. (1997) *Consciousness Lost and Found: A Neuropsychological Exploration*, Oxford: Oxford University Press.
Weiskrantz, L. (1998) Consciousness and commentaries, *International Journal of Psychology*, **33**, pp. 227–233.
Wilberg, J. (2010) Consciousness and false HOTs, *Philosophical Psychology*, **23**, pp. 617–638.
Wittgenstein, L. (2009) *Philosophical Investigations*, 4th ed., Hacker, P.M.S. & Schulte, J. (eds.), II, x, 190–192, Oxford: Wiley-Blackwell.

Jesse Prinz

Unconscious Perception and the Function of Consciousness

1. Introduction

Pre-theoretically, the divide between conscious and unconscious information processing strikes us as important, even profound. It is sometimes surmised that conscious information processing is special — capable of doing things that are impossible unconsciously. Consciousness is also said to be what makes having a mind so rewarding (or, potentially, so awful), and it has been the primary impetus for theories that reject the materialist worldview that governs mainstream scientific inquiry. When people try to specify *why* consciousness is so special, however, we often find synonyms standing in place of explanations. Conscious states feel like something — they are felt. The felt quality of experience is, indeed, interesting, and difficult to account for, but it doesn't tell us much about what consciousness adds to mental functioning. It doesn't tell us why we have conscious states and what they do. This question takes us away from the grandiose metaphysical debates about consciousness into more tractable terrain. We can put aside debates about materialism and ask, does consciousness make a difference? I want to take up this question in the area where the conscious/unconscious divide has been most intensively studied: vision science. In addressing the value of consciousness, this is a natural starting place, and a natural first question is: is there anything we can see consciously that can't also be seen unconsciously? Quite surprisingly, I think the answer to this question is no. I think it follows that conscious vision is not for seeing. What then, we may ask, is it for?

In what follows I will first briefly review the evidence for the existence of unconscious vision. Then I will review evidence that unconscious vision can represent every feature that is represented consciously. Unconscious vision can also carry out all kinds of operations on these representations, such as binding, comparison, and inference. This will lead us to our question about the function of consciousness. Conscious vision, I will suggest, adds nothing to seeing; consciousness contributes more to doing. It makes a subtle but important contribution to action. The conclusion will tell us something about the value of consciousness, but the more important lesson is really about the value of unconsciousness. In exploring the enormous power of unconscious vision, we will see that consciousness deserves less credit than it is sometimes given.

2. Is there Unconscious Vision?

2.1. Unconscious vision defined

Before recent times, the distinction between conscious and unconscious vision was not clearly drawn. Most people assumed that mental activity is conscious, and, if something could not be consciously seen, it could not be seen at all. By the mid-twentieth century, the situation had changed, and the unconscious was widely accepted. Freud had popularized the idea of unconscious motivations, and behaviourists had tried to show that consciousness itself was an illusion. It was natural, then, to assume that vision could occur without consciousness. But this hypothesis stands in need of conceptual clarification and defence. As we will see, some evidence for unconscious vision is unconvincing.

To test for the existence of unconscious vision, an operational definition is required. There must be a way to test for both features: vision and the lack of consciousness. Testing for vision involves proving that a stimulus presented to visual transducers (the eyes) has been registered by the visual system. That can be done using physiological measures, such as skin conductance or fMRI, or behavioural measures. In using physiological measures, it is not enough to show that a stimulus caused a physiological response. That would not establish that the stimulus had been *processed* by the visual system. Vision, like other senses, is in the business of interpreting inputs from the world. Mere activity could reflect preparation or even an effort to interpret without any success. What's needed is a measure of success. Thus, physiological measures of perception use physiological responses that correlate with specific kinds of information processing. For example, there is a region in the temporal cortex that is (controversially) called the fusiform face area. If this area shows increased activation after a face is presented, that is evidence that the face has been visually processed, or seen. If a stimulus with emotional content is presented, such as an expressive face, one can test for perception by testing whether it caused electrodermal changes associated with emotion.

Behavioural tests for vision include explicit measures and implicit measures. An explicit measure is one where the viewer directly reports that something was seen. In some cases people are asked to specify what they have seen, and, in other cases, they are asked to simply report that they saw something rather than nothing. The former is called 'recognition' and the latter stimulus 'detection'. Both can be accurate or inaccurate. Accuracy above chance is evidence for vision. An implicit measure is one were behaviour other than reporting is used to establish that a visually presented stimulus has been processes. For example, one can see whether the stimulus primed subsequent behaviour. In priming, the presentation of a stimulus influences performance on another task, e.g. by increasing reaction times or producing responses that are structurally or semantically related to the stimulus.

If any of these measures shows evidence of vision, we can say objective vision has occurred. Subjective vision occurs when someone reports having seen something, and purely subjective vision arises when such reports occur without behavioural or physiological evidence for accuracy. In cases where there is no evidence for objective vision and no report of having seen, we can say no vision occurred.

Testing for the lack of consciousness is trickier. One difficulty is that there is no uncontroversial measure of consciousness, either behavioural or physiological. If there were an accepted theory of consciousness, we could test to see whether the conditions specified in that theory were met. But no theory enjoys consensus, and there are many who think that consciousness is not defined with respect to any physical or functional features. We judge we are conscious (perhaps fallibly) on the basis of experience, not on the basis of behaviour or awareness of physiological changes as such. Thus, measures of consciousness will always be indirect. Of course, indirect measures are common in science, and most researchers are willing to grant that if someone says they are conscious, that report should be accepted, unless there is a special reason to deny it (such as a clinical condition known to result in failures of self-knowledge). We can say that consciousness is present, *ceteris paribus*, if there is subjective vision. This is a sufficient condition, not a necessary condition. Someone might have a conscious visual experience without realizing it was visual (for example, visual imagery in someone who is congenitally blind). Here, one might expect the person to report having had an experience, but not necessarily a visual experience. If such self-reports are correlated with an objective measure, such as behavioural success or physiology, then they should be regarded as reliable evidence for visual consciousness, even if the person having the experience does not realize the experience is visual.

Contrast this with a case where, when asked, a person claims not to have had any experience at all in response to a stimulus—not just a case where the person cannot report what was presented, but also where the person reports not having been presented with anything at all. Imagine, further, that this negative report remains even just after a stimulus has been presented, or when the person was warned in advance that a stimulus would be presented. And now restrict this to cases where there is no pathology or context that would imply any kind of memory failure. This would be evidence that there is no visual consciousness of the stimulus. We can call this an absence of subjective experience.

Putting these two operational definitions together, we can define unconscious vision as a situation where, *ceteris paribus*, objective vision occurs without subjective experience (including the experience of detection). This definition includes a *ceteris paribus* clause because there may be cases where subjective reports are rendered impossible during a conscious visual episode (imagine someone on curare). Under ordinary circumstances, however, it is safe to assume that someone who claims not to have seen anything has, in fact, seen a stimulus unconsciously if that person shows objective evidence of visually processing that stimulus.

2.2. Is there unconscious vision?

With this definition in hand, we can ask, does vision ever occur unconsciously? We are accustomed to believing the answer is obvious and affirmative, but some of the evidence that is sometimes given is less straightforward than it may initially seem. Consider what has become a textbook example: blindsight. As the name suggests, individuals with blindsight are assumed to be blind, due to lesions in the primary visual cortex, but they retain some visual acuity. When asked to guess, they are remarkably accurate, for example, at reporting the location of stimuli presented in their 'blind' visual fields. The problem is that many studies working with blindsight patients measure consciousness using an overly simple dichotomous choice: do you see the stimulus or not? Overgaard *et al.* (2008) argue that this is an insensitive measure. In a study with a blindsight patient, they deployed a four-item scale ('Did you see a clear image, an almost clear image, a weak glimpse, or nothing at all?'). Using this measure, they found that successful discrimination actually correlates with visual clarity, indicating that their patient is actually relying on visual experience. The generality of these results can be challenged (see Prinz, 2015), but a cautionary point must be conceded. With crude measures of consciousness, we may be reporting unconscious vision in cases where experience is actually present.

Take another example: change blindness. Here healthy participants are consecutively presented with two versions of the same stimulus and asked to spot the difference. Many fail to notice even dramatic changes, such as the colour of a prominent object. We are invited to think that the feature in question was not experienced when the first stimulus was presented. But failure to report a change shows only that the feature was not stored in memory. Leading models of change blindness assume that the failure has to do with how much detail gets stored in an accessible way, not a failure of representing (Simons and Levin, 1997). On the face of it, the claim that we may stare at a large object without experiencing its colour sounds odd. Do we experience it as colourless? If so, why doesn't it pop out as anomalous? Do we experience it in a way that is neutral about colour? Perhaps. But, without an explanation of why color is not consciously processed, this hypothesis seems unmotivated, and inconclusive at best.

To take one more example, consider Milner and Goodale's (1995) groundbreaking work on the distinction between dorsal and ventral streams in vision. They argue that the dorsal pathway of vision operates unconsciously. As evidence they describe a patient with damage in her ventral stream. She claims not to be able to perceive the orientation of a 'slot' placed in her visual field, but she can accurately place an object in that slot—a skill that relies on the dorsal stream. But the patient in question does not lack visual consciousness. She is visually agnosic, which means she has conscious experiences that are difficult for her to report and interpret. This shows a failure of explicit recognition, but it is possible that conscious experience guides her successful action. At least, it is a very difficult case to interpret.

There are, however, some lines of evidence that provide more decisive support for unconscious vision. Most obviously, there is the vast literature on masked priming. When a stimulus is presented for a brief duration (usually under 50 milliseconds) followed by a 'mask', people have no awareness of seeing it. The mask can be visual noise (like a scrambled object or a snow screen), something similar to the stimulus (as when a neutral face is used to mask an emotional expression), or something in the area surrounding where the stimulus was located (as in metacontrast masking, which usually uses an annulus to mask a disc). Participants in such studies fail to report that they see the masked stimulus, and accuracy would not increase if they are given a graded scale. In fact, if the trials with a masked stimulus are mixed with trials on which no stimulus is presented, people are at chance when asked to guess whether a stimulus was present. Thus, there is neither explicit recognition nor detection. But behavioural and physiological measures confirm that the stimulus can nevertheless be perceived. Masked priming is unlike change blindness in that people in change blindness studies do believe they are seeing the stimulus. Moreover, when a changing stimulus is pointed out with a cue in a change blindness paradigm, it is reliably detected, but cueing does not render masked stimuli conscious. Change blindness derives from memory limitations when we perceive complex scenes; when a stimulus is complex, only the gist and a few details get accessibly encoded in memory. With masked priming, simple stimuli are used, so there is no principled reason why they shouldn't be encoded. When unmasked, the stimuli in these studies are readily perceived and not subject to change blindness.

Further evidence for unconscious perception comes from interoccular suppression (e.g. Jiang et al., 2006). If each eye is presented with different stimuli, and these differ in contrast, the duller of the two stimuli is rendered invisible. People have no idea that they are seeing two different stimuli, and couldn't distinguish the suppression case from ones in which two copies of the same stimulus or no stimulus in the second eye were presented. Nevertheless, the suppressed stimulus impacts behaviour and impacts brain processes. Here memory is not a factor, since people fail to report the suppressed stimulus while it is still there, so this looks like a clear case of objective perception without subjective experience.

Finally, consider inattentional blindness (Mack and Rock, 1998). Here participants are presented with an attentionally demanding visual task, and then another, unexpected, stimulus is briefly flashed. The unexpected stimulus is presented above the conscious threshold (usually 200 milliseconds), but it is often missed. Participants fail to report it even when asked a series of questions akin to Overgaard et al.'s levels of awareness scale. This indicated that the stimulus is neither recognized nor detected. Nevertheless, there is evidence that these invisible stimuli are objectively seen. Mack and Rock found that words that were rendered invisible in an inattentional blindness task nevertheless had a significant impact on a subsequent word stem completion task. Inattentional blindness is easy to confuse with change blindness, but there are crucial differences. With change blindness, people are confident that they have seen everything, including the features that change (they just don't notice that they have changed). In change blindness studies, people are also asked to look for a change, so they are actively attending to everything. What they miss is not a visual feature, but the difference between visual features at two different time points. So change blindness could easily be a memory failure rather than a failure of conscious experience. Inattentional blindness is difficult to explain this way because simple stimuli are used, and the effect goes away when those stimuli are presented without being asked to perform the attentionally demanding task.

In summary, there are several lines of evidence that convincingly establish a dissociation between vision (measured by objective success) and consciousness (measured by subjective reports about having experience). Unconscious vision seems to be a real phenomenon. The next task is asking whether it differs from conscious vision.

3. Is there a Difference between Conscious and Unconscious Vision?

3.1. Representations

It is natural to suppose that conscious and unconscious vision are different in some way. Theories of the unconscious often assume that unconscious processing is different. Freud's division into Ego, Super-Ego, and Id is a case in point. Or consider dual process models, which suppose that unconscious processes are more emotionally based and make greater use of stereotypes and heuristics. There is also work in category learning, which suggests that unconscious processes are more pattern-based and less rule-governed than conscious processes. Against this background, one might expect significant differences between conscious and unconscious vision. On many models, the unconscious mind is treated as more rudimentary, even more primitive. Applied to vision, that might imply that conscious seeing is more sophisticated in terms of its representations and processes than unconscious seeing. Thus far, however, evidence has tended to disconfirm this hypothesis. Conscious and unconscious vision seem to be very much alike, and, if anything, unconscious vision is more sophisticated.

Most of the relevant research has investigated either the representational capacities of unconscious vision or the processes that can be performed on those representations. I will look at these in turn. Let's begin with representations. We can ask, is there anything represented in conscious vision that cannot also be represented unconsciously? The answer seems to be negative. The features represented in conscious vision can all be represented unconsciously.

I will start with shape. Many unconscious perception tasks depend on the successful perception of shape without consciousness. For example, some use faces (e.g. Winkielman *et al.*, 2005) and others use words (e.g. Mack and Rock, 1998). In some studies, priming carries over from the physical shape of an object. For example, Bar and Biederman (1998) found priming effects for pictures of ordinary objects, but these did not prime differently shaped objects in the same category.

In other studies, there is evidence that the semantic meaning of a shape is unconsciously perceived. Dell'Acqua and Grainger (1999) found that pictures of objects primed names of those objects, and conversely; they also found priming between pictures and conceptually associated words, such as a picture of a dog and the word 'cat'. Thus, shapes can be perceived unconsciously, and even recognized and associated. The shapes used in priming studies vary from simple polygons to complex objects, suggesting that unconscious perception of shape is not limited in any obvious way.

There is also ample evidence for unconscious colour perception. For example, Ro *et al.* (2009) used a metacontrast masking paradigm in which a coloured disc was masked by a coloured annulus. Participants had to report the colour of the consciously perceived annulus. Performance was faster when the unconscious disc had the same colour.

Ro and his collaborators have established unconscious perception of various other visual features. In another study, Persuh and Ro (2012) use metacontrast masking to show that we can unconsciously perceive brightness. They also show that unconscious perception is context-sensitive: brightness priming depends on background tone, not just brightness of the prime. Ro *et al.* (2009) also showed that there can be unconscious perception of orientation; they presented masked semi-circles at various orientations and showed facilitation on a subsequent orientation discrimination task. Rajimehr (2004) obtained further evidence for unconscious perception or orientation with Gabor patches.

Use of Gabor patches also indicates unconscious perception of texture. Rajimehr used Gabor patches of different frequencies, showing distinct effects. Other studies have investigated unconscious texture perception more directly. One particularly demanding task is texture segmentation: finding an irregularity embedded in a complex texture. Shubö and Meinecke (2007) obtained evidence for unconscious texture segmentation using a masked priming paradigm.

There is also research indicating that unconscious perception of motion is possible. Sperling *et al.* (2011) used an interocular suppression paradigm in which they presented different motion patterns in each eye. Participants consciously reported only one direction of motion, but their eye movements reflected a composite of the directions presented in both eyes. Mattler and Fendrich (2007) found that an unconsciously presented apparent motion stimulus impacted response times in a conscious motion detection task, and also biased interpretation when participants were consciously presented with ambiguous motion.

From these findings and countless others like them it seems safe to infer that the features familiar from conscious vision can also be perceived unconsciously. In fact, unconscious vision can perceive much more. There is strong evidence that most vision is unconscious, and that consciousness can only arise at very specific stages of visual processing. Elsewhere, I have surveyed evidence in support of an 'intermediate-level' theory of consciousness (Prinz, 2012; Jackendoff, 1997). According to this theory, sensory systems are organized into a three-stage hierarchy. At the low level, local features are detected, at the intermediate level they are integrated into coherent wholes and scenes, and, at the high level, stimuli are represented more categorically, in a way that abstracts away from size, position, orientation, and other highly specific characteristics. The intermediate-level theory states that consciousness rises only at the second of these three stages: we are not conscious of discrete disconnected features, nor of abstract sensory representations. If this is right, then unconscious vision encompasses two levels of representation that are never conscious.

One might be tempted to suppose that intermediate-level representations are always conscious while the other two stages are never conscious. This would establish that conscious and unconscious perception use different kinds of representations. But this is unsustainable. For example, Daems and Verfaillie (1999) found that priming with masked images of human body postures depends on viewpoint and left/right orientation—two features associated with intermediate-level vision. Eddy and Holcomb (2011) found evidence for both viewpoint-specific and viewpoint-invariant priming using masked pictures of familiar objects rotated in space. Such work suggests that different stages of vision can be processed unconsciously, and specific features of the experimental design will determine which stages have priming effects.

In summary, there is evidence that unconscious vision can represent every kind of stimulus feature represented in conscious vision and much more. Unconscious vision does not appear to be a primitive counterpart to conscious vision. It has a claim to be more representationally sophisticated.

3.2. Processes

It may be objected that unconscious vision is representationally sophisticated but deeply impoverished when it comes to operations or processes carried out on those representations. Here too, however, evidence suggests otherwise. Many complex processes can be carried out on visual representations outside of consciousness.

Consider binding, which was once hypothesized to be the hallmark of consciousness (Crick and Koch, 1990; Engel and Singer, 2001). A visual representation is said to be bound with different feature dimensions, such as colour and shape, which are treated by the visual system as belonging to the same object. The supposition that binding requires consciousness has roots in Treisman and Gelade's (1980) feature integration theory, which claims that binding is achieved by attention (as we will see below, attention is thought to be related to consciousness). Now, however, there is ample evidence that such binding can occur without consciousness. For example, Rajimehr (2004) found that unconscious priming of orientation with Gabor patches can depend on the colour of those patches. There is also recent work suggesting that Gabor patch priming can depend on location (Keizer *et al.*, 2015). Such work shows that two different visual features can be linked to each other unconsciously. There is also ample evidence that objects with multiple spatial parts can be linked together unconsciously. Consider the vast literature on masked priming with words. The letters that form a word must be treated as parts of an organized whole, which encodes the relative location and shape of each letter.

Other authors have suggested that consciousness is co-instantiated with information integration (Tononi and Edelman, 1998). Information integration is not the same as binding. Binding involves treating two features as belonging to the same object, and integration involves bringing different stimuli or information channels together. For example, integration is required to locate a visually presented object in a scene. Mudrik *et al.* (2011) designed a clever study to prove that this can happen without consciousness. Using interocular suppression, it has been shown that an invisible stimulus can overcome suppression and enter conscious awareness if it has some striking feature. Murdrik *et al.* used interocular suppression to mask images of objects in various background settings. They found that the suppressed stimuli were able to break into awareness when the object pictured was incongruous with its background, such as a person placing a chessboard in an oven, or a group of people playing basketball with a watermelon. This shows that the object and background were being integrated unconsciously.

There is also evidence that unconscious vision can transfer information to another sense modality. For example, Chen and Spence (2011) showed that sounds can increase the visibility of otherwise invisible masked pictures, when the two belong to the same category (e.g. a barking sound and a picture of a dog). Here an unconscious visual stimulus is integrated with a conscious sound and brought into awareness. In another study, Lamy *et al.* (2008) use an unconscious auditory prime (a masked spoken word) to facilitate a visual task (word stem completion). Thus information can transfer from unconscious sound to vision or from unconscious vision to sound. Either way, consciousness is not required for linking vision to another modality.

Unconscious vision can also be used to compare two stimuli. For example, in one of Rajimehr's (2004) experiments, Gabor patches changed orientation unconsciously and this change is detected. Change requires detecting difference in time. Difference can also be detected unconsciously in space. Lin and Murray (2014) presented two masked shapes side by side, and obtained evidence that people can unconsciously assess whether the shapes are the same or different.

There is also extensive evidence for unconscious visual learning and memory. For example, Esteves *et al.* (1994) used masked emotional expressions in a Pavlovian conditioning study. A follow-up study replicated the effect with pictures of snakes and spiders (Öhman and Soares, 1998). Merickle and Smith (2005) tested the impact of unconscious perception on memory by using a temporally extended replication of Mack and Rock's (1998) inattentional blindness paradigm. They presented participants with words at the same time as an attentionally demanding task, and then asked them to perform a word stem completion task, using a word other than any they had seen in the first part of the study. Those who consciously perceived words during the attention task were able to provide different words during the stem completion task, but those who experienced inattentional blindness showed a tendency to use the previously presented words for stem completion. This much replicated Mack and Rock. But Merikle and Smith extended these findings by presenting people with the stem completion task more than 30 minutes after the attention task. Even after this very long interval, the effect remained, suggesting long-term retention of unconsciously presented words.

In another striking example of unconscious performance, Sklar *et al.* (2012) used interocular suppression to unconsciously present mathematical statements, such '9 − 4 − 3 ='. Then, they consciously presented integers and asked participants to pronounce them. The found facilitation when the integer corresponded to the solution of the mathematical statement (in this case '2'). Using methods of this kind they showed that people unconsciously perform arithmetic. This is particularly striking in the case of three number problems, like the one given here, since other work shows that these are considerably more difficult than problems with two numbers.

The arithmetic example illustrates how unconscious vision can contribute to thinking. Unconscious vision can also contribute to action. Earlier we saw that there is some controversy about the interpretation of patients with lesions in their ventral visual streams. These patients may have disorganized conscious experiences rather than none at all. In the lab, however, one can show that an invisible stimulus can guide motor response. Schmidt (2002) used a metacontrast masking paradigm in which red discs were rendered invisible by green annuli and conversely. They then presented two annuli at once, one red masking a green disc, and one green masking a red disc. When participants were asked to point to the red annulus, their fingers first began to point towards the green one, and conversely. This shows that hand movements were being lured by an unconscious stimulus. In another study, Binstead et al. (2007) used masking to conceal circles that were either small or large. Fitt's law predicts that people will move more slowly when reaching for small targets. When asked to point in the location of the invisible circle, movements were indeed slower for the small circles than for the large.

Invisible stimuli influence not only how we move, but also decisions about whether to act or inhibit action. Ocampo et al. (2015) asked people to perform a reaching task, and then presented masked instruction to either stop or continue. The unconscious instruction to stop interfered with task performance. In a similar spirit, Wokke et al. (2011) conducted an experiment in which every trial was preceded by an instruction indicating whether they should press a key upon seeing a triangle or a square. In some cases, a square required the key press and in others the triangle. The target (square or triangle) was presented consciously, but, just beforehand, a square or triangle was also presented unconsciously. When the unconscious shape was different from the shape that required a key press, it caused response inhibition. This shows that unconscious stimuli can affect conscious control in an adaptive way, sensitive to stimulus–response rules that are newly assigned at each trial.

The foregoing examples show that unconscious vision can influence executive processing (e.g. the decision to execute or inhibit action). The reverse is true as well; executive decisions can influence unconscious vision. For example, Enns and Oriet (2007) gave participants a task in which they measured the influence of unconsciously presented faces. The faces varied along three dimensions: emotion, ethnicity, and sex. They found that the degree of influence of these three factors depended on relevance to the task being performed. Thus, we adaptively use unconscious information.

There is also a sizeable literature exploring the impact of attention on unconscious vision. The interpretation of this research is somewhat contentious, since the tasks that have been described as involving attention may actual involve related processes that precede attention or correlate with attention (Prinz, 2011). Without getting embroiled in that debate, it can be shown without controversy that unconscious vision allows for such things as lateral inhibition and selection. Thus, for example, Hsieh et al. (2011) used interocular suppression to unconsciously present people with a pop-out display—a red patch mixed among many green patches. Then they consciously presented a target either in the same location as the red dot had occupied or a new location. Location congruence improved performance even though the pop-out display had been unconscious.

Collectively, all these findings indicate that unconscious vision is rich in both representational capacity and in the processes that can be carried out on these representations. Efforts to find a kind of representation or process that is unique to conscious vision have met with frustration. It looks like unconscious vision can do anything that conscious vision can do.

I don't mean to imply that differences will never be found. Perhaps there are a few tricks reserved for conscious vision. I also don't mean to imply that conscious vision isn't better at doing some of the things we can do unconsciously. For example, we can do arithmetic with three digits unconsciously, but what about four or five or ten? Keeping track of many items or many steps is something that consciousness is good for, and that is important. But note that this appears to be a quantitative difference, not a qualitative one. It is not an example of a process that cannot be achieved with unconscious vision (multi-step arithmetic), but rather a difference in how much of that process can be carried out. The point of this review is to show the surprising degree to which unconscious vision can carry out operations of the kind that has been associated with consciousness. Processes that are unique to consciousness have been elusive, and many proposals, such as binding, have been refuted.

So we are left with a question: if consciousness involves no distinct processes, what function does it serve? This is the question I want to turn to now. The interim moral is that unconscious vision is very impressive. So impressive, in fact, that the value of consciousness needs to be established. In taking up that task, in the next section, I don't want to lose sight, so to speak, of the lessons learned here. Unconscious vision is remarkably powerful. It seems that anything we can see consciously can also be seen unconsciously, and many things can only be seen unconsciously. We can also perform complex, multi-step, multi-modal, adaptive operations on unconscious visual representations. Consciousness seems to add very little to vision. It doesn't help us see anything. It is a puzzle why we need consciousness at all.

4. The Nature and Function of Conscious Vision

4.1 Consciousness and attention

The upshot of the foregoing review is that, in terms of representations and operations on representations, unconscious vision seems capable of doing everything that conscious vision can do and more. If so, consciousness does not contribute to visual function. It follows that consciousness is not for seeing. This is a surprising conclusion since, in folk psychology, we tend to say that consciousness makes the difference between seeing and not seeing. It also flies in the face of other work on the conscious/unconscious distinction, which emphasizes distinctive rules and representations on either side of the divide. The evidence summarized above suggests that consciousness does not provide a distinctive way of seeing. All this leaves us wondering what consciousness has to contribute.

To address this question, it is necessary to say a bit about the nature of consciousness. That question is among the most contested in all of cognitive science and the philosophy of mind. I cannot defend a theory of consciousness here, but I will briefly outline the theory I take to be most plausible in light of the evidence. I offer a lengthy defence elsewhere (Prinz, 2012). To minimize reliance on controversial claims, I will try to extrapolate the aspect of this theory that is most widely shared by other contributors to the debate about consciousness. I will use this to set up a proposal about the function of consciousness that could be accepted by those coming from various different theoretical perspectives.

In a nutshell, I think consciousness is attention. More precisely, I think that the modulation of sensory inputs that are properly referred to as attention are precisely the same ones that constitute conscious perception. Attention is necessary and sufficient for making a perceptual state conscious. I will briefly state some of the evidence for this, and then I will indicate why it is controversial and how we can bypass the controversy here.

Four lines of evidence can be marshalled in favour of the equation of attention and consciousness. First, items that attract attention enter into consciousness. If you attend to a stimulus (top-down attention), or happen to see a striking stimulus (bottom-up attention), then, *ceteris paribus*, it will be consciously perceived (I will come back to the *ceteris paribus* clause shortly). Second, if attention is withdrawn from a stimulus, it will be rendered unconscious. We saw this with inattentional blindness. It is also true in cases of pathology, such as unilateral neglect: when attention centres are damaged, deficits of consciousness arise. Third, neural correlates of attention seem to coincide with neural correlates of consciousness. At the systems level, conscious vs. unconscious sensory processes are associated with activity in sensory areas, inferior parietal cortex, and lateral prefrontal cortex—all areas that activate in attention tasks. At the neurocomputational level, both are associated with relatively fast, phase-locked neural activity. Fourth, the equation of attention and consciousness can explain a variety of behavioural results. Most obviously, it explains inattentional blindness. Attention has also been posited as a mechanism to explain backwards masking—the time it takes for attention to modulate a visual state coincides with the temporal parameters within which masking is possible, and masks, whether they cover a stimulus or appear in its surround, are effective attention lures. Interocular suppression shows a similar pattern. One eye suppresses another when it is presented with features that capture low-level attention, and stimuli in the suppressed eye can creep into consciousness if they are capable of recapturing attention.

Despite these converging lines of evidence, many authors reject the hypothesis that consciousness is attention. The main reason is that there is empirical evidence purporting to show both attention without consciousness and consciousness without attention. To illustrate the former, recall the study by Hsieh *et al.* (2011), which was mentioned above. Here visual pop-out effects—a classic example of bottom-up attention—are shown to occur in a stimulus that has been rendered invisible by interocular suppression. To illustrate an alleged case of consciousness without attention, consider a study by Reddy *et al.* (2006). Here participants were given an attentionally demanding letter task (finding an L in a group of Ts), and then faces were briefly flashed in the periphery; the letter task seems to have no impact on conscious face detection.

I think the Reddy *et al.* study, and others like it, are unconvincing. They use a paradigm very much like inattentional blindness, but get opposite results. Why? The reason is their choice of briefly flashed stimuli. Faces (as well as emotionally charged words, animals, and various other ecologically exciting objects) capture attention. The letter task is fairly demanding, but not so demanding that it prevents faces from causing participants to briefly allocate some attention to the periphery without significant loss of performance. This also works with other stimuli known to capture attention, but, crucially, it doesn't work for less exciting stimuli like simple colours and shapes. When such neutral stimuli are used, inattentional blindness results. Thus, the Reddy study actually helps to prove that attention dictates the contents of consciousness, not the opposite.

The Hsieh et al. result is harder to accommodate, and the vast majority of efforts to dissociate attention and consciousness take this form. Many tasks have been designed to show selective enhancement within an unconscious visual state. Here, I think the best response is to concede that the debate is verbal. 'Attention' can be used in different ways, and the best use depends on the overall use in ordinary language, scientific practice, and theoretical utility. I think there are better and worse ways to define attention, but, in the present context, that doesn't matter. The substantive question concerns the actual mechanisms at play in the Hsieh study. Here, it is safe to say that selection is achieved through lateral inhibition. Contrast between one patch and surrounding patches results in the selection of the distinctive patch. Selection can involve a number of changes, both functional and physiological. The cells corresponding to the distinctive patch may exhibit increased excitation, and the surround may be suppressed; receptive fields may warp so that the distinctive patch occupies more visual space; changes may increase luminance, contrast, sharpness, and other aspects of acuity; and saccadic eye movements and motor plans may shift towards the distinctive stimulus. All of these changes can take place unconsciously. In addition, given the way executive function can modulate unconscious visual states, such changes can occur as a result of top-down effects, not just lateral inhibition. So, there is quite a lot that takes place when an unconscious stimulus is selected, and selection can be caused in various different ways. Many authors are inclined to call this 'attention'.

At the same time, we must grant that there is one crucial thing that doesn't happen to an unconscious stimulus when it is selected in all these ways. The distinctive stimulus does not enter the systems that control reporting and deliberation. Consequently, participants in the Hsieh task do not realize it is there. They do not report its presence, and, though it may affect their actions, they do not make explicit plans on the basis of having seen it. The systems that allow reporting and deliberation depend on encoding in what is called 'working memory'. Only stimuli that enter working memory can guide reports and deliberation. The distinctive stimulus in the Hsieh study never gets access to working memory. Interocular suppression prevents that from happening. Putting this all together, it seems to follow that access to working memory makes the difference between visibility and invisibility. It so happens that I think the term 'attention' is best defined as access to working memory. On this use, participants in the Hsieh et al. study exhibit pre-attentive selection, but they cannot attend—i.e. access the stimulus in working memory—because the stimulus in the other eye has occupied attention too fully. But this is just a verbal preference.

We can avoid unnecessary debates by defining consciousness as working memory access and avoiding the word attention.

The claim that consciousness arises with access to working memory would not be accepted by all consciousness researchers, but shares considerable support. For example, it is consistent with the global workspace theory of consciousness, which is the most popular theory in cognitive neuroscience and psychology. I will assume that this theory is true, and explore some implications for the present discussion. One immediate implication is that the difference between conscious and unconscious vision is not a difference in the representations involved, nor even the operations on those representations. Access to working memory is not a representational property, like binding, but rather a change in how a representation relates to other neurocognitive systems. It is the becoming-available of a representation.

4.2. The function of consciousness

If visual consciousness arises when visual states become available to working memory, we can say something about the function of consciousness. What consciousness adds is not representational features, but information flow. Consciousness allows visual representations to get encoded in working memory. This is its function. Now we can ask, what good is that?

Here it is tempting to answer by saying something about the function of working memory. That would be a mistake. Working memory is very valuable, to be sure. It is the gateway to reporting and deliberation, for example. But, the function of consciousness is to make information available to working memory, and availability is different from encoding. Working memory has a very small capacity — perhaps three items — and consciousness is quite rich. This shows that, of all the information available to working memory, very little gets in. So the value of consciousness must be expressed in terms of the availability relation, not encoding.

Elsewhere I speculate that consciousness earns its keep by providing a 'menu for action' (Prinz, 2012). The things that are available to working memory become, thereby, candidates for deliberative action. They present a choice space from which an organism can then engage in further processes with the full resources of executive systems — systems that are, themselves, unconscious. Here I want to offer a slightly different formulation, since the menu metaphor now strikes me as somewhat too passive.

I propose that the function of consciousness, as achieved by availability to working memory, is to 'cue and confirm'. When we see things unconsciously, we usually do nothing. Aside from some very automatic or overlearned responses, an unconscious visual stimulus is likely to cause little more than a few fluctuations in motor preparation and eye movements. Without consciousness, we don't *know* that a stimulus is present, and do not realize there is anything upon which we can act or reflect. Cueing changes that. It tells us that something is present, and this knowledge creates the possibility of deliberate response. Now suppose that one directs a deliberative action towards a stimulus and suddenly it blinks out of consciousness. Without continued consciousness, we cannot confirm that the action was successful. If that were to occur, we wouldn't know what to do next. Do we repeat the action or go on to a new task? Cueing and confirmation work in tandem. If you see a road before you, it may cue you to take a step forward, and if you see that the step has succeeded in advancing your position, you may take another step. We need both the cue and the confirmation. Consciousness gives us both.

The helps to explain why consciousness allows us to do things that would be impossible without it. Unconscious vision may be able to modulate actions, but it doesn't prompt us to act, and it doesn't give us a progress report that allows us to perform actions that are complex, such as actions that are carried out in a sequence of steps. Unconscious vision is also very limited with respect to multi-step cognitive actions. Recall that in unconscious arithmetic we can subtract three digits. But we can't subtract four. One explanation is that we do not get confirmation without consciousness. With consciousness, we can subtract one number, then another, and and then another *ad infinitum*, because each step can be confirmed before moving on. This is the cognitive equivalent of walking down a path. The unconscious arithmetic problem doesn't even serve as a cue. It elicits an automatic associative response, but it does not invite us to deliberate about what to do. If an unconscious stimulus does not have a potentiated response, or if it has too many, then it will not be effective. It can cause an effect, but can't cue a response. Cueing makes something available for choice, and confirmation tells us when our choices pan out.

5. Conclusions

If the cue-and-confirm story is right, then consciousness makes a massive difference in the life of an organism. It allows flexible decision making, and actions that have a potentially unlimited number of steps. So clearly consciousness is important. My focus here, however, has been on the enormous power of unconscious vision. That point can be extended to all forms of unconscious mental activity. The vast majority of what we do is unconscious. Recognizing objects, constructing sentences, and, I would argue, all of higher cognition is carried out unconsciously. Consciousness gives us (limited) access to the inputs of these processes and the output, by the action, is hidden from view. The case of vision is instructive because no domain has been more thoroughly investigated in consciousness studies. The main lesson there is that consciousness does not contribute to visual function. This suggests that the visual unconscious deserves full credit for seeing. Generalizing, almost everything the mind can do, it can do unconsciously. Consciousness merely cues us to do things and confirms when we have.

Acknowledgments

I am deeply indebted to Zdravko Radman. His friendship, encouragement, inspiration, and extraordinary patience made this chapter possible.

References

Bar, M. & Biederman, I. (1998) Subliminal visual priming, *Psychological Science*, **9**, pp. 464–469.

Binsted, G., Brownell, K., Vorontsova, Z., Heath, M. & Saucier, D. (2007) Visuomotor system uses target features unavailable to conscious awareness, *Proceedings of the National Academy of Sciences*, **104**, pp. 12669–12672.

Chen, Y.-C. & Spence, C. (2011) Crossmodal semantic priming by naturalistic sounds and spoken words enhances visual sensitivity, *Journal of Experimental Psychology: Human Perception and Performance*, **37**, pp. 1554–1568.

Crick, F. & Koch, C. (1990) Towards a neurobiological theory of consciousness, *Seminars in Neuroscience*, **2**, pp. 263–275.

Daems, A. & Verfaillie, K. (1999) Viewpoint-dependent priming effects in the perception of human actions and body postures, *Visual Cognition*, **6**, pp. 665–693.

Dell'Acqua, R. & Grainger, J. (1999) Unconscious semantic priming from pictures, *Cognition*, **73**, pp. B1–B15.

Eddy, P.J. & Holcomb, M.D. (2011) Invariance to rotation in depth measured by masked repetition priming is dependent on prime duration, *Brain Research*, **1424**, pp. 38–52.

Engel, A.K. & Singer, W. (2001) Temporal binding and the neural correlates of sensory awareness, *Trends in Cognitive Science*, **5**, pp. 16–25.

Enns, J.T. & Oriet, C. (2007) Visual similarity in masking and priming: The critical role of task relevance, *Advances in Cognitive Psychology*, **3**, pp. 211–216.

Esteves, F., Parra, C., Dimberg, U. & Öhman, A. (1994) Nonconscious associative learning: Pavlovian conditioning of skin conductance responses to masked fear-relevant facial stimuli, *Psychophysiology*, **31**, pp. 375–385.

Hsieh, P., Colas, J.T. & Kanwisher, N. (2011) Unconscious pop-out: Attentional capture by unseen feature singletons only when top-down attention is available, *Psychological Science*, **22**, pp. 1220–1226.

Jackendoff, R. (1987) *Consciousness and the Computational Mind*, Cambridge, MA: MIT Press.

Jiang, Y., Costello, P., Fang, F., Huang, M. & He, S. (2006) A gender- and sexual orientation-dependent spatial attentional effect of invisible images, *Proceedings of the National Academy of Sciences*, **103**, pp. 17048–17052.

Keizer, A.W., Hommel, B. & Lamme, V.A.F. (2015) Consciousness is not necessary for visual feature binding, *Psychonomic Bulletin and Review*, **22**, pp. 453–460.

Lamy, D., Mudrik, L. & Deouell, L.Y. (2008) Unconscious auditory information can prime visual word processing: A process-dissociation procedure study, *Consciousness and Cognition*, **17**, pp. 688–698.

Lin, Z. & Murray, S.O. (2014) Unconscious processing of an abstract concept, *Psychological Science*, **25**, pp. 296–298.

Mack, A. & Rock, I. (1998) *Inattentional Blindness*, Cambridge, MA: MIT Press.

Mattler, U. & Fendrich, R. (2007). Priming by motion too rapid to be consciously seen, *Perception & Psychophysics*, **69**, pp. 1389–1398.

Merikle, P.M. & Smith, S.D. (2005) Memory for information perceived without awareness, in Ohta, N., MacLeod, C.M. & Uttl, B. (eds.) *Dynamic Cognitive Processes*, Tokyo: Springer.

Milner, D. & Goodale, M. (1995) *The Visual Brain in Action*, Oxford: Oxford University Press.

Mudrik, L., Breska, A., Lamy, D. & Deouell, L.Y. (2011) Integration without awareness: Expanding the limits of unconscious processing, *Psychological Science*, **22**, pp. 764–770.

Ocampo, B., Al-Janabi, S. & Finkbeiner, M. (2015) Direct evidence of cognitive control without perceptual awareness, *Psychonomic Bulletin & Review*, **22**, pp. 1083–1088.

Öhman, A. & Soares, J. (1998) Emotional conditioning to masked stimuli: Expectancies for aversive outcomes following non-recognized fear-relevant stimuli, *Journal of Experimental Psychology: General*, **127**, pp. 69–82.

Overgaard, M., Fehl, K., Mouridsen, K., Bergholt, B. & Cleeremans, A. (2008) Seeing without seeing? Degraded conscious vision in a blindsight patient, *PLoS One*, **3**, e3028.

Persuh, M. & Ro, T. (2013) Unconscious priming requires primary visual cortex at specific temporal phases of processing, *Journal of Cognitive Neuroscience*, **25**, pp. 1493–1503.

Prinz, J.J. (2011) Is attention necessary and sufficient for consciousness?, in Mole, D., Smithies, D. & Wu, W. (eds.) *Attention: Philosophical and Psychological Essays*, Oxford: Oxford University Press.

Prinz, J.J. (2012) *The Conscious Brain: How Attention Engenders Experience*, New York: Oxford University Press.

Prinz, J.J. (2015) Unconscious perception, in Matthan, M. (ed.) *Oxford Handbook of Philosophy of Perception*, Oxford: Oxford University Press.

Rajimehr, R. (2004) Unconscious orientation processing, *Neuron*, **41**, pp. 663–673.

Reddy, L., Wilken, P. & Koch, C. (2004) Face-gender discrimination is possible in the near-absence of attention, *Journal of Vision*, **4**, pp. 106–117.

Ro, T., Singhal, N., Breitmeyer, B. & Garcia, J. (2009) Unconscious processing of color and form in metacontrast masking, *Attention, Perception, & Psychophysics*, **71**, pp. 95–103.

Schmidt, T. (2002) The finger in flight: Real-time motor control by visually masked color stimuli, *Psychology of Science*, **13**, pp. 112–118.

Schubö, A. & Meinecke, C. (2007) Automatic texture segmentation in early vision: Evidence from priming experiments, *Vision Research*, **47**, pp. 2378–2389.

Simons, D.J. & Levin, D.T. (1997) Change blindness, *Trends in Cognitive Science*, **1**, pp. 261–267.

Sklar, A., Levy, N., Goldstein, A., Mandel, R., Maril, A. & Hassin, R.R. (2012) Reading and doing arithmetic nonconsciously, *Proceedings of the National Academy of Sciences*, **109**, pp. 19614–19619.

Spering, M., Pomplun, M. & Carrasco, M. (2011) Tracking without perceiving: A dissociation between eye movements and motion perception, *Psychological Science*, **22**, pp. 216–225.

Tononi, G. & Edelman, G.M. (1998) Consciousness and complexity, *Science*, **282**, pp. 1846–1851.

Treisman, A. & Gelade, G. (1980) A feature integration theory of attention, *Cognitive Psychology*, **12**, pp. 97–136.

Winkielman, P., Berridge, K. & Wilbarger, J. (2005) Unconscious affective reactions to masked happy versus angry faces influence consumption behavior and judgments of value, *Personality and Social Psychology Bulletin*, **1**, pp. 121–135.

Wokke, M.E., van Gaal, S., Scholte, H.S., Ridderinkhof, K.R. & Lamme, V.A.F. (2011) The flexible nature of unconscious cognition, *PLoS ONE*, **6**, e25729.

Shaun Gallagher

Prenoetic Effects on Perception and Judgment

> We feel things differently according as we are sleepy or awake, hungry or full, fresh or tired; differently at night and in the morning, differently in summer and in winter, and above all things differently in childhood, manhood, and old age. (James, 1890, 1, p. 232)

The notion of a prenoetic effect attempts to capture the idea of certain embodied processes that happen non-consciously but that nonetheless have an effect on one's conscious experience of the world. Some of these processes can be described in terms of neuronal processes. Some of them related to non-conscious aspects of movement and action may be captured in descriptions of body schematic processes (Gallagher, 2005; see Cushing *et al.*, this volume). Others are less determinate and may involve constraints or affordances provided in certain environments. I'll argue that the concept of prenoetic effect fits well with enactivist accounts of experience, and does not fit well with representationalist accounts.

I'll start by providing some examples of prenoetic (specifically affective, non-conscious) processes that shape perception and judgment. I'll then try to clarify how prenoetic effects relate to brain function and how all of this fits with, and indeed enriches the enactivist model.

1. Some Examples of Prenoetic Effects

Prenoetic effects include affective processes (in the broadest sense); not only emotional processes, but processes that are associated with autonomic and endocrine systems—things like hunger, fatigue, changes in homeostasis. We can start with a simple example from William James (1890)—that an apple appears larger and more intensely red when the perceiver is hungry than when satiated. Hunger can also affect our perception of food-related words, allowing us to pick them out faster compared to other non-food related words (Radel and Clément-

Guillotin, 2012). A study by Danziger, Levav and Avnaim-Pesso (2011) shows that hunger distorts cognitive processes. The study shows that the rational application of legal reasons does not sufficiently explain the decisions of judges. Whether the judge is hungry or satiated may play an important role.

> The percentage of favorable rulings drops gradually from ≈65% to nearly zero within each decision session [e.g. between breakfast and lunch] and returns abruptly to ≈65% after a [food] break. Our findings suggest that judicial rulings can be swayed by extraneous variables that should have no bearing on legal decisions. (Dansiger, Levav and Avnaim-Pesso, 2011, p. 6889)

Such an effect is extraneous to the legal process, but not extraneous to the jurist's embodied cognition. The embodied affect (hunger) has an effect on the jurist's judgments, on on the weighing of evidence, and so forth, and doesn't appear out of nowhere just when the judicial decision is made. It is rather something that has been emerging in the background, with the judge unaware of it.

Fatigue too can have an effect on perception. In a set of well-known experiments, Proffitt has shown that subjects estimate the grade of an incline to be steeper when wearing a heavy backpack or when fatigued, in comparison to wearing none or when rested (Proffitt et al., 1995). Durgin et al. (2009) have challenged Proffitt's results, but in a way that nicely makes the same point that I am after. Durgin et al. showed that steeper estimates of an incline while wearing a backpack 'are judgmental biases that result from the social, not physical, demands of the experimental context' (p. 1). Without awareness of this bias, subjects who sense the aim of the experiment estimate a steeper incline than subjects who are given an explicit reason for wearing the heavy backpack (e.g. that it contains electromyographic equipment to measure muscle tension). In other words, those subjects who had a nonconscious sense of the experimenters' intentions were biased in favour of those intentions, without knowing it. But this too is a prenoetic effect — one that is social in nature. Other studies show that the behaviour of others can lead to a priming effect on object evaluation. Subjects presented with a face looking towards (or away from) an object are primed to evaluate the object as more (or less) likeable than those objects that don't receive attention from others, specifically when they're asked, although they are not conscious of the influence. If one adds an emotional expression to the face, one get's a stronger effect (Bayliss et al., 2006; 2007). Seeing another person act with ease (or without ease) toward an object will influence observers' feelings about the object (Hayes et al., 2008).

If Proffitt is right (see Proffitt, 2009; 2013, for further discussion), his experimental results point to the embodied-affective nature of cognition. But if Durgin *et al.* (2009; 2012) are right, their results still point to an embodied phenomenon—namely, the significant prenoetic effects that others (or other people's intentions) have for our perceptual experiences.

Indeed, this helps to point to the fact that typically, in any situation, there is not one simple, isolated affect—there is rather a cocktail, a mélange of aspects that make up an affective state (Gallagher and Bower, 2014). For example, my hike up the mountain results in a perception or judgment that is shaped by a combination of my fatigue, my troubled respiration, the grittiness of the dust on my body, my hunger, my pain, and the kinaesthetic difficulty involved in climbing. The mountain path will look quite different and less challenging after a good night's sleep, not because of certain objective qualities that belong to the path, but because of my affective state. My affective state will prenoetically condition my perception of the path. Likewise, to bring the social back into it, the climb may look less difficult if I am with a group of friends and they are enjoying the hike or we are absorbed in conversation. These are prenoetic qualifications on my perceptions, actions, and judgments, and they can equally have an effect on my phenomenal consciousness even if I am not aware of the reason that I am feeling one way or the other (Gallagher and Aguda, 2015). There's a difference in *what it is like* to be on the mountain path in the morning after a good night's rest, and *what it is like* to be on the very same mountain path at the end of a long day of hiking. My experience is modulated by my situation, which includes my bodily condition. Depending also on my practical interests, my physical state may be felt as an overwhelming fatigue that is a barrier to any further climbing, or it may contribute to a feeling of satisfaction as I sip a glass of wine in front of the fire at the end of the day.

2. Embodied Cognition and the Brain

Prenoetic effects can involve neural and extra-neural processes. A system's satiety and energy level depend on peripheral metabolic processes that involve hormonal regulation, e.g. insulin and leptin circulate in the brain in proportion to recent energy intake and established body adiposity (fat levels) (Havel, 2000). In such cases, explanations in terms of representations, even B[ody]-formatted representations (Gallese and Sinigaglia, 2011; Goldman, 2012; 2014) seem inadequate. B-formatted (neural) representations specifically represent states of the subject's own body from an internal perspective; they include interoceptive or motoric representations 'of one's own bodily states and activities' (Goldman, 2012, pp. 71, 73). Information about one's body may originate peripherally, but becomes B-formatted only when represented centrally in somatosensory and motor cortexes, but also in the 'interoceptive cortex' (Craig, 2002) registering 'pain, tickle, temperature, itch, muscular and visceral sensations, sensual touch, and other feelings from (and about) the body' (Craig, 2012, p. 74). On such an account, however, the actual body plays a marginal and perhaps even trivial role in cognition. It ignores the fact that the brain does what it does in tandem with the body. When the bodily system is fatigued or hungry, these conditions influence brain function, not through the mediation of representations of fatigue or hunger, but through real physical effects. Parts of the brain, e.g. the hypothalamus, operate on homeostatic principles rather than anything that can be construed as representational principles. Perner and Ogden (1988), for example, suggest that hunger is a 'non-representational internal state' — supposedly just the opposite of what Clark and Toribio (1994) would call a representation-hungry state.

Indeed, the body regulates the brain as much as the brain regulates the body. Autonomic, peripheral, and immune systems communicate with the brain in a bidirectional manner (Silverman, Sternberg et al., 2010; Sternberg and Gold, 1998). Homeostatic regulation happens via mutual (largely chemical) influences between parts of the endocrine system, and signals from the autonomic system. Low glucose levels (hypoglycemia, which is a biological condition not caused by the brain) may mean slower or weaker brain function, or some things turning off. At the extreme, it can mean brain death. In cases of hypoglycemia, perception is not modulated because the brain *represents* hunger or fatigue, but because the perceptual system (brain and body) is chemically (materially) affected by the actual hunger and fatigue. There are real physical connections here in the complex chemistry of the body-brain system but also in its perceptual and action coupling with the environment.

The environment includes physical, social, and cultural aspects that can have an affect on, and can be affected by, my embodied system. My hunger may not affect my perception so much if a sexually attractive person walks into the room. This has something to do, not only with who is entering the room, but also with processes in the hypothalamus. A similar example involves changes in the perception of sexual attractiveness due to testosterone depletion (hypogonadism) or across menstrual phase (Alexander and Sherwin, 1991). Such effects are not a case of confused hypothalamic body-formatted representations. They're matters of bodily chemistry.

The body's *affective* life regulates brain function via hormonal and neurotransmitter levels. Together with peripheral and autonomic systems (including heart function [Garfinkle et al., 2013] and respiration [Liu et al., 2014]), such prenoetic effects shape cognitive function. For example, the functioning of the circulatory system has a prenoetic effect on perception of fear. Circulation and heartbeat influence how and whether fear-inducing stimuli (images of fearful faces, in the reported experiments) are processed (Garfinkel et al., 2013). When the heart contracts in its systole phase, fearful stimuli are more easily recognized, and they tend to be perceived as more fearful than when presented in a diastole phase. Accordingly, cognition is not just the work of the brain, but the effect of a dynamic interplay between brain, endocrine, autonomic, and peripheral systems that are closely connected with the body's environment.

To fully understand the cognitive event that constitutes something on the order of the judge's legal ruling, moreover, one has to consider not only non-representational, unruly processes complicated by autonomic and endocrine responses, but also social and institutional environments (Gallagher, 2013). Despite the 'extraneous' effects of prenoetic bodily processes, the legal ruling is framed by social practices and institutional factors that make it the legal ruling that it is. In the full picture, affective phenomena related in part to body chemistry, together with social and institutional practices that define the situated, contextual conditions of such events, are good examples of how cognition can be the result of complex dynamical brain–body–environment couplings that simply can't be captured in purely representational terms.

One may be able to explain how such affective, and more generally prenoetic, subpersonal, non-conscious processes, affect perception and judgment in terms of predictive coding models. With respect to affect, for example, Barrett and Bar's *affective prediction hypothesis* 'implies that responses signaling an object's salience, relevance or value do not occur as a separate step after the object is identified. Instead, affective responses support vision from the very moment that visual stimulation begins' (Barrett and Bar, 2009, p. 1325). Even neuronal activity in the earliest of perceptual processing areas, such as visual area V1, reflects more than simple feature detection. For example, V1 neurons are activated in ways that anticipate reward if they have been tuned by prior experience (Shuler and Bear, 2006). But this means that priors are not necessarily top-down assumptions or beliefs; they may find specificity in the tuning of brain processes via neural plasticity, and may even involve these broad, bottom-up affective aspects of embodied life embedded in the larger brain–body–environment system. Thus, along with the earliest visual processing, the medial orbital frontal cortex is activated initiating a train of muscular and hormonal changes throughout the body, generating 'interoceptive sensations' from organs, muscles, and joints associated with prior experience, and integrated with current sensory information that helps to guide response and subsequent actions (Barrett and Bar, 2009).

We can think, then, that visual stimulation generates specific bodily affective changes that are integrated with sensory-motor processes tied to the current situation. This is part of early processing, so that before the perceiver fully recognizes an object or other person or their situation, the perceptual system is already dynamically adjusting into overall peripheral and autonomic patterns based on current status and prior associations.

3. How this Fits with an Enriched Enactivist Model

On the enactivist view, brains play an important part in the dynamical attunement of organism to environment. Cognition (perception, action, judgment) involves the integration of brain processes with fully embodied and situated processes—a complex mix of transactions that involve stomach, heart, respiration, movement, gesture, engagement with others, and that incorporate artefacts, tools, and technologies (Clark, 2008; Malafouris, 2010), situated in various physical environments, and defined by diverse social roles and institutional practices (Gallagher, 2013). Brains, which have evolved with their bodies, are fully integrated into the larger system along with all these other factors. The brain would work differently if its embodiment were different—if hormones were different, if we never got hungry or fatigued or emotional, etc. Our experience of the world and our affordances would be different if the brain evolved in a body without arms or legs.

I refer to this as an enriched enactivist model only in the sense that prenoetic effects include more than just sensory-motor contingencies, and sometimes enactivist approaches are thought of simply in terms of such body-schematic contingencies (e.g. Noë, 2004; O'Regan and Noë, 2001). They do include such non-conscious body-schematic factors, but clearly all of these affective aspects, in the broad sense including hunger, fatigue, autonomic factors, etc., also have an effect on what counts as an affordance or action possibility (e.g. Colombetti, 2013; Di Paolo, 2005; Gallagher and Bower, 2014; Thompson, 2007).

Changes or adjustments to brain chemistry, to its electrical rhythms, or to neural processing will accompany changes in these other extraneural factors, not because the brain represents such changes and responds to them in central command mode, but because the brain is part of the larger embodied system that is coping with its changing environment. Just as the hand adjusts to the shape of the object to be grasped, so the brain adjusts to the circumstances of organism-environment.

A standard philosophical question at this point is: are such things merely causally coupled or constitutive of cognition (Adams and Aizawa, 2001; 2010a,b)? One way to respond to this question is to think that if something is merely causal then one should be able to substitute some other mechanism and get the same effect. If constitutive, no substitution will work without changing the nature of the phenomenon. On this simple test it's not clear what kind of substitution can be made for hunger or fatigue. On some models of cognition, for example, the claim has always been that the causal mechanisms can be brain-based neural mechanisms or can be swapped out for silicon-based computer chips (at least in principle), which seems to suggest that the neural processes underpinning cognition are merely causal. Issues concerning functionalism and causality are too complex to take up here, but in any case these issues all concern causal efficacy rather than constitution. For Adams and Aizawa, for something to be a constitutive part of a cognitive process it must involve an intrinsic mark of the mental, which they identify as non-derived representational content. If things like hunger and fatigue are non-representational, on this view they would be merely causal factors that influence cognition, but not constitutive aspects of cognition.

In contrast to this classical metaphysical conception of a synchronic material or compositional notion of constitution, where something's intrinsic nature just is what makes it what it is, and where facts about external causal relations tell us nothing about constitution (e.g. Bennett, 2011; 2014) — a conception that lends itself to an account of static properties or states, but not of processes — an alternative conception of constitution involves diachronic processes that may in fact be causal (Kirchhoff, 2015). On this view, embodied mental processes distributed across different factors/levels (neural, behavioural, environmental) at different timescales are constituted in a temporally integrated dynamical system. The constituent elements may very well be in complex, reciprocal causal relations with each other, but just these reciprocal causal relations make the mental process what it is. In a dynamic gestalt composed of processes that unfold over time, and characterized by recursive reciprocal causality relations, changes in any processual part will lead to changes in the whole, and changes in the whole will imply changes in the processual parts. In contrast to a synchronic, compositional notion of constitution, these kinds of causal relations are diachronically constitutive of the phenomenon. The notion of a causal/coupling-constitution fallacy, where constitution is defined synchronically and independently of causation, does not apply to the type of diachronic processes described in dynamic patterns (Kirchhoff, 2015; Gallagher, in press). Of course one will still need a way to distinguish between non-constitutive causal factors and diachronic constitutive causal factors in any particular case. This can be a test by intervention: if an intervention (e.g. the removal of a particular factor, X, or the substitution of a different factor, Y) changes the system as a whole, or causes a significant change in cognitive outcome, X would count as a constitutive element (see Campbell, 2007; Woodward, 2003, for the interventionist notion of causality).

The various prenoetic factors that I have discussed here are, in this sense, constitutive of cognition. They point to a more complete account than any account that can be given in narrow or 'weak' embodiment terms of neural B-formatted representations (Alsmith and Vignemont, 2012). They also fit well with a more complete enactivist account that takes into consideration more than just sensory-motor contingencies.

References

Adams, F. & Aizawa, K. (2001) The bounds of cognition, *Philosophical Psychology*, **14** (1), pp. 43–64.

Adams, F. & Aizawa, K. (2010a) Defending the bounds of cognition, in Menary, R. (ed.) *The Extended Mind*, Cambridge, MA: MIT Press.

Adams, F. & Aizawa, K. (2010b) The value of cognitivism in thinking about extended cognition, *Phenomenology and the Cognitive Sciences*, **9** (4), pp. 579–603.

Alexander, G.M. & Sherwin, B.B. (1991) The association between testosterone, sexual arousal, and selective attention for erotic stimuli in men, *Hormones and Behavior*, **25** (3), pp. 367–381.

Alsmith, A.J.T. & De Vignemont, F. (2012) Embodying the mind and representing the body, *Review of Philosophy and Psychology*, **3** (1), pp. 1–13.

Barrett, L.F. & Bar, M. (2009) See it with feeling: Affective predictions during object perception, *Philosophical Transactions of the Royal Society of London B*, **364** (1521), pp. 1325–1334.

Barrett, L.F. & Bliss-Moreau, E. (2009) Affect as a psychological primitive, *Advances in Experimental Social Psychology*, **41**, pp. 167–218.

Bayliss, A.P., Paul, M.A., Cannon, P.R. & Tipper, S.P. (2006) Gaze cueing and affective judgments of objects: I like what you look at, *Psychonomic Bulletin & Review*, **13**, pp. 1061–1066.

Bayliss, A.P., Frischen, A., Fenske, M.J. & Tipper, S.P. (2007) Affective evaluations of objects are influenced by observed gaze direction and emotional expression, *Cognition*, **104**, pp. 644–653.

Bennett, K. (2004) Spatio-temporal coincidence and the grounding problem, *Philosophical Studies,* **118** (3), pp. 339–371.

Bennett, K. (2011) Construction area (no hard hat required), *Philosophical Studies,* **154**, pp. 79–104.

Campbell, J. (2007) An interventionist approach to causation in psychology, in Gopnik, A. & Schulz, L. (eds.) *Causal Learning: Psychology, Philosophy, and Computation,* Oxford: Oxford University Press.

Clark, A. (2008) *Supersizing the Mind: Reflections on Embodiment, Action, and Cognitive Extension,* Oxford: Oxford University Press.

Clark, A. & Toribio, J. (1994) Doing without representing?, *Synthese,* **101** (3), pp. 401–431.

Colombetti, G. (2013) *The Feeling Body: Affective Science Meets the Enactive Mind,* Cambridge, MA: MIT Press.

Craig, A.D. (2002) How do you feel? Interoception: The sense of the physiological condition of the body, *Nature Reviews Neuroscience,* **3**, pp. 655–666.

Danziger, S., Levav, J. & Avnaim-Pesso, L. (2011) Extraneous factors in judicial decisions, *Proceedings of the National Academy of Sciences,* **108** (17), pp. 6889–6892.

Di Paolo, E.A. (2005) Autopoiesis, adaptivity, teleology, agency, *Phenomenology and the Cognitive Sciences,* **4** (4), pp. 429–452.

Durgin, F.H., Baird, J.A., Greenburg, M., Russell, R., Shaughnessy, K. & Waymouth, S. (2009) Who is being deceived? The experimental demands of wearing a backpack, *Psychonomic Bulletin & Review,* **16**, pp. 964–969.

Durgin, F.H., Klein, B., Spiegel, A., Strawser, C.J. & Williams, M. (2012) The social psychology of perception experiments: Hills, backpacks, glucose, and the problem of generalizability, *Journal of Experimental Psychology: Human Perception and Performance,* **38** (6), p. 1582.

Gallagher, S. (2005) *How the Body Shapes the Mind,* Oxford: Oxford University Press.

Gallagher, S. (2011) Interpretations of embodied cognition, in Tschacher, W. & Bergomi, C. (eds.) *The Implications of Embodiment: Cognition and Communication,* Exeter: Imprint Academic.

Gallagher, S. (2013) The socially extended mind, *Cognitive Systems Research,* **25/26**, pp. 4–12.

Gallagher, S. (2014) Pragmatic interventions into enactive and extended conceptions of cognition, *Nous – Philosophical Issues,* **24**, pp. 110–126.

Gallagher, S. (in press) *Enactivist Interventions: Rethinking Mind and Cognition,* Oxford: Oxford University Press.

Gallagher, S. & Aguda, B. (2015) The embodied phenomenology of phenomenology, *Journal of Consciousness Studies*, **22** (3-4), pp. 93-107.

Gallagher, S. & Bower, M. (2014) Making enactivism even more embodied?, *AVANT: Trends in Interdisciplinary Studies* (Poland), **5** (2), pp. 232-247.

Gallese, V. & Sinigaglia, C. (2011) What is so special about embodied simulation?, *Trends in Cognitive Sciences*, **15** (11), pp. 512-519.

Garfinkel, S., Minati, L. & Critchley, H. (2013) Fear in your heart: Cardiac modulation of fear perception and fear intensity, Poster presented at the *British Neuroscience Association Festival of Neuroscience*, 8 April.

Goldman, A.I. (2012) A moderate approach to embodied cognitive science, *Review of Philosophy and Psychology*, **3** (1), pp. 71-88.

Goldman, A.I. (2014) The bodily formats approach to embodied cognition, in Kriegel, U. (ed.) *Current Controversies in Philosophy of Mind*, New York and London: Routledge.

Goldman, A. & De Vignemont, F. (2009) Is social cognition embodied?, *Trends in Cognitive Sciences*, **13** (4), pp. 154-159.

Havel, P.J. (2000) Role of adipose tissue in body-weight regulation: Mechanisms regulating leptin production and energy balance, *Proceedings of the Nutrition Society*, **59** (3), pp. 359-371.

Hayes, A.E., Paul, M.A., Beuger, B. & Tipper, S.P. (2008) Self produced and observed actions influence emotion: The roles of action fluency and eye gaze, *Psychological Research*, **72** (4), pp. 461-472.

James, W. (1890) *Principles of Psychology*, 2 vols., New York: Dover.

Kirchhoff, M.D. (2015) Extended cognition and the causal-constitutive fallacy: In search for a diachronic and dynamical conception of constitution, *Philosophy and Phenomenological Research*, **90** (2), pp. 320-360.

Liu, L., Papanicolaou, A.C. & Heck, D.H. (2014) *Visual Reaction Time Modulated by Respiration*, (Working paper), University of Tennessee Medical Center, Memphis.

Malafouris, L. (2010) *How Things Shape the Mind*, Cambridge, MA: MIT Press.

Noë, A. (2004) *Action in Perception*, Cambridge, MA: MIT Press.

O'Regan, J.K. & Noë, A. (2001) A sensorimotor account of vision and visual consciousness, *Behavioral and Brain Sciences*, **24** (5), pp. 939-973.

Perner, J. & Ogden, J.E. (1988) Knowledge for hunger: Children's problem with representation in imputing mental states, *Cognition*, **29** (1), pp. 47-61.

Proffitt, D.R. (2009) Affordances matter in geographical slant perception, *Psychonomic Bulletin & Review*, **16**, pp. 970–972.

Proffitt, D.R. (2013) An embodied approach to perception by what units are visual perceptions scaled?, *Perspectives on Psychological Science*, **8** (4), pp. 474–483.

Proffitt, D., Bhalla, M., Gossweiler, R. & Midgett, J. (1995) Perceiving geographical slant, *Psychonomic Bulletin & Review*, **2** (4), pp. 409–428.

Proffitt, D., Stefanucci, J., Banton, T. & Epstein, W. (2003) The role of effort in perceiving distance, *Psychological Science*, **14** (2), pp. 106–112.

Radel, R. & Clément-Guillotin, C. (2012) Evidence of motivational influences in early visual perception hunger modulates conscious access, *Psychological Science*, **23** (3), pp. 232–234.

Shuler, M.G. & Bear, M.F. (2006) Reward timing in the primary visual cortex, *Science*, **311** (5767), pp. 1606–1609.

Silverman, M.N., Heim, C.M., Nater, U.M., Marques, A.H. & Sternberg, E.M. (2010) Neuroendocrine and immune contributors to fatigue, *Journal of Physical Medicine and Rehabilitation*, **2** (5), pp. 338–346.

McCann, S.M., Lipton, J.M., Sternberg, E.M., Chrousos, G.P., Gold, P.W. & Smith, C.C. (1998) *Molecular Aspects, Integrative Systems, and Clinical Advances: Third International Congress of the Society for Neuroimmunomodulation, 1996*, New York: New York Academy of Sciences, 840.

Thompson, E. (2007) *Mind in Life: Biology, Phenomenology and the Sciences of Mind*, Cambridge, MA: Harvard University Press.

Woodward, J. (2003) *Making Things Happen: A Theory of Causal Explanation*, Oxford: Oxford University Press.

Ben R. Newell

The Anatomy of an (Unconscious) Decision

In 2014 Newell and Shanks published a review critiquing evidence of unconscious influences on decision making (Newell and Shanks, 2014a). The review generated comments running the gamut from outright rejection of the approach and conclusions (Dijksterhuis *et al.*, 2014; Hassin and Milyavsky, 2014) to praise and broad agreement (Baumeister, Vohs and Masicampo, 2014; Hogarth, 2014; Huizenga *et al.*, 2014). These reactions exemplify the enduring interest, controversy, and sheer difficulty in making theoretical progress when it comes to considering how our behaviour is shaped by factors outside our awareness.

The aim of this chapter is to pursue two threads of research emerging from the discussion of the Newell and Shanks review (hereafter N&S). The first thread focuses on an approach regarded by N&S as offering clear potential for demonstrating unconscious influences on decision making: the biasing of decisions via eye-gaze-contingent prompts (e.g. Richardson, Spivey and Hoover, 2009). The second thread picks up on the criticism that N&S failed to consider evidence of the potential for unconscious biases in the decisions of legal practitioners (Konečni, 2015).

At first glance these threads appear unrelated. The first is concerned with fleeting, moment-to-moment changes in the direction of gaze and how they might affect decision making (e.g. Pärnamets *et al.*, 2015). The second concerns situations traditionally thought to involve deep deliberation, careful consideration of facts, and explicitly reasoned arguments and judgments (e.g. Dhami, 2003). And yet, an analysis of these seemingly polar opposite decision processes helps in revealing the complex anatomy of purportedly unconscious decisions. The anatomy of the decision (a term adopted from Abercrombie's 1960 text) emphasizes the need to dissect the structure of these decisions and to scrutinize the empirical and theoretical bases for claiming unconscious influences.

The chapter proceeds as follows. In Section 1 a very brief review of the N&S 'lens-model' approach to assessing unconscious influences on decision making is presented. Section 2 discusses the idea that one can 'track eyes to change minds' and presents new data questioning recent bold claims about the impact of passive monitoring of eye-gaze on decision making. In Section 3 the arguments raised by Konečni (2015) regarding legal practitioners are discussed. Section 4 draws out some general conclusions within the context of the differences between intuitive and unconscious judgment.

1. A Lens Model for Assessing Unconscious Influence

Following N&S's approach, the term decision making is used here to refer to the mental processing that leads to the selection of one amongst several actions (choices). This interpretation renders consciousness a property of individuals and excludes from consideration questions about how areas of the brain 'make decisions'. For example, the visual system's computation of low-level properties — such as motion detection by area V5 — is not decision making under this definition. It is, however, perfectly reasonable to ask whether an individual's judgment of motion is conscious.

N&S propose a framework for thinking about how decisions could be influenced by unconscious processes; this framework lends itself to evaluating the eye-movement (Section 2) and legal examples (Section 3) considered here. The framework is based loosely on the *lens model* (Brunswik, 1952), popularized in the judgment and decision making field by Hammond, Stewart, and many others (see Hammond and Stewart, 2001).

The basic premise of the lens model is that a decision maker views the world through a 'lens of cues' that mediates between a stimulus in the environment and the internal perceptions of the decision maker, as shown in Figure 1. The double convex lens in the centre of the diagram shows a constellation of cues that diverge from a criterion or event in the environment (left side of figure). The decision maker uses these cues to achieve (e.g. correctly estimate) the criterion and so these cues are shown as converging (right side of figure) on a point of response or judgment in the mind of the decision maker.

A more detailed account of how unconscious influences can arise at each of the points (A to E in the diagram) can be found in N&S; here the focus is on Points C and D. The former refers to situations in which a decision maker is simply unaware of the influence that a particular cue has on a subsequent decision. As we will see, this type of influence appears to capture the effect that eye-movements might have on our decisions. Point D refers to unawareness of the way in which cues are weighted or utilized when arriving at a judgment. This kind of influence appears to map onto the claims made for unconscious influences in legal decisions.

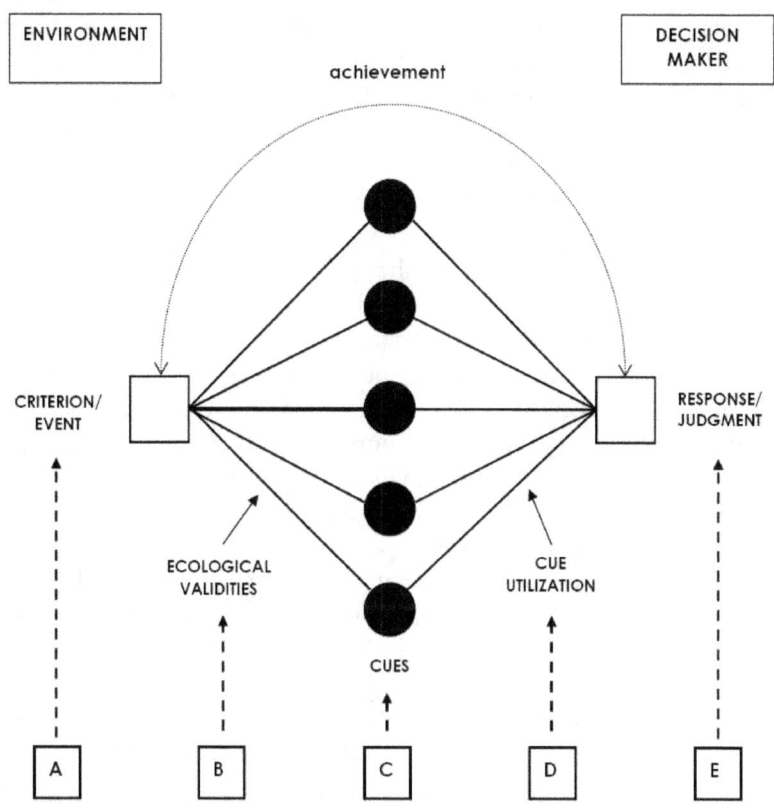

Figure 1. A lens model framework illustrating possible loci of unconscious influences on decision making. (Reprinted with permission from Newell and Shanks, 2014a.)

2. Eye-Gaze and Choice

Many of the 'classic' experimental tasks claimed to provide evidence for unconscious influences on decision making are highly reflective in nature. Take, for example, the infamous Iowa Gambling Task (IGT; e.g. Bechara *et al.*, 1997): participants are given explicit instructions to make a particular decision (maximize reward), they are fully aware of the independent variables (reward magnitudes of different options) and that they have multiple trials in which to learn about the structure of the task. It seems unsurprising, then, that when given a proper opportunity to introspect about their own mental processes and the influences of the independent variables on their behaviour, participants' reports can be quite insightful (e.g. Konstantinidis and Shanks, 2014; Maia and McClelland, 2004).

N&S proposed that a potentially more promising approach to finding unconscious influences is to look at situations in which attention is diverted *away* from the experimenter's hypothesis. As a paradigmatic example, they singled out a demonstration of the influence of eye-gaze on choice (Richardson *et al.*, 2009).

Eye-gaze duration is known to predict subsequent preference. For example, when selecting between snack foods pre-rated for equivalent appetitive value, participants are more likely to choose the food that they have looked at for longer (Krajbich, Armel and Rangel, 2010).

Experimenter-controlled manipulations of gaze duration have also been shown to bias subsequent choice for foods (Armel, Beaumel and Rangel, 2008), as well as faces (Shimojo *et al.*, 2003). Such data have led some researchers to conclude, 'one's own gaze bias may be interpreted as preference at subconscious levels' (*ibid.*, p. 1321). Indeed, the fast food giant Pizza Hut has recently begun trials of the world's first 'subconscious menu': pizza-lovers simply glance through the tablet-based menu, and before they 'consciously' make a decision the software infers topping-preferences from gaze duration and places an order (Henderson, 2014).

But what if Pizza Hut could have a direct, causal influence on pizza choice simply by prompting patrons to order at a particular point during their perusal of the menu? Pärnamets *et al.* (2015), building on Richardson *et al.* (2009), reported an arresting demonstration of just this type of causal effect—not on (trivial) pizza preference but on substantive moral choices. Participants' gaze was monitored and information about the location of their gaze at particular time-points in a decision trial was used to prompt responses. Each decision trial comprised a two-alternative choice between answers to a morally complex question (e.g. Murder is sometimes justifiable?). At the start of each trial one of the two potential answers ('Sometimes' or 'Never') was *randomly predetermined* as the 'target' response. The striking finding was that participants' decisions could be biased towards this randomly determined target simply by prompting participants to make a response as soon as the target option had been looked at for a specified duration. Specifically, participants chose the predetermined target on 58.2% of trials—a proportion significantly above the chance responding expected if the prompt had had no influence.

Moreover, this influence on choice appeared to occur even though participants were apparently unaware that the appearance of the response prompt was based on their own eye-movements. In terms of the lens model depicted in Figure 1, the prompt-bias is a seemingly clear example of people being unaware of the influence of a cue (the prompt) on a subsequent decision (i.e. point C). Pärnamets *et al.* (2015) assessed awareness of the prompt's influence via post-test questioning. (Such awareness questionnaires are often fraught with interpretational difficulties (N&S). However, the relatively straightforward nature of the design and the transparency of the connection between eye-movements and the appearance of the prompt render their use in this instance defensible.)

Pärnamets *et al.*'s findings have substantial theoretical implications because they suggest that the process of making a (moral) decision is not only *reflected* by a participant's eye-gaze but that gaze can also *determine* that decision. This causal interpretation implies a complex intertwining of abstract moral values and sensory-motor mechanisms that goes beyond our current understanding of how evidence accumulates in the formation of decisions (Pärnamets *et al.*, 2015).

A full understanding of the implication of these findings—both for theories of decision making and for the role of awareness—requires a detailed dissection of the methods employed by Pärnamets *et al.* As with many surprising effects, it turns out that the devil may be in the details. Participants in the Pärnamets *et al.* study (*ibid.*, Experiment 2) sat in front of computer monitors wearing headphones. Each trial started with a central fixation point while a statement (e.g. Murder is sometimes justifiable?) was played over the headphones. Immediately after the recording had finished playing, two response options ('Sometimes', 'Never') were shown on the screen, one on the right-hand side and one on the left-hand side. The left-hand response option was designated as the target on (a randomly selected) half of the trials, and the right-hand option on the other half.

Eye-gaze was defined as falling on an item if it fell in the half of the screen containing that option; the response prompt was triggered when the target option on the current trial had accumulated at least 750 ms gaze time, and the non-target had accumulated at least 250 ms of gaze time. At this point the response options disappeared and were replaced by masks for 167 ms, before the response prompt appeared. If the trigger condition was not met within 3000 ms from the onset of the options, then the trial timed out, and the mask and response prompt appeared.

In the primary analysis of their results, Pärnamets et al. chose only to focus on the subset of trials that satisfied the trigger conditions—i.e. those in which a response had been prompted within 3000 ms. At first glance this seems like a sensible analysis decision, but it brings with it the potential to skew the data in favour of finding the desired effect. To illustrate why, consider what could happen on a trial involving an easier question than the murder one. For example, consider the statement 'Australia is larger than Iceland' with the response options 'True' and 'False'. In this case presumably most participants would readily realize the statement was true and thus would not gaze (for long) at the 'False' option. A failure to gaze for long enough at 'False' would mean that trials in which that response (i.e. False) had been predetermined as the target would time out (according to the trigger conditions defined above) and be omitted from the analysis. Trials for which 'True' had been predetermined as the target would, in contrast, be included in the analysis—assuming that participants at least glanced (250 ms) at the 'False' response in addition to dwelling (for at least 750 ms) on 'True'. Any such bias could result in a greater proportion of target-chosen trials appearing in the prompted relative to the timed-out trials.

Although the Australia-Iceland question is illustrative (Parnamets et al. did not use this question), it seems fair to assume that the questions they used had a distribution of difficulty levels, thus contributing to a possible asymmetry in the distribution of trials. Indeed Pärnamets et al. were aware of this potential asymmetry. In an analysis reported after their main results they stated that the biasing effect fell from 58.2% target selection to 53.8% when all trials were included (i.e. timed-out and triggered trials); a value still significantly above chance. Such a drop is consistent with the claim that when time-outs are excluded the analysis is skewed in favour of showing the response-prompt bias (target-selected trials) at the expense of a more accurate reflection of how self-directed gaze biases choice.

This speculation is also supported by new data that illustrate how excluding timed-out trials systematically and falsely inflates support for the response-prompt effect. Newell et al. (2016) replicated the Pärnamets et al. study—using the same set of questions and the same presentation and trigger procedures. They found—consistent with Pärnamets et al.—that when time-out trials were excluded a response-prompt bias effect was found: participants chose the target on 53.3% of trials (significantly above chance). However, when all trials were included this dropped to 49.6% target selection—in other words no evidence that the prompt had influenced choice. Notably, Newell et al.

had almost twice as many participants in their experiment (N = 37) as Pärnamets et al. (N = 20).

An additional study reported by Newell et al. provides an even more straightforward demonstration of the impact of excluding timed-out trials. Rather than moral questions, participants were given a perceptual discrimination task in which they had to decide which of two squares presented on screen was more dense (in terms of pixel concentration). The trial structure and the trigger conditions were identical to the moral choice experiments. On some trials the discrimination task was objectively (and presumably subjectively) easy because the dot-density of the paired patterns differed by 10%. On other trials the discrimination was objectively impossible because paired patterns of *equal* density were presented.

Newell et al. expected that in the cases in which stimulus discrimination was possible, and relatively easy, the gaze-contingent appearance of the response prompt would be unlikely to enhance participants' choice of the randomly determined target stimulus. This is because any effect of prompting would be over-ridden by the participant's certainty in knowing which option is correct. In contrast, they expected to find a bias towards the randomly determined target in conditions where the discrimination was objectively impossible. Under these conditions participants have no perceptual evidence on which to base a decision, so it seems plausible that their choices may be more susceptible to external influences, such as the time at which a decision is prompted.

The analysis when timed-out trials were *excluded* was surprising. Newell et al. found significant biasing effects both for the impossible (53.4%) *and the easy trials* (56.5%). What was puzzling about the latter effect was that participants were almost 100% accurate (98.7%) in choosing the correct (denser) alternative on these easier trials. This juxtaposition of a biasing effect (participants chose the target on more than 50% of trials) and near-100% accuracy on the *possible* trials seems paradoxical. How can both of these be true, if the target is randomly determined on each trial (such that the target will be the incorrect option on 50% of trials)?

The answer again lies in the excluded time-out trials. Specifically, participants showed a strong, systematic bias to choose the *non*-target on trials that timed out. On *easy* trials, when the target is the low-density pattern, it is presumably obvious to participants that it is not the correct answer; consequently participants will often not look at this pattern for long enough (750 ms) to trigger the prompt. Hence the trial will time out and be excluded from the analysis. The more accurate reflection comes from analyses including *all trials*. Here the findings matched Newell et al.'s predictions. On *easy* trials, there was no evidence for an effect of the gaze-contingent prompt (target selection = 49.8%), but on the impossible trials a small but reliable effect of the prompt remained (52.3% target selection).

This re-examination of the decision-prompt bias could be dismissed as hair-splitting and clarification of methodological niceties. Newell *et al.* (2016) argue against this interpretation strongly: it is only via a careful dissection of the design and the data that assumptions can be tested adequately. In general these new results are consistent with the idea that participants' choices are reflected in their eye-gaze (a well-established finding; e.g. Shimojo *et al.*, 2003), but do not support the suggestion of a causal link wherein the timing of the gaze-contingent response prompt influences the choice that is subsequently made. The one exception to this pattern is the above-chance selection of the target in the analysis of all trials for the impossible perceptual stimuli. This result suggests that, when there is no objective perceptual information available to a decision maker, the timing of the prompt can have a small but detectable effect on choice. Moreover, participants did not appear to be aware of the influence of the prompt in these trials. Section 4 returns to the implications of these findings for the larger question regarding unconscious influences on decision making; first Section 3 considers the potential for unconscious influences on much more serious and consequential moral decisions—those made by legal professionals.

3. Are there Unconscious Influences on Judicial Decisions?

In 2015 Konečni published a critique of N&S which, although largely favourable, highlighted what he perceived to be an 'unfortunate analytical omission of judicial decision making' (p. 1) from the N&S review. Konečni laments this omission for two reasons: 1) decisions made by legal practitioners have considerable moral, social, and political importance, and can lead to enormous human costs if errors are made; 2) databases detailing many relevant legal decisions are readily available and could be mined for evidence of unconscious influences (e.g. Konečni and Ebbesen, 1984). Konečni goes on to speculate that the 'most compelling possible reason' for the omission is the observation that 'public scrutiny and the demands of high office allegedly largely preclude (or even prohibit) unconscious influences' (p. 2) in judicial decision making. The actual reason for the omission was a practical one driven by constraints on article length. Nonetheless, it is highly instructive to examine the issues raised by Konečni and to explore the implications for understanding the anatomy of an unconscious decision.

Konečni argues that the notion of 'individualized justice' (broadly the idea that every case is different) behooves judges to take all factors into account and weigh them appropriately in a conscious and explicit manner. In reality, Konečni claims that individualized justice permits 'a comparatively free rein being given to intuition and to an "instinctive" (in the sense of unconscious) synthesis of information' (p. 2).

The complete story is, however, perhaps more complex and more interesting. The central point of Konečni's critique is that when the databases he cites are consulted carefully, they reveal a striking mismatch between judges' decisions and their introspections about how those decisions are made (e.g. Konečni and Ebbesen, 1984). This apparent mismatch emphasizes the importance of distinguishing among three components: 1) what judges publicly say they do; 2) what they privately think they do; and 3) what they actually do (Konečni, 2015). To illustrate the distinction, Konečni provides the following example:

> A sentencing judge may believe that he or she is taking—prejudically, intentionally—a defendant's race into account; the judge would of course never publicly admit to these prejudices and behavioral intentions; but the actual sentencing decision, which usually relies on very few factors, may completely disregard race (as it should)! In other words, the closet racist and the public civil libertarian may behaviorally be neither. (p. 2)

There are several different elements to unpack here. First let's examine the idea that a judge may be a 'closet racist'. In the only brief foray N&S made into the legal sphere, they did in fact consider this type of potential unconscious influence (N&S, 2014, footnote 4). Specifically, they discussed work by Blair and colleagues purporting to demonstrate the unconscious influence of Afrocentric features on sentencing decisions. For example, Blair, Judd and Chapleau (2004) found that an analysis of black and white US prison inmates with approximately equivalent criminal histories received roughly equivalent sentences. Crucially, however, when sentences within-race were examined, Blair et al. found that inmates whose faces had been independently rated as more Afrocentric (e.g. darker skin, wide nose, full lips) received harsher sentences than those rated as less Afrocentric. Eberhardt et al. (2006) extended this work and showed that in cases involving a white victim, the more stereotypically black a defendant was perceived to be the more likely that person was to receive a death sentence.

The authors' interpretation of these effects is that the perceived Afrocentricity of the faces is the direct cause of the harsher sentencing. Moreover, they argue that because these effects operate outside of awareness, people are unable to control their influence on decisions (e.g. Blair, Judd and Fallman, 2004). However, N&S argued that there is no clear evidence that Afrocentric features are the direct—or *proximal*— cause of behaviour. The notion of a proximal cause is important here and it strikes at the heart of what N&S termed the *relevance criterion* (see Table 1 in N&S). This criterion states simply that when assessing awareness one should target only information that is relevant to the observed behaviour. In the Blair and Eberhardt studies it is relatively easy to imagine that another—perhaps correlated—attribute drove the sentencing behaviour, rather than Afrocentricity *per se*. In other words, the number of Afrocentric features is not the direct cause of the behaviour and thus assessing its influence represents a failure to adhere to the *relevance criterion*. As N&S noted:

> [S]uppose that number of Afrocentric features in faces is correlated, in the minds of judges, with some other attribute such as hostility or low intelligence. Use of this correlated attribute might be entirely conscious (though of course deeply unjust). Moreover, it would not be surprising on this alternative hypothesis that participants are unable to control the influence of Afrocentric features on their judgments. (p. 19)

This interpretation is strikingly similar to the one offered by Konečni of a judge who is consciously aware of potential prejudices and thus attempts to take them into account when sentencing. Nonetheless, as Konečni notes, the 'public civil libertarian' would never admit to taking such information into account.

This contrast between the closet racist and the public libertarian echoes the vast amount of research on so-called 'unconscious bias'; the idea that our explicit attitudes towards different segments of society (e.g. women, ethnic minorities, LGBTIQ) are divorced from our more authentic implicit attitudes (e.g. Nosek, 2007). The most popular interpretation of the dissociation between implicit and explicit attitudes is that the latter are unconscious in the sense of being inaccessible to introspection (Tierney, 2008). However, the example of the judges provides pause for thought: the closet racist 'knows' he's a racist—he is aware of his implicit attitude—he is just not willing to express it publicly (perhaps a form of motivated self-presentation). How often might people be in a similar state? How often are 'unconscious biases' actually available to consciousness?

The principal experimental tool used to demonstrate the apparent dissociation between implicit and explicit attitudes is the Implicit Association Test—IAT (Greenwald, McGhee and Schwartz, 1989). Details of the exact procedure are beyond the scope of this chapter, but in essence the task measures participants' accuracy and response times in associating different categories of objects. For example, participants might have to categorize photographs of white people with positive words and black people with negative words and vice versa. The differences in accuracy and response times between trials perceived as 'compatible' (e.g. white and good, for US university undergraduates) and incompatible (e.g. black and good) is taken as a measure of implicit attitudes. These implicit attitudes are then often contrasted with explicit attitudes elicited by asking participants to indicate how 'warmly' or 'coolly' they feel towards the different groups tested in the IAT.

An impressive body of research illustrates the low correlation between explicit attitudes (measured by the 'thermometer scale') and implicit attitudes (measured by the IAT) (e.g. Hoffman *et al.*, 2005). This low correlation is taken as evidence that implicit attitudes — although an apparently more accurate reflection of true beliefs — are inaccessible to conscious report. However, in an ingenious set of studies Hahn *et al.* (2014) challenge this widely held conclusion. Hahn *et al.*'s analysis is predicated on the idea espoused by the *relevance criterion*: if you want to assess a person's awareness of the basis of an observed behaviour then you need to ask them for information that is directly relevant to that behaviour. This is not what is done typically in IAT studies. The attitude information elicited in the explicit assessment may well not be the same as the information underlying the IAT performance; thus the latter might not be unconscious, it might just be different.

In a series of four experiments Hahn *et al.* found that this 'not unconscious, just different' hypothesis garnered a great deal of support. The support was revealed via a simple — yet clever — methodological change: participants were asked to *predict* their performance on the IAT before taking the test. That is, participants were asked, for example, how easy it would be to sort pictures of black people with the category 'good' and white with the category 'bad', relative to the reverse arrangement. The key result was that these predicted ratings correlated much more strongly with actual performance on the IAT than did the explicit ratings, elicited using the standard 'thermometer' scale. Moreover, this strong correlation held irrespective of whether participants were given detailed instructions about the IAT procedure before making their predictions.

The clear implication of these results is that a lack of correspondence between implicit and explicit attitudes — the result commonly reported in IAT studies — tells us very little about people's awareness of their implicit reactions (*ibid.*). If the lack of correspondence emerges from a failure to apply N&S's relevance criterion then any conclusions about the unconscious nature of particular biases appear unfounded. Hahn *et al.*'s results remain rather troubling, however. The patterns of responses in the IAT tests still revealed biases in the sense that participants were less accurate in categorizing black with good than white with good, for example, but crucially this pattern cannot be attributed to the operation of an unconscious process. This is important because it suggests we cannot abdicate responsibility for responding in a particular way to some hardwired/culturally embedded reaction that is outside of awareness and conscious control. However unpleasant and unjustified these 'gut reactions' might be, Hahn *et al.*'s data suggest that people have unique insight into them—and therefore a responsibility to address any inherent biases. Indeed, one of the participants in Hahn *et al.*'s study commented in debriefing: 'I feel guilty because I think I am an intuitive person. Yet, based on this test, it shows that if I go with my initial gut reaction about race and value judgments, I am actually quite judgmental' (p. 1388).

This participant thus provides another illustration of the conflict Konečni identified in the judge trying to reconcile the inner racist with the external civil libertarian. Konečni's final claim, however, was that the decisions often handed down by judges take none of this information into account but rather are based on a very small subset of cues. The principal evidence for this claim comes from statistical analyses of sentencing decisions made in courts in the San Diego area in the 1970s and 1980s. The details of these analyses need not concern us, but in essence they show that in decisions about bail, for example, judges tend to follow a single cue: the prosecutor's recommendation. Similarly, in sentencing decisions, the probation officer's recommendation appeared to be the *only* cue used in 90% of the decisions analysed by Konečni and Ebbesen (1984). In more recent analyses of UK courts' bail decisions, Dhami (2003) makes similar claims about judges using simple heuristics that reflect 'passing the buck'—relying on decisions made previously by police, prosecution, or previous members of the bench.

If these statistical analyses are to be believed and the judges in question claimed to have used more than a single cue then this would be evidence for an unconscious influence located at Point D on Figure 1 —unawareness of the way in which cues are integrated to make a decision. The Dhami data do not speak directly to this issue because she reports no independent assessments of what judges said they did. (In fact such an analysis would be very difficult to perform because, in fitting heuristic models, Dhami averaged over decisions made by multiple different judges within a particular court—over a series of days—rather than focusing on individual judges.) Konečni and Ebbesen (1984), on the other hand, do have such data and claim a stark discrepancy between the reported and actual cues used. What remains unresolved, however, is whether this is a genuine unconscious influence, a failure to elicit the relevant explicit knowledge, or a deliberate attempt by judges not to reveal their private beliefs and insight into their own decision processes in interviews and questionnaires. Konečni and Ebbesen (1984) appear to favour the last interpretation.

4. Conclusion: Unconscious or Intuitive?

What conclusions can be drawn from our dissection of two very different examples of purported unconscious influence on decision making? The re-examination of the eye-gaze experiments discussed in Section 2 raised doubts about the idea that moral judgments made in the blink of an eye could be biased by factors outside participants' awareness. A small but reliable effect of the gaze-contingent prompt did remain, however, for the 'impossible' perceptual judgments. In terms of the lens model (Figure 1), this latter effect could be taken as evidence for unawareness at Point C: participants simply do not realize that the time at which they are prompted influences their subsequent choice. The effect is small (barely above the 50% chance level) and in need of further examination, not least with the addition of potentially more sensitive measures of awareness. Nonetheless, such a result provides partial support to N&S's contention that, in contrast to often rather heavy-handed (and controversial) behavioural priming procedures (e.g. Doyen et al., 2012; Newell and Shanks, 2014b; Shanks et al., 2013), eye-gaze offers a window onto (unconscious) cognitive processing that is potentially less susceptible to demand characteristics or inference of the experimenter's hypothesis.

The literature discussed in Section 3 concerning much weightier and more consequential real-world moral decisions provides suggestive evidence of a mismatch between what judges say they do and what they actually do, but there is no conclusive proof that this mismatch is a result of unconscious biases. That is, echoing N&S's conclusion, there is little evidence of unawareness at Point D (Figure 1) of the decision process. Indeed, Konečni's own critique hints at the discrepancy being due to judges' reluctance to divulge, publicly, what they know to be externally invalid reasons for passing judgments. This accurate, although potentially unpalatable, insight into the bases of decisions appears to converge with Hahn et al.'s re-evaluation of the IAT literature. Low correlations between explicit and implicit attitudes do not necessarily indicate inaccessibility of the latter. Participants appear to be able to predict their gut reactions while at the same time realizing that they are not valid bases for explicit attitudes (Hahn et al., 2014).

Questioning the influence on decision making of factors outside our awareness is not synonymous with questioning the role of intuition or instinctive judgment. There are clearly phenomenological and other differences between intuitive and deliberative decisions (Hammond,

1996; Hogarth, 2001) — the latter involve intermediate, inferential steps, whereas the former can be driven by feelings of fluency, familiarity, and recognition (Kahneman and Klein, 2009; Simon, 1992). But such intuitive judgment does not imply the influence of unconscious information processing, nor reliance on information that is inaccessible to introspection. Thus a potentially very fruitful question for future research is to ask when our gut reactions should be relied upon and when they should be ignored. In other words, what are the characteristics for developing accurate 'intuitive expertise' (Kahneman and Klein, 2009)?

Recent developments in understanding how our cognitive machinery interacts with the structure of environments in which decisions are made are beginning to shed more light on this issue. For example, how do we become calibrated to potential sources of bias in the samples of information we receive (e.g. Fiedler, 2012); how do feedback and environment structure interact to improve judgment (e.g. Hawkins *et al.*, 2014; Hogarth and Soyer, 2014; Newell, Lagnado and Shanks, 2015); and does awareness of the potential for biases improve our ability to know when to listen to or ignore our 'gut' (e.g. Soll, Milkman and Payne, 2015)? These and many other questions are ripe for research. Moreover, they could be readily addressed if we resist the temptation to relegate our decision making to a black-box labelled 'the unconscious' (e.g. Dijksterhuis and Strick, 2016),[1] realize we cannot abdicate responsibility to mysterious machinations, and concentrate on the learning and knowledge acquisition that precedes decisions and the feedback which follows (Newell *et al.*, 2015). In the words of Abercrombie (1960, p. 18), a detailed understanding of the anatomy of a decision (unconscious or otherwise) requires a focus on 'how it comes to be as it is, how it is adapted to its function and in what conditions in can develop differently'.

---

[1] Perhaps the most stinging critique of N&S came from Dijksterhuis *et al.* (2014), who accused us of cherry-picking data in our evaluation of Unconscious Thought Theory (the idea that superior decisions are made via active, offline, unconscious weighting and integration of information). We note that in the most recent treatment of this theory (Dijksterhuis and Strick, 2016) the authors failed to address an important study by Nieuwenstein *et al.* (2015) which reported a large-scale replication study that yielded no evidence for an unconscious thought advantage, and a meta-analysis showing that previous reports of an effect were confined to underpowered studies that used relatively small sample sizes.

Acknowledgment

The author thanks David Shanks and Emmanouil Konstantinidis for very helpful comments on an earlier draft of this chapter and the Australian Research Council (DP 140101145) for research support.

References

Abercrombie, M.L.J. (1960) *The Anatomy of Judgment*, London: Hutchinson & Co.

Armel, K.C., Beaumel, A. & Rangel, A. (2008) Biasing simple choices by manipulating relative visual attention, *Judgment and Decision Making*, **3**, pp. 396–403.

Baumeister, R.F., Vohs, K.D. & Masicampo, E.J. (2014) Maybe it helps to be conscious, after all, *Behavioral & Brain Sciences*, **37**, pp. 20–21.

Bechara, A., Damasio, H., Tranel, D. & Damasio, A.R. (1997) Deciding advantageously before knowing the advantageous strategy, *Science*, **275**, pp. 1293–1295.

Blair, I.V., Judd, C.M. & Chapleau, K.M. (2004a) The influence of Afrocentric facial features in criminal sentencing, *Psychological Science*, **15**, pp. 674–679.

Blair, I.V., Judd, C.M. & Fallman, J.L. (2004b) The automaticity of race and Afrocentric facial features in social judgments, *Journal of Personality and Social Psychology*, **87**, pp. 763–778.

Brunswik, E. (1952) *The Conceptual Framework of Psychology*, Chicago, IL: University of Chicago Press.

Dhami, M.K. (2003) Psychological models of professional decision-making, *Psychological Science*, **14**, pp. 175–180.

Dijksterhuis, A., van Knippenberg, A., Holland, R.W. & Veling, H. (2014) Newell and Shanks' approach to psychology is a dead end, *Behavioral & Brain Sciences*, **37**, pp. 25–26.

Dijksterhuis, A. & Strick, M. (2016) A case for thinking without consciousness, *Perspectives on Psychological Science*, **11**, pp. 117–132.

Doyen, S., Klein, O., Pichon, C.-L. & Cleeremans, A. (2012) Behavioral priming: It's all in the mind, but whose mind?, *PLoS One*, **7**, art. e29081.

Eberhardt, J.L., Davies, P.G., Purdie-Vaughns, V.J. & Johnson, S.L. (2006) Looking Deathworthy: Perceived stereotypicality of black defendants predicts capital sentencing outcomes, *Psychological Science*, **17**, pp. 383–386.

Fiedler, K. (2012) Meta-Cognitive Myopia and the dilemmas of inductive-statistical inference, *Psychology of Learning and Motivation*, **57**, pp. 1–55.

Greenwald, A.G., McGhee, D.E. & Schwartz, J.L.K. (1998) Measuring individual differences in implicit cognition: The Implicit Association Test, *Journal of Personality and Social Psychology*, **74**, pp. 1464–1480.

Hahn, A., Judd, C.M., Hirsh, H.K. & Blair, I.V. (2014) Awareness of implicit attitudes, *Journal of Experimental Psychology: General*, **143**, pp. 1369–1392.

Hammond, K.R. (1996) *Human Judgment and Social Policy: Irreducible Uncertainty, Inevitable Error, Unavailable Injustice*, New York: Oxford University Press.

Hammond, K.R. & Stewart, T.R. (eds.) (2001) *The Essential Brunswik: Beginnings, Explications, Applications*, Oxford: Oxford University Press.

Hassin, R.R. & Milyavsky, M. (2014) But what if the default is defaulting?, *Behavioral & Brain Sciences*, **37**, pp. 29–30.

Hawkins, G., Hayes, B.K., Donkin, C., Pasqualino, M. & Newell, B.R. (2015) A Bayesian latent mixture model analysis shows that informative samples reduce base rate neglect, *Decision*, **2**, pp. 306–318.

Henderson, J.M. (2014) *Eyetracking Technology Knows Your Subconscious Pizza Desires... or not*, [Online], https://theconversation.com/eyetracking-technology-knows-your-subconscious-pizza-desires-or-not-35132.

Hofmann, W., Gawronski, B., Gschwendner, T., Le, H. & Schmitt, M. (2005) A meta-analysis on the correlation between the Implicit Association Test and explicit self-report measures, *Personality and Social Psychology Bulletin*, **31**, pp. 1369–1385.

Hogarth, R.M. (2001) *Educating Intuition*, Chicago, IL: University of Chicago Press.

Hogarth, R.M. (2014) Automatic processes, emotions and the causal field, *Behavioral & Brain Sciences*, **37**, pp. 31–32.

Hogarth, R.M. & Soyer, E. (2011) Sequentially simulated outcomes: Kind experience versus nontransparent description, *Journal of Experimental Psychology: General*, **140**, pp. 434–463.

Huizenga, H.M., van Duijvenvoorde, A.C.K., van Ravenzwaaij, D., Wetzels, R. & Jansen, B.R. (2014) Is the unconscious, if it exists, a superior decision maker?, *Behavioral & Brain Sciences*, **37**, pp. 32–33.

Kahneman, D. & Klein, G. (2009) Conditions for intuitive expertise: A failure to disagree, *American Psychologist*, **64**, pp. 515–526.

Konečni, V.J. (2015) Are there unconscious influences on judicial decisions?, *Abnormal Behavioral Psychology*, **1**, pp. 101–102.

Konečni, V.J. & Ebbesen, E.B. (1984) The mythology of legal decision-making, *International Journal of Law and Psychiatry*, **7**, pp. 5–18.

Konstantinidis, E. & Shanks, D.R. (2014) Don't bet on it! Wagering as a measure of awareness in decision making under uncertainty, *Journal of Experimental Psychology: General*, **143**, pp. 2111-2134.

Krajbich, I., Armel, K.C. & Rangel, A. (2010) Visual fixations and the computation and comparison of value in simple choice, *Nature Neuroscience*, **13**, pp. 1292-1298.

Maia, T.V. & McClelland, J.L. (2004) A reexamination of the evidence for the somatic marker hypothesis: What participants really know in the Iowa gambling task, *Proceedings of the National Academy of Sciences*, **102**, pp. 16075-16080.

Nieuwenstein, M.R., Wierenga, T., Morey, R.D., Wicherts, J.M., Blom, T.N., Wagenmakers, E.-J. & van Rijn, H. (2015) On making the right choice: A meta-analysis and large-scale replication attempt of the unconscious thought advantage, *Judgment and Decision Making*, **10**, pp. 1-17.

Newell, B.R. & Shanks, D.R. (2014a) Unconscious Influences on decision making: A critical review, *Behavioral and Brain Sciences*, **37**, pp. 1-63.

Newell, B.R. & Shanks, D.R. (2014b) Prime numbers: Anchoring and its implications for theories of behavior priming, *Social Cognition*, **32**, pp. 88-108.

Newell, B.R., Lagnado, D.A. & Shanks, D.R. (2015) *Straight Choices: The Psychology of Decision Making*, 2nd ed., Hove: Psychology Press.

Newell, B.R., Torgerson, T., Saranu, C. & Le Pelley, M.E. (2016) The eyes have it? Can perceptual and moral decisions be influenced by eye movements?, Manuscript submitted for publication.

Nosek, B.A. (2007) Implicit-explicit relationships, *Current Directions in Psychological Science*, **16**, pp. 65-69.

Pärnamets, P., Johansson, P., Hall, L., Balkenius, C., Spivey, M.J. & Richardson, D.C. (2015) Biasing moral decisions by exploiting the dynamics of eye gaze, *Proceedings of the National Academy of Sciences*, **112**, pp. 4170-4175.

Richardson, D.C., Spivey, M.J. & Hoover, M.A. (2009) How to influence choice by monitoring gaze, in Taatgen, N., van Rijn, H., Nerbonne, J. & Schomaker, L. (eds.) *Proceedings of the 31st Annual Conference of the Cognitive Science Society*, p. 2244.

Shanks, D.R., Newell, B.R., Lee, E.H., Balakrishnan, D., Ekelund, L., Cenac, Z., Kavvadia, F. & Moore, C. (2013) Priming intelligent behavior: An elusive phenomenon, *PLoS One*, **8** (4), art. e56515.

Shimojo, S., Simion, C., Shimojo, E. & Scheier, C. (2003) Gaze bias both reflects and influences preference, *Nature Neuroscience*, **12**, pp. 1317-1322.

Simon, H.A. (1992) What is an explanation of behavior?, *Psychological Science*, **3**, pp. 150–161.

Soll, J.B., Milkman, K.L. & Payne, J.W. (2015) A user's guide to debiasing, in Keren, G. & Wu, G. (eds.) *Wiley-Blackwell Handbook of Judgment and Decision Making*, Chichester: John Wiley & Sons, Ltd.

Tierney, J. (2008) A shocking test of bias, *The New YorkTimes*, November 18, [Online], http://tierneylab.blogs.nytimes.com/2008/11/18/a-shocking-test-of-bias/.

Thomas P. Reber

Memory, Consciousness, and the Hippocampus

1. Memory and Consciousness

Memory can be described as past experience shaping future thinking and behaviour. Nowadays most researchers agree that memory is not a unitary phenomenon but can be subdivided into several distinct memory systems and/or processes, each responsible for the storage and retrieval of a different kind of knowledge (Cabeza and Moscovitch, 2013). On a neural level, a memory system has been defined as 'the minimal neural network required to record, retain, and retrieve a form of knowledge' (Gabrieli et al., 1990). It feels like something rather than nothing as we go through our waking life. This phenomenon is what one usually refers to when speaking of consciousness. The neural correlate of consciousness (NCC) has been defined as the minimal set of neuronal events that are sufficient for a certain conscious experience (Crick and Koch, 1995; Mormann and Koch, 2007). On a subjective level, consciousness of mnemonic processes can manifest as, for example, what it feels like to mentally place oneself back in time, and recall a personal experience. Thus there is likely some overlap of the NCC and the memory system required for recollection of personally experienced episodes. Consciousness may even be a prerequisite for a memory system to operate (e.g. Squire, 1992), or, vice versa, memory systems could form a precondition for conscious experience (Crick and Koch, 1990). Of course, memories can affect our behaviour in the absence of conscious awareness of mnemonic operations. Research on the extent by which unconscious or implicit memories impact upon our behaviour has been an area of active research and lively debate, and has been instrumental in devising current models of human long-term memory.

In fact, demonstrations of intact implicit memory in an otherwise severely memory-impaired neurological patient, H.M., have been at the very heart of the idea that there are multiple forms of memory (Squire, 2009). In an attempt to cure H.M. from his pharmacologically intractable epilepsy, parts of his brain, namely the hippocampus in the medial temporal lobe, were resected. While speech, general knowledge, short-term memory, and most of his cognitive functioning remained intact, H.M. displayed a severe deficit in the ability to form new autobiographical memories (Scoville and Milner, 1957). Strikingly however, H.M. was able to acquire new sensory-motor skills such as tracing a star-shape with a pencil while watching his drawing hand only through a mirror (Corkin, 1968). After several days of training, H.M. even showed surprise at how well he performed at this task as he had no conscious recollection of earlier training sessions whatsoever. Thus, skill acquisition and conscious recollection of the episodes in which skill acquisition happened appear to rely on different structures in the brain. H.M.'s surprise at his good performance hints at profound lack of awareness of his recently acquired sensory-motor memories, and hence suggests that there are forms of memory which can be accessed by conscious experience, i.e. explicit memory, or not, i.e. implicit memory.

Further studies on healthy participants revealed that implicit versus explicit memories not only dissociate regarding their neural correlate but also regarding various other factors (reviewed in Tulving and Schacter, 1990). For example, a seminal experiment by Jacoby (1983) entailed that participants study a list of words either by reading single words aloud (e.g. 'cold'), by reading a word in the context of its antonym (e.g. 'cold – hot'), or participants were required to generate study words ('cold') from a context word (e.g. 'hot –?'). The explicit memory test required participants to classify words into studied or new. Explicit memory was indicated by correct recognition of studied words, and was best for words which were studied in the generation-in-context condition ('hot –?'). The implicit test entailed very brief presentations of studied and new words on a computer screen (e.g. 40 milliseconds), and participants were instructed to merely try to identify the words if possible. Identification was set to be rather difficult as only 50% of correct identifications were achieved overall. Strikingly, the proportion of correct identifications was higher for studied versus new words, suggesting that reactivation of implicit memories preceded and facilitated the (conscious) perception of briefly presented words – a phenomenon referred to as priming. Maybe even more striking was the finding that implicit memory was best for words studied as single items. Thus, the study format had opposing effects on implicit and explicit memory performance (Blaxton, 1989; Jacoby, 1983; Tulving and Schacter, 1990).

2. Models of Long-Term Memory

One immediate conclusion from these implicit-explicit dissociations is that memory is not one single faculty (although see Berry, Shanks and Henson, 2008). Standard models account for these dissociations by introducing consciousness as a decisive feature to distinguish certain kinds of memory (Squire, 1992; Tulving, 2002). On a more computational level, these initial studies suggest that implicit memory is rather limited as it promotes establishment of new memory representations only through repeated learning trials, as in the case of procedural memory (Corkin, 1968), or seems confined to encoding of single items rather than associations between items, as in the case of priming (Jacoby, 1983). More generally, implicit memory has been described as rigid and slow because knowledge seems to be represented on a perceptual or procedural level, and can hardly be used to generalize to new situations.

Explicit memory, on the other hand, has been described as much more elaborate. Autobiographical memories are formed rapidly in just one episode, and contain information on a conceptual level. Such episodic memories consist of associations of multiple elements (e.g. 'what—where—when'), which can be accessed via multiple retrieval cues. Furthermore, generalization of knowledge to new situations is feasible as elements of an episode can be flexibly recombined across distinct memory representations (Eichenbaum, 2004; Henke, 2010).

Theoretical models of memory can be broadly assigned to two categories (Cabeza and Moscovitch, 2013), namely memory systems models and processing modes models. Memory systems models (Squire, 1992; Tulving, 2002) seek to explain implicit-explicit dissociations by postulating the existence of memory systems that require consciousness to operate, and memory systems that can operate without consciousness. In the tradition of categorizing memory systems, further differentiation of declarative (explicit) and non-declarative (implicit) memory systems have been made. Within the realm of declarative memory (Squire, 1992), episodic memories are distinguished from semantic memory (Tulving, 2002). Episodic memories have been defined as the conscious memories for personally experienced events in time and place (e.g. 'Yesterday I ate a doughnut for breakfast at the coffee shop around the corner'). Episodic memories crucially depend on an intact hippocampus in the medial temporal lobe and enable the fast acquisition of novel relations between elements. Semantic memory is the memory for general world knowledge, facts, concepts, and meaning (e.g. 'A doughnut is a ring-shaped deep-fried confectionery'), and depends on prefrontal and lateral temporal structures in the brain

(Patterson, Nestor and Rogers, 2007). Non-declarative or implicit memory, on the other hand, is a heterogeneous set consisting of procedural memory (e.g. the skill to form a nicely shaped doughnut with your hands), classical conditioning (e.g. when you always listen to a specific song when eating a doughnut, listening to the song alone makes you salivate in anticipation of the doughnut), or priming (e.g. faster reading of the word 'doughnut' on repeated occurrences).

Another group of models proposes so called 'processing modes'. Rather than consciousness, processing modes mainly differ in computational aspects. Thus, implicit and explicit memory rather differ in the extent to which they require processing on a perceptual or conceptual level (Blaxton, 1989; Roediger, 1990), or e.g. learning of single items versus associations between items (Henke, 2010). Put differently, if tasks to measure implicit and explicit memory both impose the same demands with regard to the processing mode (e.g. perceptual or conceptual retrieval), no dissociations in implicit and explicit memory should be observed.

A recent processing modes model has been formulated by Henke (2010). She distinguishes between three modes of processing, each with its own set of computational features and neuroanatomical correlates. Neocortical structures in the brain are shared among all processing modes but each processing mode is associated with specific additional structures. The first processing mode enables fast encoding of associations on a conceptual level, which can be flexibly expressed in novel situations. This mode of processing maps on to episodic memory according to systems views and critically depends on an intact hippocampus. Incremental learning of associations that are rather limited in flexibility constitutes the second mode of processing, which is reminiscent of procedural and semantic memory according to systems views, and relies on basal ganglia and the cerebellum. Finally, the mode for rapid encoding of single items requires parahippocampal regions, and maps onto priming.

3. Deciding between Models of Memory

3.1. Implicit memory for novel associations

Research on the limits of implicit memory has been instrumental to decide between models of long-term memory. Decisive evidence can be gathered by investigating the computational features of implicit memory, which are believed to be limited according to systems views (Squire, 1992; Tulving, 2002) but equally as elaborate as explicit memories according to processing modes views (Henke, 2010). One such feature is the formation of novel associations on a conceptual level requiring an intact hippocampus.

An early demonstration of implicit association formation has been conducted by Graf and Schacter (1985). Here, the study phase entailed reading unrelated pairs of words such as 'kindly—stick'. Testing consisted of word stem completion, i.e. saying the first word that comes to mind when presented with three letters, which form the beginning of multiple words. Word stem completion was performed on the second word 'sti__', which was either presented in its original pairing ('kindly—sti__') or in a pair with another word from the study list ('dryer—sti__'). Implicit memory for novel associations was indicated in healthy participants as they used studied words more frequently to complete stems in original versus rearranged test pairings, which was not the case in amnesics (Graf and Schacter, 1985; Schacter and Graf, 1986; Shimamura and Squire, 1989). Advocates of systems views explain this finding by asserting that putative implicit processes are mediated by explicit processes in healthy subjects (Shimamura and Squire, 1989). Furthermore, using an identification task on brief presentations (50–200 ms) of test items, amnesics identified intact pairs ('kindly—stick') more frequently than rearranged pairs ('dryer—still'). As identification may tap more into perceptual processing than word stem completion, this result has been taken as evidence that amnesics are able to implicitly learn new perceptual rather than conceptual associations (Gabrieli *et al.*, 1997).

Further evidence for implicit learning of novel associations has been gathered using the so-called contextual cueing task (Chun and Phelps, 1999). Here, visual search for the rotated letter T was tested when shown in context of rotated versions of the letter L in different colours. Visual search was facilitated during repeated trials when the context was identical versus different in healthy participants but not amnesics (*ibid.*). As both groups performed normally on a perceptual single-item priming task, this study suggested that the hippocampus is crucial also for implicit learning of novel associations. However, a following study suggested that contextual cueing does not depend on the hippocampus but adjacent structures (Manns and Squire, 2001).

Another line of research uses eye-movements as an index of implicit memory (Hannula and Ranganath, 2009; Ryals *et al.*, 2015; Ryan *et al.*, 2000). For example, participants freely viewed images of scenes for study (Ryan *et al.*, 2000). During test, eye-movement trajectories for old, new, and manipulated scenes are recorded. Manipulations entail, for example, removing people from a studied scene. Healthy participants but not amnesics displayed intact memory for scenes as indexed by more frequent fixations to regions on the image where manipulations occurred (*ibid.*). Similar to the above stem completion of word pairs tasks (e.g. Graf and Schacter, 1985), memory for image manipulations entail associations of items in context. The implicit status of this form of memory, however, has been questioned. Although most studies argue that memory expressed in eye-movements is implicit (e.g. Hannula and Ranganath, 2009; Ryals *et al.*, 2015; Ryan *et al.*, 2000), Smith, Hopkins and Squire (2006) found the opposite, namely that viewing patterns were altered only when participants were aware of the changes to the scene.

That explicit retrieval of studied items may have been responsible for putative implicit effects has been a major issue in implicit memory research (Mulligan, 1997; Shimamura and Squire, 1989). A very conservative way to exclude conscious processing is to present encoding stimuli very briefly (tens of milliseconds) and pattern masked. Masking renders stimuli invisible and hence prevents awareness of their mere existence. Implicit effects are measured on ensuing stimuli presented for conscious perception. Task instructions entail judgments which make no reference to the subliminal stimulus *per se* but on the ensuing stimulus (e.g. word/non-word decision). Implicit effects are indicated if subliminal stimuli nevertheless impact on behaviour in this so-called 'indirect test'. Whether such subliminal presentations bypass conscious awareness completely is usually established in a separate test in which stimuli are presented with the same masking procedure as in the main priming experiment, but subjects perform a direct test on the masked stimuli (e.g. presence/absence decision). To establish claims of 'pure' unconscious processing one seeks effects on the indirect test in the absence of significant effects in the direct, threshold test (Greenwald, Klinger and Schuh, 1995; Snodgrass and Shevrin, 2006).

This dissociation procedure has been considered the gold-standard for claims of unconscious processing, and there is general agreement that subliminal stimuli can impact on perceptual processes. The question whether subliminal stimuli can reach conceptual stages of processing has been intensely debated (for a review see Kouider and Dehaene, 2007). Although recent studies (Dehaene *et al.*, 1998; Greenwald, Draine and Abrams, 1996; Klauer *et al.*, 2007) have led to renewed acceptance, some researchers still question the feasibility of conceptual processing of subliminal stimuli (Kunde, Kiesel and Hoffmann, 2003).

Subliminal stimulation procedures have also been used to investigate the unconscious formation of novel associations into long-term memory. In a study performed by Henke *et al.* (2003), subliminal encoding entailed twelve-fold repetition of a subliminal stimulus (16.7 ms presentation time, pattern masked) within a six-second time window. This 'subliminal episode' consisted of an image of a face and job description (e.g. 'chemist' or 'musician') captioned underneath. After a five-minute study test delay, participants were presented with the faces for conscious inspection alone, and were asked to guess whether the depicted person works in academia or as an artist. Thus successful implicit memory not only afforded the learning of novel face-occupation associations but also required representations on a conceptual level. Implicit memory was indicated by shorter reaction latencies for correct versus incorrect choices at test. Furthermore, encoding of face-occupation associations was correlated with increased activity in the hippocampus as evidenced by functional magnetic resonance imaging (fMRI). Thus, although consciousness was excluded from encoding, Henke *et al.* (2003) reported evidence of hippocampal involvement in learning of novel associations on a conceptual level (see also Degonda *et al.*, 2005; Duss *et al.*, 2011; Zust *et al.*, 2015).

Further studies using this subliminal stimulation procedure investigated the feasibility of implicit conceptual associative learning using perceptually dissimilar retrieval cues (Reber and Henke, 2011). Here, subliminal encoding stimuli consisted of word pairs such as 'table—car' or 'apple—glass'. Retrieval words were semantically related to encoding words and were either arranged such that conceptual relations of encoding word pairs were kept intact ('desk—bus', analogues) or not ('counter—banana', broken analogues). The indirect retrieval task entailed judging whether two words in a pair fit together or not. Implicit memory for conceptual associations was indicated by more frequent fit judgments for analogues versus broken analogues (*ibid.*; Reber *et al.*, 2014). Brain imaging suggested that measures of implicit memory were correlated with the fMRI signal in hippocampal and prefrontal structures (Reber *et al.*, 2014). Furthermore, amnesic patients were impaired on this task, while subliminal conceptual single-item priming appeared intact (Duss *et al.*, 2014).

3.2. Flexibility of implicit memory

Generalization of knowledge to novel situations is held to be confined to explicit and particularly episodic memories according to systems views. Transitive inference and related tasks are widely used tests to gain a measure of this flexibility of memory representations (e.g. Greene *et al.*, 2001; Smith and Squire, 2005). In transitive inference, participants learn a set of new associations between previously unrelated stimuli in repeated learning trials through feedback. For example, the choice of stimulus B in context of stimulus A is punished, while the choice of B in context of C is rewarded. Reward and punishment during learning is given in such a manner that a reward hierarchy across all stimuli in the set can be inferred (A > B > C > D > E > F). Flexibility of memory representations is indicated if participants choose the correct stimulus according to this hidden relational hierarchy on test items that have never been encountered before (e.g. choose C over E).

While numerous studies demonstrate that an intact hippocampus is required for flexible memory expression (e.g. D'Angelo, Rosenbaum and Ryan, 2016; Smith and Squire, 2005), the question whether inference can also occur implicitly is being discussed (Greene, 2007). Structured interviews on whether participants gained insight of the reward hierarchy was used to assess whether above-chance performance on inference can also be achieved implicitly. Initial evidence of implicit inference (Greene *et al.*, 2006; 2001) was dismissed, as putative inference has rather been due to learning of unequal reward histories of single stimuli ('choosing C was often rewarded'; Frank, O'Reilly and Curran, 2006; 2008).

Studies conducted in response to this criticism eliminated the possibility of this 'pseudo-inference', and still found evidence for implicit inference (Leo and Greene, 2008; Reber *et al.*, 2016). For example, one study entailed one-trial encoding of word pairs ('winter—red', 'red—cat'), and subjects had to indicate whether they thought two words in a pair fit together or not. The same task was used on test word pairs ('winter—cat') which either were indirectly related, as they were presented in context of the same encoding ('red'), or not (Reber *et al.*, 2016). This study was conducted with patients implanted with electrodes in their medial temporal lobes for epilepsy monitoring. Here, encoding of overlapping associations ('winter—red', 'red—cat') was associated with increased event related potentials in hippocampal contacts. Furthermore, evidence that relational inference may also be possible under subliminal encoding conditions (Henke, Reber and Duss, 2013; Reber and Henke, 2012; Reber *et al.*, 2012) disagrees with notions that implicit inference is contaminated by explicit processes.

4. Conclusions

Taken together, there is mounting evidence to suggest that implicit memory may be more elaborate than initial studies with amnesic patients suggest (e.g. Corkin, 1968). These findings challenge a distinction of memory systems based on consciousness (Squire, 1992; Tulving, 2002). Although processing modes views have been refuting consciousness as a criterion to distinguish between forms of memory from their beginning (Blaxton, 1989; Henke, 2010; Roediger, 1990), it has been rather recently that more convincing empirical evidence was gathered to support these views. One reason for this rather late revival might be that evidence favouring processing modes views relies on convincing demonstrations of elaborate but unconscious processing. Methods to establish these claims have been refined in the last decades but have also been questioned repeatedly. Nevertheless, more recent notions in the memory systems tradition are increasingly accepting of the idea that awareness of memory does not always go hand in hand with the underlying memory system (Dew and Cabeza, 2011; Hannula and Greene, 2012; Park and Donaldson, 2016; Reder, Park and Kieffaber, 2009).

As an intermediate conclusion, awareness of memory could be seen as an aspect that is not primarily linked with mnemonic operations *per se* but rather as a modulating factor of memory performance. Here, it seems noteworthy that memory for subliminal stimuli are usually considerably weaker (Pessiglione *et al.*, 2008) or even absent (Reber and Mormann, 2016) as compared to when conscious perception of the stimuli and explicit strategies are used. The profit from conscious processing seems larger for more elaborate versus rather simple forms of memory as more elaborate forms of memory likely depend on cooperative recruitment of distant brain areas (e.g. King *et al.*, 2015), which has also been reported as a neural correlate of consciousness in general (e.g. Gaillard *et al.*, 2009). Furthermore, even though overlaps in brain regions recruited by implicit and explicit memory have been found (Turk-Browne, Yi and Chun, 2006), explicit conditions often also recruit additional regions (Buckner and Koutstaal, 1998; Schott *et al.*, 2005). Thus, more research is needed not only to replicate findings of elaborate unconscious processing but also to compare neural correlates and connectivity of explicit and implicit memory.

Acknowledgment

I thank K. Henke and L. Chaieb for careful reading of the manuscript. Correspondence should be addressed to treber@live.com. T.P.R. is supported by the Swiss National Science Foundation (Grant P300P1_161178).

References

Berry, C.J., Shanks, D.R. & Henson, R.N.A. (2008) A unitary signal-detection model of implicit and explicit memory, *Trends in Cognitive Sciences*, **12** (10), pp. 367–373.

Blaxton, T.A. (1989) Investigating dissociations among memory measures: Support for a transfer-appropriate processing framework, *Journal of Experimental Psychology: Learning, Memory, and Cognition*, **15** (4), pp. 657–668.

Buckner, R.L. & Koutstaal, W. (1998) Functional neuroimaging studies of encoding, priming, and explicit memory retrieval, *Proceedings of the National Academy of Sciences USA*, **95** (3), pp. 891–898.

Cabeza, R. & Moscovitch, M. (2013) Memory systems, processing modes, and components functional neuroimaging evidence, *Perspectives on Psychological Science*, **8** (1), pp. 49–55.

Chun, M.M. & Phelps, E.A. (1999) Memory deficits for implicit contextual information in amnesic subjects with hippocampal damage, *Nature Neuroscience*, **2** (9), pp. 844–847.

Corkin, S. (1968) Acquisition of motor skill after bilateral medial temporal-lobe excision, *Neuropsychologia*, **6** (3), pp. 255–265.

Crick, F. & Koch, C. (1990) Towards a neurobiological theory of consciousness, *Seminars in the Neurosciences*, **2**, pp. 263–275.

Crick, F. & Koch, C. (1995) Are we aware of neural activity in primary visual cortex?, *Nature*, **375** (6527), pp. 121–123.

D'Angelo, M.C., Rosenbaum, R.S. & Ryan, J.D. (2016) Impaired inference in a case of developmental amnesia, *Hippocampus* (ahead of print).

Degonda, N., Mondadori, C.R.A., Bosshardt, S., Schmidt, C.F., Boesiger, P., Nitsch, R.M., Hock, C. & Henke, K. (2005) Implicit associative learning engages the hippocampus and interacts with explicit associative learning, *Neuron*, **46** (3), pp. 505–520.

Dehaene, S., Naccache, L., Le Clec'H, G., Koechlin, E., Mueller, M., Dehaene-Lambertz, G., van de Moortele, P. & Le Bihan, D. (1998) Imaging unconscious semantic priming, *Nature*, **395** (6702), pp. 597–600.

Dew, I.T. & Cabeza, R. (2011) The porous boundaries between explicit and implicit memory: Behavioral and neural evidence, *Annals of the New York Academy of Sciences*, **1224** (1), pp. 174–190.

Duss, S.B., Oggier, S., Reber, T.P. & Henke, K. (2011) Formation of semantic associations between subliminally presented face-word pairs, *Consciousness and Cognition*, **20** (3), pp. 928–935.

Duss, S.B., Reber, T.P., Hänggi, J., Schwab, S., Wiest, R., Müri, R.M., Brugger, P., Gutbrod, K. & Henke, K. (2014) Unconscious relational encoding depends on hippocampus, *Brain*, awu270.

Eichenbaum, H. (2004) Hippocampus: Cognitive processes and neural representations that underlie declarative memory, *Neuron*, **44** (1), pp. 109–120.

Frank, M.J., O'Reilly, R.C. & Curran, T. (2006) When memory fails, intuition reigns: Midazolam enhances implicit inference in humans, *Psychological Science*, **17** (8), pp. 700–707.

Frank, M.J., O'Reilly, R.C. & Curran, T. (2008) Midazolam, hippocampal function, and transitive inference: Reply to Greene, *Behavioral and Brain Functions*, **4** (1), 5.

Gabrieli, J., Milberg, W., Keane, M.M. & Corkin, S. (1990) Intact priming of patterns despite impaired memory, *Neuropsychologia*, **28** (5), pp. 417-427.

Gabrieli, J., Keane, M.M., Zarella, M.M. & Poldrack, R.A. (1997) Preservation of implicit memory for new associations in global amnesia, *Psychological Science*, **8** (4), pp. 326-329.

Gaillard, R., Dehaene, S., Adam, C., Clémenceau, S., Hasboun, D., Baulac, M., Cohen, L. & Naccache, L. (2009). Converging intracranial markers of conscious access, *PLoS Biol*, **7** (3), e1000061.

Graf, P. & Schacter, D.L. (1985) Implicit and explicit memory for new associations in normal and amnesic subjects, *Journal of Experimental Psychology, Learning, Memory, and Cognition*, **11** (3), pp. 501-518.

Greene, A.J. (2007) Human hippocampal-dependent tasks: Is awareness necessary or sufficient?, *Hippocampus*, **17** (6), pp. 429-433.

Greene, A.J., Spellman, B.A., Levy, W.B., Dusek, J.A., & Eichenbaum, H.B. (2001) Relational learning with and without awareness: Transitive inference using nonverbal stimuli in humans, *Memory & Cognition*, **29** (6), pp. 893-902.

Greene, A.J., Gross, W.L., Elsinger, C.L. & Rao, S.M. (2006) An fMRI analysis of the human hippocampus: Inference, context, and task awareness, *Journal of Cognitive Neuroscience*, **18** (7), pp. 1156-1173.

Greenwald, A.G., Klinger, M.R. & Schuh, E.S. (1995) Activation by marginally perceptible ('subliminal') stimuli: Dissociation of unconscious from conscious cognition, *Journal of Experimental Psychology: General*, **124** (1), pp. 22-42.

Greenwald, A.G., Draine, S.C. & Abrams, R.L. (1996) Three cognitive markers of unconscious semantic activation, *Science*, **273** (5282), pp. 1699-1702.

Hannula, D.E. & Ranganath, C. (2009) The eyes have it: Hippocampal activity predicts expression of memory in eye movements, *Neuron*, **63** (5), pp. 592-599.

Hannula, D.E. & Greene, A.J. (2012) The hippocampus reevaluated in unconscious learning and memory: At a tipping point?, *Frontiers in Human Neuroscience*, **6**, art. 80.

Henke, K. (2010) A model for memory systems based on processing modes rather than consciousness, *Nature Reviews Neuroscience*, **11** (7), pp. 523-532.

Henke, K., Mondadori, C.R.A., Treyer, V., Nitsch, R.M., Buck, A. & Hock, C. (2003) Nonconscious formation and reactivation of semantic associations by way of the medial temporal lobe, *Neuropsychologia*, **41** (8), pp. 863–876.

Henke, K., Reber, T.P. & Duss, S.B. (2013) Integrating events across levels of consciousness, *Frontiers in Behavioral Neuroscience*, **7**, art. 68.

Jacoby, L.L. (1983) Remembering the data: Analyzing interactive processes in reading, *Journal of Verbal Learning and Verbal Behavior*, **22** (5), pp. 485–508.

King, D.R., de Chastelaine, M., Elward, R.L., Wang, T.H. & Rugg, M.D. (2015) Recollection-related increases in functional connectivity predict individual differences in memory accuracy, *Journal of Neuroscience*, **35** (4), pp. 1763–1772.

Klauer, K.C., Eder, A.B., Greenwald, A.G. & Abrams, R.L. (2007) Priming of semantic classifications by novel subliminal prime words, *Consciousness and Cognition*, **16** (1), pp. 63–83.

Kouider, S. & Dehaene, S. (2007) Levels of processing during nonconscious perception: A critical review of visual masking, *Philosophical Transactions of the Royal Society B: Biological Sciences*, **362** (1481), pp. 857–875.

Kunde, W., Kiesel, A. & Hoffmann, J. (2003) Conscious control over the content of unconscious cognition, *Cognition*, **88** (2), pp. 223–242.

Leo, P.D. & Greene, A.J. (2008) Is awareness necessary for true inference?, *Memory & Cognition*, **36** (6), pp. 1079–1086.

Manns, J.R. & Squire, L.R. (2001) Perceptual learning, awareness, and the hippocampus, *Hippocampus*, **11** (6), pp. 776–782.

Mormann, F. & Koch, C. (2007) Neural correlates of consciousness, *Scholarpedia*, **2** (12), 1740.

Mulligan, N.W. (1997) Attention and implicit memory tests: The effects of varying attentional load on conceptual priming, *Memory & Cognition*, **25** (1), pp. 11–17.

Park, J.L. & Donaldson, D.I. (2016) Investigating the relationship between implicit and explicit memory: Evidence that masked repetition priming speeds the onset of recollection, *NeuroImage*, **139**, pp. 8–16.

Patterson, K., Nestor, P.J. & Rogers, T.T. (2007) Where do you know what you know? The representation of semantic knowledge in the human brain, *Nature Reviews Neuroscience*, **8** (12), pp. 976–987.

Pessiglione, M., Petrovic, P., Daunizeau, J., Palminteri, S., Dolan, R.J. & Frith, C.D. (2008) Subliminal instrumental conditioning demonstrated in the human brain, *Neuron*, **59** (4), pp. 561–567.

Reber, T.P. & Henke, K. (2011) Rapid formation and flexible expression of memories of subliminal word pairs, *Frontiers in Psychology*, **2**, art. 343.

Reber, T.P. & Henke, K. (2012) Integrating unseen events over time, *Consciousness and Cognition*, **21** (2), pp. 953-960.

Reber, T.P., Luechinger, R., Boesiger, P. & Henke, K. (2012) Unconscious relational inference recruits the hippocampus, *Journal of Neuroscience*, **32** (18), pp. 6138-6148.

Reber, T.P., Luechinger, R., Boesiger, P. & Henke, K. (2014) Detecting analogies unconsciously, *Frontiers in Behavioral Neuroscience*, **8**, art. 9.

Reber, T.P., Do Lam, A., Axmacher, N., Elger, C., Helmstaedter, C., Henke, K. & Fell, J. (2016) Intracranial EEG correlates of implicit relational inference within the hippocampus, *Hippocampus*, **26** (1), pp. 54-66.

Reber, T.P. & Mormann, F. (2016) Cue visibility predicts instrumental conditioning (in preparation).

Reder, L.M., Park, H. & Kieffaber, P.D. (2009) Memory systems do not divide on consciousness: Reinterpreting memory in terms of activation and binding, *Psychological Bulletin*, **135** (1), pp. 23-49.

Roediger, H.L. (1990) Implicit memory: Retention without remembering, *American Psychologist*, **45** (9), pp. 1043-1056.

Ryals, A.J., Wang, J.X., Polnaszek, K.L. & Voss, J.L. (2015) Hippocampal contribution to implicit configuration memory expressed via eye movements during scene exploration, *Hippocampus*, **25** (9), pp. 1028-1041.

Ryan, J.D., Althoff, R.R., Whitlow, S. & Cohen, N.J. (2000) Amnesia is a deficit in relational memory, *Psychological Science*, **11** (6), pp. 454-461.

Schacter, D.L. & Graf, P. (1986) Preserved learning in amnesic patients: Perspectives from research on direct priming, *Journal of Clinical and Experimental Neuropsychology*, **8** (6), pp. 727-743.

Schott, B.H., Henson, R.N., Richardson-Klavehn, A., Becker, C., Thoma, V., Heinze, H.-J. & Düzel, E. (2005) Redefining implicit and explicit memory: The functional neuroanatomy of priming, remembering, and control of retrieval, *Proceedings of the National Academy of Sciences USA*, **102** (4), pp. 1257-1262.

Scoville, W.B. & Milner, B. (1957) Loss of recent memory after bilateral hippocampal lesions, *Journal of Neurology, Neurosurgery, and Psychiatry*, **20** (1), pp. 11-21.

Shimamura, A.P. & Squire, L.R. (1989) Impaired priming of new associations in amnesia, *Journal of Experimental Psychology: Learning, Memory, and Cognition*, **15** (4), pp. 721-728.

Smith, C. & Squire, L.R. (2005) Declarative memory, awareness, and transitive inference, *The Journal of Neuroscience*, **25** (44), pp. 10138-10146.

Smith, C., Hopkins, R.O. & Squire, L.R. (2006) Experience-dependent eye movements, awareness, and hippocampus-dependent memory, *The Journal of Neuroscience*, **26** (44), pp. 11304–11312.

Snodgrass, M. & Shevrin, H. (2006) Unconscious inhibition and facilitation at the objective detection threshold: Replicable and qualitatively different unconscious perceptual effects, *Cognition*, **101** (1), pp. 43–79.

Squire, L.R. (1992) Declarative and nondeclarative memory: Multiple brain systems supporting learning and memory, *Journal of Cognitive Neuroscience*, **4** (3), pp. 232–243.

Squire, L.R. (2009) The legacy of patient H.M. for neuroscience, *Neuron*, **61** (1), pp. 6–9.

Tulving, E. (2002) Episodic memory: From mind to brain, *Annual Review of Psychology*, **53** (1), pp. 1–25.

Tulving, E. & Schacter, D.L. (1990) Priming and human memory systems, *Science*, **247** (4940), pp. 301–306.

Turk-Browne, N.B., Yi, D.-J. & Chun, M.M. (2006) Linking implicit and explicit memory: Common encoding factors and shared representations, *Neuron*, **49** (6), pp. 917–927.

Zust, M.A., Colella, P., Reber, T.P., Vuilleumier, P., Hauf, M., Ruch, S. & Henke, K. (2015) Hippocampus is place of interaction between unconscious and conscious memories, *PLoS One*, **10** (3), e0122459.

Penka Hristova

Can We Think Unconsciously Via Analogy?

Analogies are defined as the process of mapping entities which share a similar structure but not necessarily similar elements (i.e. objects, events, categories) (Gentner, 1983). The Rutherford analogy is the textbook example for analogy, based on the relational similarity established between dissimilar elements (i.e. planets revolving the sun and electrons revolving the atomic nucleus) (see Gentner, 1983). Since relational, but not superficial (semantic, associative, or perceptual) similarity is the prerequisite for analogies (*ibid.*), people may connect and transfer knowledge across domains. Novel things, which seem unfamiliar and unidentifiable at the beginning, can be assimilated if they resemble the relational structure of specifically stored past experience (often called the source analogue or the base).

The question of whether and under what circumstances people may draw conclusions based on analogies without awareness, effort, or intention turned out to be important for analogy making research and our understanding of cognition in general. As Douglas Hofstadter had argued, people can't use anything but analogy in high-level perception, categorization, and comprehension (Hofstadter, 2001), i.e. analogy may be considered as a 'core' mechanism of human cognition instead of a solitary mark for human reasoning (Holyoak, Gentner and Kokinov, 2001). In fact, many research groups had tried to explore that possibility and to investigate analogical mapping as a fundamental mechanism taking part in perception (Hofstadter, 1995; Mitchell, 1993; French, 1995; Petkov *et al.*, 2007), judgment (Kokinov, Hristova and Petkov, 2004), decision making (Markman and Moreau, 2001), categorization (Goldwater, Markman and Stilwell, 2011; Medin, Goldstone and Gentner, 1993), memory (Kokinov and Petrov, 2001; Feldman and Kokinov, 2009; Pavlova and Kokinov, 2014), comprehension of ambiguous statements (Day and Gentner, 2007; Popov and Hristova,

2014). The assumption underlying most of those attempts, however, was that analogies would be drawn whenever two relationally similar structures are processed or the target stimulus reminds one of a relationally similar one independent of the task at hand and despite a lack of explicit instruction for searching, verifying, or completing a given analogy. Hence, analogies were considered to be unconscious (being able to be processed without deliberation and awareness), autonomous (i.e. being able to start when given circumstances are present), and effortless, too (i.e. being able to be processed without depleting working memory resources). This standpoint was, however, seriously attacked on the grounds of the accumulated empirical evidence in the field (Holyoak, 2012), most of which was based on research with an explicit instruction for making, finding, or verifying analogies (for a similar argument, Day and Gentner, 2007).

Since it was found that explicit analogies are effortful, cognitively exhaustive, and computationally costly (Halford, 1992; Cho, Holyoak and Cannon, 2007), attempts to study the impact of analogies without an explicit instruction (i.e. Day and Goldstone, 2011; Schunn and Dunbar, 1996) were reconciled as being a result of relational priming instead of analogical mapping (Holyoak, 2012). The transfer of knowledge in those studies in turn was suggested to result from a piecemeal and partial, rather than one-to-one, mapping between analogous tasks (Holyoak, 2012). Likewise, neuroimaging studies on explicit analogies reported a consistent activation of anterior regions of the prefrontal cortex (PFC) (Christoff et al., 2001; Bunge et al., 2005; Green et al., 2006; Luo et al., 2003; Qiu et al., 2008; Green et al., 2010). Since those regions were associated with stimulus-independent but explicit processing of internally generated information (Christoff and Gabrieli, 2000; Christoff and Keramatian, 2007; Christoff et al., 2003; Smith et al., 2006), analogies in turn were associated with explicit information processing too. It was argued, however, that automatic analogies (i.e. frequently used analogies), if such exist, may also be processed by the same PFC regions as the explicit and effortful ones, but be less detectable by fMRI since automatic analogies recruit a smaller number of cells (Speed, 2010). Comparing, however, explicit analogy with neutral (i.e. another task which does not require analogy making) conditions may not clarify the existence and mechanisms of implicit analogies. Finally, the very fact that the PFC is recruited for analogy making but also for decision making, reasoning, and problem solving might indicate that analogies are the fundamental component of cognition (Speed, 2010). However, the general view, or at least the one maintained by some researchers in the field, is that implicit, i.e. unintentional, unconscious and effortless analogies do not exist (Holyoak, 2012). And the effects resembling automatic analogies (i.e. Day and Goldstone, 2011; Schunn and Dunbar, 1996) were indeed enacted by different mechanisms such as relational priming (Holyoak, 2012).

1. Relational Priming vs. Analogical Mapping

Relational priming is the facilitation in the processing of a word pair by the previous word pair that instantiates the same relation (e.g. COMPUTER PASSWORD is comprehended faster after PADLOCK KEY, compared to STEEL KEY) (Estes, 2003; Estes and Jones, 2006; Spellman, Holyoak and Morrison, 2001; Popov and Hristova, 2015). It was also obtained with pictorial stimuli (Raffray, Pickering and Branigan, 2007) and with mathematical expressions (Bassok, Chase and Martin, 1998; Bassok, Pedigo and Oskarsson, 2008), and even across domains (Fisher, Bassok and Osterhout, 2010) and between modalities (Livins, Doumas and Spivey, 2015). In line with the empirical findings for unintentional and uncontrollable relational integration (Mather *et al.*, 2014), relational priming was observed when the encoding of relations was not necessitated by the task (Bassok, Chase and Martin, 1998; Bassok, Pedigo and Oskarsson, 2008; Estes and Jones, 2006; Popov and Hristova, 2015; Hristova, 2009a; Raffray *et al.*, 2007).

Some researchers, such as Goswami (1991) and Leech, Mareschal and Cooper (2008), suggested that relational priming alone can explain many of the key findings in the analogy making domain (Leech, Mareschal and Cooper, 2008), and depending on the task demands analogies may tap different processes. For example, a:b::c:d analogical completion may be underlined by mechanisms distinct from the ones used in more complex problem solving tasks, which may require a more explicit and purposeful mechanism of structure mapping instead of relational priming. In addition, relational as well as conceptual priming was given as an example of a mechanism that substantially determines the content and processing speed of information that take part in analogical mapping and retrieval, even without awareness (Kokinov and Hristova, 2012). Hence, even analogies that we draw deliberately and with awareness take advantage of such unconscious and unintentional processes and thus can hardly be defined as being entirely conscious. Therefore, conscious-unconscious analogies do not necessarily imply different mechanisms and most probably lay on a continuum rather than as a dichotomy as the two terms imply in some dual-system accounts (Kokinov and Hristova, 2012).

To sum up, although some researchers suggested that relational priming is the developmentally plausible vehicle for analogical mapping (Leech, Mareschal and Cooper, 2008), others admit that some piecemeal transfer can possibly happen due to relational priming

(Holyoak, 2012; Schunn and Dunbar, 1996), but contrasted it to analogical mapping (French, 2008; Holyoak, 2012; Holyoak and Hummel, 2008). Hence, even though relational priming can happen unintentionally and effortlessly (Popov and Hristova, 2015), resembling in that way the defining characteristics of Type 1 processing in the terminology of Evans and Stanovich (2013), analogical mapping may not necessarily follow the same rules. The debatable ground is not whether people may transfer relational knowledge unconsciously and unintentionally across domains but rather how they do so, in terms of with what mechanisms such transfer happens. And to be more specific, can analogies be autonomous and unconscious mechanisms for relational transfer?

2. Prerequisites of Unintentional and Unconscious Analogies

Traditionally, analogies are divided into the sub-processes of: representation-building (encoding of the relevant relations); retrieval of analogical base/source, when needed; mapping base onto the target; transfer of unmapped base elements/relation; and learning (Kokinov and French, 2003). Two of those sub-processes, i.e. mapping and representation-building, were considered to be crucial for analogies, while the others (i.e. retrieval of a source analogy, transfer, and learning) usually build upon them. On one hand, mapping of the base onto the target structure is, by definition, the necessary and sufficient process for analogy making (Gentner, 1983). On the other hand, representation-building in terms of relations was considered to be an important gateway to spontaneous analogies, by explaining the so-called 'analogical paradox'[1] and the superior performance of experts compared to novices on tasks within their domain of expertise[2] (Dunbar, 2001). Both analogical mapping and relational encoding of AB pairs are associated with different event-related potentials during simple A:B::C:D proportional analogies, compared to a control task (Maguire *et al.*, 2012; Krawczyk *et al.*, 2010). If those two defining sub-processes cannot be

---

[1] The surprising finding that people spontaneously and easily generate analogies in a naturalistic setting, whilst at the same time having severe problems in solving problems by analogy in psychological laboratories. The explanation of the paradox was connected to relational encoding, since the generation of analogies encourages a relation-based memory search, while analogical problem solving is usually based on an object-based memory search.

[2] Experts outperform novices because of the superior encoding of higher-order relations between problems' elements, since they rely on well-structured relational knowledge in their domain of expertise.

carried out unintentionally and unconsciously, thus, most probably, analogies can only be made deliberately.

2.1. Encoding relations does not require awareness and intentions

The common position in the field of analogy making is that analogies require argument-free representation (Doumas, Hummel and Sandhofer, 2008; Hummel and Holyoak, 2003). In other words, relations need to be represented independently of their arguments, since the same relation can bind different objects and simultaneously bind the same objects to different arguments.[3] Hence, processes of relational encoding should represent relational information in an abstract enough form so as to be mapped and transferred across relationally similar structures, independent of the specific objects that they link.

Lin and Murray (2014) showed that encoding of the same–different relation does not require awareness. Participants were able to extract the relation between two masked geometrical figures presented briefly on the screen (i.e. for 16.7 or 33.3 ms) and slowed down their responses for DIFFERENT compared to SAME non-visible go-trials, because they associated them with no-go- and respectively go-visible trials. As participants were not subjectively aware of detecting any relation on the non-visible trials, the same–different discrimination was interpreted as an unconscious one.

Adult participants were quicker at indicating that two words can be meaningfully integrated if they instantiate a frequent (i.e. PAPER NOTE) instead of a less frequent relation (i.e. PAPER CUT) (Gagné and Shoben, 1997; Gagné and Spalding, 2004), depending on the frequency of pairing the two semantic categories, denoted by the two nouns (Maguire, Maguire and Cater, 2010). Such relational integration (i.e. the process of relating two nouns) was also reported with 4- to 5-year-old children (Krott, Gagné and Nicolades, 2009), with the (structurally different from English) Indonesian language (Storms and Wisniewski, 2005) and also when controlled processes were unlikely[4] and, in addition, the two words were unassociated, dissimilar, and low in co-occurrence (Mather et al., 2014). It was argued that the mechanism of complementary role activation accounts better for the behavioural observations than mechanisms requiring strategic control (e.g.

---

[3] For example, 'John financially supports Mary' (supports (John, Mary)) and 'Mary financially supports Tom' (supports (Mary, Tom)) instantiate the same relation between supporter and supported and, also, the same entities fulfil different roles within the same relations (i.e. Mary).

[4] Since the tasks were for perceptual identification and colour naming (Mather, Jones and Estes, 2014).

expectancy generation, semantic matching) or semantic, associative, and compound cue similarity (i.e. spreading activation, activation of distributed representations, episodic retrieval, and compound cue models) (Estes *et al.*, 2011; Mather *et al.*, 2014). Complementary role activation predicts that the two words unintentionally and uncontrollably activate their typical relational roles and, respectively, their common relation if such exists. For example, 'wood' and 'chair' can be easily integrated via the roles 'material' and, respectively, 'furniture', which in turn are meaningfully related to the 'made of' relation (the example is from the stimuli used in the Mather *et al.*'s, 2014, study). The activated relation facilitates the processing of the two words and, in turn, reduces the lexical decision time (Estes and Jones, 2009) or the time for sensicality judgment (i.e. whether the word pair make sense as a phrase) (Gagné and Shoben, 1997; Gagné and Spalding, 2004; Maguire, Maguire and Cater, 2010). Thus, complementary role activation from stimuli, which can fulfil specific roles within a given context, seems to be compatible with the requirements for argument-free representations for further analogical mapping, even without any effort (Maguire *et al.*, 2014). The usage of those automatically encoded relations seems, however, to depend crucially on the frequency of the instantiated relation (Gagné and Shoben, 1997; Gagné and Spalding, 2004; Maguire *et al.*, 2010). Hence, at least under some circumstances, i.e. when analogies involve mapping of frequently used relations between base and target elements, those relations can be taken for granted through the effortless and unintentional mechanism of complementary role activation.

Observations based on explicit analogy tasks, however, suggest deliberate, context sensitive, and dynamic relational encoding. Representation-building of the A:B terms during explicit A:B::C:D analogy making in some neuroimaging (Maguire *et al.*, 2012; Krawczyk *et al.*, 2010; Qiu *et al.*, 2008) and eye-tracking (Thibaut *et al.*, 2011) studies seems to require more than automatic complementary role activation, since an excessive processing of the AB relation was observed in the analogy compared to a control condition. Similarly it was argued (Chalmers, French and Hofstadter, 1992; French, 2008) that the CD relation determines the choice of AB relation, since every two things can be related in different ways. For example, the *offspring* relation between DOG PUPPY in an AB pair is appropriate when the CAT KITTEN is the CD pair, but the *bigger than* relation would be more appropriate if the WATCH CLOCK is the CD pair in a simple proportional analogy task (French, 2008). Some computational models of analogy making use incremental mechanisms for the encoding of AB in the context of the CD relation, where the AB relation can be instantiated several times until a meaningful mapping between AC and BD is not established (Yan, Forbus and Gentner, 2003; Keane, 1990). Others use relational similarity between the AB and CD pair as a constraint for analogical mapping (Holyoak and Thagard, 1989; Kokinov and Petrov, 2001). In addition, Cho, Holyoak and Cannon (2007) found that explicit instruction for integrating several relations in the base over-taxes working memory resources. Hence, it may well be the case that the extensive processing of AB relations during explicit analogies attests to an additional processing due to the heightened precision in choosing the relevant AB relation or integration of several AB relations when needed.

In sum, although people may encode relations between two entities without a prompt, awareness, and cognitive effort (Mather *et al.*, 2014), they seem to process them extensively during explicit analogy making compared to a control task (Maguire *et al.*, 2012; Krawczyk *et al.*, 2010; Qiu *et al.*, 2008, Thibaut *et al.*, 2011). The explicit prompt for finding and verifying analogies, however, requires explicit relational processing. Possibly, indeed probably, the reported extensive processing builds upon the same relations, which were unintentionally instantiated between the analogy making terms, even without awareness and effort.

2.2. Analogical mapping does not necessarily require intentions and awareness

Evidence for unintentional and unconscious analogical mapping has been found in studies on text comprehension (Day and Gentner, 2007), problem solving (Day and Goldstone, 2011), and also instances where analogies have been found to have a detrimental effect on a main colour naming task (Hristova, 2009a,b). Participants in those studies were neither explicitly instructed to search the analogy, nor that an analogy between the latter and the former stimuli exist. Despite the fact that they were asked to focus on the main task, they nevertheless transferred knowledge from previous relationally similar episodes, facilitating (Day and Gentner, 2007; Day and Goldstone, 2011; Hristova, 2009a,b) or hindering in this way the performance on the main task (Experiment 2, Hristova, 2009b). Since analogical transfer follows the sub-process of analogical mapping, one might draw the conclusion that mapping via analogy proceeds without deliberation or awareness too (for similar argument, see Day and Gentner, 2007).

Day and Gentner (2007) showed that analogy but not priming predicts how people will comprehend an ambiguous passage. Participants transferred more from analogous stories, although the same facts were presented three times more often and were equally memorable when presented within the control text. Most of them (80%), however, claimed they didn't need and didn't use older passages to comprehend newer ones (*ibid.*). Similarly, both production (Goldwater *et al.*, 2011; Popov and Hristova, 2014) and comprehension (Popov and Hristova, 2014) of ambiguous sentences were found to depend significantly on thematic similarity with a syntactically different sentence. Previous presentation of the key role (i.e. 'The doctor and the patient watched by using glasses' or 'The doctor and the patient, who wore glasses, watched') was not enough to facilitate its use in the comprehension of the subsequent sentences, but the presentation of the whole thematic structure (i.e. 'The doctor watched the patient by using glasses' or 'The doctor watched the patient who wore glasses') significantly determined the way people preferred to disambiguate the prepositional phrase attachment ambiguity (i.e. 'The hunter watched the alpinist with binoculars') (Popov and Hristova, 2014). The sources and targets, however, shared high surface similarity in those studies, which may constrain the effect of unintentional and unconscious transfer between structurally similar episodes only to those conditions.

Day and Goldstone (2011), however, reported analogical transfer between two superficially dissimilar domains. They demonstrated a robust improvement in performance for a target problem when it was preceded immediately by a source problem that was consistent with the goal and structure of the target. Those participants who noticed the analogy did not perform analogical transfer better by applying the appropriate strategy. Moreover, explicit instruction focusing participants' attention on the specific analogical correspondences between the tasks significantly impeded successful problem solving on a second task. Although the authors admit that the obtained effect may be due to relational priming instead of unconscious and unintentional analogical transfer, they argue that the latter can happen even between superficially dissimilar analogous cases, given that their representations are compatible in terms of their spatial and perceptual characteristics.

Analogical mapping and transfer between superficially distinct elements were also reported with a modified Stroop colour-naming paradigm (Hristova, 2009a,b). A series of experiments with the same stimuli and experimental question studied whether participants would start to map and transfer knowledge between analogically related subsequent pairs even though this might be detrimental to the main colour identification task. Participants were shown a pair of words on each trial and were asked to judge as quickly and accurately as possible the colour of the ink of one of the words (i.e. green or red). Not surprisingly, a main effect of colour congruence was obtained: the naming of the colour of the word in a word pair was faster following a word pair containing a word of a congruent, compared to an incongruent, colour (i.e. a red word in a target pair was processed faster after a red and slower after a green word and, respectively, a green target word was processed faster after a green compared to a red word). Importantly, however, if two subsequent pairs shared an analogous relation, the colour congruency effect was even greater. It was found that response times were significantly slower for incongruent and significantly faster for congruent analogical trials (e.g. FILTER WATER proceeded by CENSORSHIP TEXT) compared to non-analogically related ones (e.g. FILTER WATER proceeded by RING GOLD). Since the modified colour-naming paradigm and the relational priming predict only a delayed, or respectively faster, response for both congruent and incongruent trials in colour word pairs, the data were interpreted as showing unintentional analogical mapping and transfer. Participants anticipated the colour of words, based on spontaneous unintentional and unconscious analogy with the previous word pair. Thus, they reacted slower or faster when the target (e.g. FILTER WATER) followed an analogically related word pair (e.g. CENSORSHIP TEXT) with different, compared to same, colour. If the common relational structure was violated by changing only one of the words in the proceeding pair (e.g. SHIP TEXT instead of CENSORSHIP TEXT), the effect of the colour anticipation disappeared, probably because the analogy itself was demolished (Experiment 3, Hristova, 2009b). Hence, neither semantic priming nor priming of a common role explains the obtained unintentional analogy with the proceeding pair. Within the whole series of experiments with the colour-naming paradigm, participants were not subjectively aware that some of the stimuli were analogous, thus indicating unconscious analogical processing in a task that does not necessitate such.

3. Conclusion

Assuming that analogy making is a complex, cognitively demanding, and deliberate process, the empirical evidence from studies explicitly requiring participants to do analogies may lead to the misleading generalization that all analogies, including implicit ones, share these characteristics. However, this is wrong because it was found that participants readily transferred knowledge from analogous past situations without an explicit prompt (Day and Gentner, 2007; Day and Goldstone, 2011; Hristova, 2009a,b; Popov and Hristova, 2014; Shun and Dunbar, 1997), without being aware of doing so and even when analogies may hamper the main task (Hristova, 2009a,b). Relational priming can hardly explain the transfer obtained in those studies, since at least in some of them it was controlled for (e.g. Day and Gentner, 2007; Popov and Hristova, 2014) and it predicts results different from the ones that were indeed obtained (Day and Gentner, 2007; Hristova, 2009a,b). If relational priming was found only to facilitate subsequent processing of relationally similar information, the implicit analogies, which people were found to draw without prior intentions and conscious recollections, were able to slow responding down if it contradicted the expectations generated from analogical mapping (Day and Gentner, 2007, Experiment 3; Hristova, 2009b). Hence, analogical mapping, instead of relational priming, seems to underlie knowledge transfer in the reviewed studies.

Importantly, the introspective self-reports of participants who took part in those studies of implicit and unintentional analogies indicated low or no awareness of any relation between the analogous passages (i.e. 20% of participants in Day and Gentner's study, 2007), between the analogous tasks (i.e. 29.3% of Day and Goldstone's Experiment 2, 2011), and between the word pairs (i.e. none of the participants in Hristova, 2009a,b). Furthermore, those who reported awareness of between-task structural similarities were not superior in transferring knowledge, but rather were associated with better performance in general (Experiment 2, Day and Goldstone, 2011). Therefore, this favours the conclusion that awareness may arise out of some confounding individual differences, such as intelligence, rather than being considered as a reason for better analogical transfer. Similarly, Evans and Stanovich (2013) have pointed out that Type 2 reasoning (i.e. reflective, effortful, working memory resource-dependent processing), which usually correlates with deliberate processing and meta-awareness, correlates in turn with individual differences such as the ability for cognitive decoupling.

Likewise, the fact that analogical mapping was found to operate even when it may hamper the main task (Hristova, 2009a,b) supports more strongly than the reported self-awareness the possibility for analogies to be processed unconsciously, unintentionally, and in parallel, despite the fact that they may interfere with the main cognitive mechanisms required by the current task. Such autonomous unconscious analogical mapping is consistent with the idea that analogies are a fundamental cognitive mechanism that take part in a number of cognitive processes such as perception, categorization, comprehension, etc. (Holyoak et al., 2001). The studies reviewed in that paper, however, seem to suggest that such unintentional and unconscious analogies may readily link simple structures of familiar relations. Familiarity of the mapped structures may also require different levels of deliberate processing, since unfamiliar relations impose difficulties for autonomous relational encoding (Mather et al., 2014) and probably require re-representations (Gentner and Forbus, 2011).

Last but not least, most computational models of analogy, such as SME (Falkenhainer, Forbus and Gentner, 1989), LISA (Hummel and Holyoak, 1997), COPICAT (Hofstadter, 1984), and AMBR (Kokinov and Petrov, 2001), do not distinguish between the mechanism of implicit and explicit analogies, or conscious and unconscious analogies, suggesting in that way that they are probably the same. Indeed, it was argued that analogies need autonomous decentralized mechanisms, most of them processing information outside awareness, in order for the dynamic, context sensitive, and emergent properties of analogies to be preserved (for a review, see Kokinov and Hristova, 2012). Perhaps, the explicit prompt for making analogies reinforces deliberate representation building and probably the thorough evaluation of the final analogy, but still people may not be aware how, in terms of what mechanisms are used, they have reached the final answer. Therefore, if explicit analogies cannot be adequately explained only in terms of deliberate and intentional mechanisms as was argued by Kokinov and Hristova (2012), and implicit analogies themselves seem to use the same key mechanisms as the explicit ones as is argued here, the unimodal (Kruglanski and Gigerenzer, 2011) rather than the dual-process view on thinking (Evans and Stanovich, 2013) may turn out to be the cognitively plausible explanatory framework for analogy making.

References

Bassok, M., Chase, V.M. & Martin, S.A. (1998) Adding apples and oranges: Alignment of semantic and formal knowledge, *Cognitive Psychology*, **35**, pp. 99–134.

Bassok, M., Pedigo, S.F. & Oskarsson, A.T. (2008) Priming addition facts with semantic relations, *Journal of Experimental Psychology: Learning, Memory, and Cognition*, **34**, pp. 343–352.

Bethell-Fox, C.E., Lohman, D.F. & Snow, R.E. (1984) Adaptive reasoning: Componential and eye movement analysis of geometric analogy performance, *Intelligence*, **8**, pp. 205–238.

Blanchette, I. & Dunbar, K. (2002) Representational change and analogy: How analogical inferences alter target representations, *Journal of Experimental Psychology: Learning, Memory, and Cognition*, **28**, pp. 672–685.

Bunge, S.A., Wendelken, C., Badre, D. & Wagner, A.D. (2005) Analogical reasoning and prefrontal cortex: Evidence for separable retrieval and integration mechanisms, *Cerebral Cortex*, **15**, pp. 239–249.

Catrambone, R. (2002) The effects of surface and structural feature matches on the access of story analogs, *Journal of Experimental Psychology: Learning, Memory, & Cognition*, **28**, pp. 18–34.

Chalmers, D.J., French, R.M. & Hofstadter, D.R. (1992) High-level perception, representation, and analogy: A critique of artificial intelligence methodology, *Journal of Experimental and Theoretical Artificial Intelligence*, **4**, pp. 185–211.

Cho, S., Holyoak, K.J. & Cannon, T.D. (2007) Analogical reasoning in working memory: Resources shared among relational integration, interference resolution, and maintenance, *Memory & Cognition*, **35** (6), pp. 1445–1455.

Christoff, K. & Gabrieli, J.D.E. (2000) The frontopolar cortex and human cognition: Evidence for a rostrocaudal hierarchical organization within the human prefrontal cortex, *Psychobiology*, **28** (2), pp. 168–186.

Christoff, K., Gordon, A.M., Smallwood, J., Smith, R. & Schooler, J.W. (2009) Experience sampling during fMRI reveals default network and executive system contributions to mind wandering, *Proceedings of the National Academy of Sciences USA*, **106** (21), pp. 8719–8724.

Christoff, K. & Keramatian, K. (2007) Abstraction of mental representations: Theoretical considerations and neuroscientific evidence, in Bunge, S.A. & Wallis, J.D. (eds.) *Perspectives on Rule-Guided Behavior*, Oxford: Oxford University Press.

Christoff, K., Prabhakaran, V., Dorfman, J., Zhao, Z., Kroger, J.K., Holyoak, K.J., *et al*. (2001) Rostrolateral prefrontal cortex involvement in relational integration during reasoning, *Neuroimage*, **14**, pp. 1136–1149.

Christoff, K., Ream, J.M., Geddes, L.P.T. & Gabrieli, J.D.E. (2003) Evaluating self-generated information: Anterior prefrontal contributions to human cognition, *Behavioral Neuroscience*, **117** (6), pp. 1161–1168.

Cole, M. & Scribner, S. (1974) *Culture and Thought*, New York: John Wiley & Sons, Inc.

Day, S. & Gentner, D. (2007) Nonintentional analogical inference in text comprehension, *Memory and Cognition*, **35**, pp. 39–49.

Day, S. & Goldstone, R. (2011) Analogical transfer from a simulated physical system, *Journal of Experimental Psychology: Learning, Memory, and Cognition*, **37**, pp. 551–567.

Dixon, J. & Dohn, M. (2003) Redescription disembeds relations: evidence from relational transfer and use in problem solving, *Memory & Cognition*, **31** (7), pp. 1082–1093.

Doumas, L.A.A., Hummel, J.E. & Sandhofer, C.M. (2008) A theory of the discovery and predication of relational concepts, *Psychological Review*, **115**, pp. 1–43.

Dunham, P. & Dunham, F. (1995) Developmental antecedents of taxonomic and thematic strategies at 3 years of age, *Developmental Psychology*, **31**, pp. 483–493.

Estes, Z. (2003) Attributive and relational processes in nominal combination, *Journal of Memory and Language*, **48**, pp. 304–319.

Estes, Z. & Jones, L.L. (2006) Priming via relational similarity: A copper horse is faster when seen through a glass eye, *Journal of Memory and Language*, **55**, pp. 89–101.

Estes, Z. & Jones, L.L. (2009) Integrative priming occurs rapidly and uncontrollably during lexical processing, *Journal of Experimental Psychology: General*, **138**, pp. 112–130.

Estes, Z., Golonka, S. & Jones, L.L. (2011) Thematic thinking: The apprehension and consequences of thematic relations, in Ross, B. (ed.) *Psychology of Learning and Motivation*, vol. 54, Burlington, MA: Academic Press.

Evans, J.St.B.T. & Stanovich, K.E. (2013) Dual-process theories of higher cognition: Advancing the debate, *Perspectives on Psychological Science*, **8**, pp. 223–241, 263–271.

Falkenhainer, B., Forbus, K. & Gentner, D. (1989) The structure-mapping engine: Algorithm and examples, *Artificial Intelligence*, **41**, pp. 1–63.

Feldman, V. & Kokinov, B. (2009) Analogical episodes are more likely to be blended than superficially similar ones, *Proceedings of the 31st Annual Conference of the Cognitive Science Society*, Hillsdale, NJ: Erlbaum.

Fisher, K.J., Bassok, M. & Osterhout, L. (2010) When two plus two does not equal four: Event-related potential responses to incongruous arithmetic word problems, in Ohlsson, S. & Catrambone, R. (eds.) *Proceedings of the 32nd Annual Conference of the Cognitive Science Society*, Austin, TX: Cognitive Science Society.

French, R.M. (1995) *The Subtlety of Same-ness*, Cambridge, MA: MIT Press.

French, R. (2008) Relational priming is to analogy-making as one-ball juggling is to seven-ball juggling. Commentary on R. Leech, D. Mareschal and R.P. Cooper, Analogy as relational priming: A developmental and computational perspective on the origins of a complex skill, *Behavioral and Brain Sciences*, **31**, pp. 386–387.

Gagné, C.L. (2001) Relation and lexical priming during the interpretation of noun–noun combinations, *Journal of Experimental Psychology: Learning, Memory, and Cognition*, **27**, pp. 236–254.

Gagné, C.L. & Shoben, E. (1997) The influence of thematic relations on the comprehension of modifier-noun combinations, *Journal of Experimental Psychology: Learning, Memory, and Cognition*, **23**, pp. 71–87.

Gagné, C.L. & Spalding, T.L. (2004) Effect of discourse context and modifier relation frequency on conceptual combination, *Journal of Memory and Language*, **50**, pp. 444–455.

Gentner, D. (1983) Structure-mapping: A theoretical framework for analogy, *Cognitive Science*, **7** (2), pp. 155–170.

Gentner, D., Rattermann, M.J. & Forbus, K.D. (1993) The roles of similarity in transfer: Separating retrievability from inferential soundness, *Cognitive Psychology*, **25**, pp. 524–575.

Gentner, D. & Forbus, K. (2011) Computational models of analogy, *WIREs Cognitive Science*, **2**, pp. 266–276.

Gick, M. & Holyoak, K. (1983) Schema induction and analogical transfer, *Cognitive Psychology*, **15**, pp. 1–38.

Glady, Y., French., R.M., Thibaut, J.P., et al. (2014) Adults' eye tracking search profiles and analogy difficulty, in Bello, P., Guarini, M., McShane, M. & Scassellati, B. (eds.) *Proceedings of the 36th Annual Meeting of the Cognitive Science Society*, Austin, TX: Cognitive Science Society.

Goldwater, M.B., Markman, A.B. & Stilwell, C.H. (2011) The empirical case for role-governed categories, *Cognition*, **118**, pp. 359–376.

Goldwater, M.B., Tomlinson, M.T., Echols, C.H. & Love, B.C. (2011) Structural priming as structure-mapping: Children use analogies from previous utterances to guide sentence production, *Cognitive Science*, **35** (1), pp. 156–170.

Gordon, P.C. &, Moser, S. (2007) Insight into analogies: Evidence from eye movements, *Visual Cognition*, **15**, pp. 20–35.

Goswami, U. (1991) Analogical reasoning: What develops? A review of research and theory, *Child Development*, **62**, pp. 1–22.

Green, A.E., Fugelsang, J.A., Kraemer, D.J., Shamosh, N.A. & Dunbar, K.N. (2006) Frontopolar cortex mediates abstract integration in analogy, *Brain Research*, **22**, pp. 125–137.

Green, A.E., Fugelsang, J.A., Kraemer, D.J.M., Gray, J.R. & Dunbar, K.N. (2010) Connecting long distance: Semantic distance in analogical reasoning modulates frontopolar cortex activity, *Cerebral Cortex*, **20**, pp. 70–76.

Green, A.E., Fugelsang, J.A., Kraemer, D.J.M., Gray, J.R. & Dunbar, K.N. (2012) Neural correlates of creativity in analogical reasoning, *Journal of Experimental Psychology: Learning, Memory, and Cognition*, **38**, pp. 264–272.

Halford, G.S. (1992) Analogical reasoning and conceptual complexity in cognitive development, *Human Development*, **35**, pp. 193–217.

Hofstadter, D.R. (1995) *Fluid Concepts and Creative Analogies: Computer Models of the Fundamental Mechanisms of Thought* (together with the Fluid Analogies Research Group), New York: Basic Books.

Hofstadter, D.R. (2001) Analogy as the core of cognition, in Gentner, D., Holyoak, K.J. & Kokinov, B.N. (eds.) *The Analogical Mind: Perspectives from Cognitive Science*, Cambridge, MA: MIT Press.

Hofstadter, D.R. (1984) The Copycat project: An experiment in nondeterminism and creative analogies, *MIT AI Memo*, p. 755.

Holyoak, K.J. (2008) Relations in semantic memory, in Gluck, M.A., Anderson, J.R. & Kosslyn, S.K. (eds.) *Memory and Mind: A Festschrift for Gordon H. Bower*, New York: Erlbaum.

Holyoak, K.J. (2012) Analogy and relational reasoning, in Holyoak, K.J. & Morrison, R.G. (eds.) *The Oxford Handbook of Thinking and Reasoning*, pp. 234-259, Oxford: Oxford University Press.

Holyoak, K.J. & Thagard, P. (1997) The analogical mind, *American Psychologist*, **52**, pp. 35-44.

Holyoak, K.J. & Thagard, P. (1989) Analogical mapping by constraint satisfaction, *Cognitive Science*, **13**, pp. 295-355.

Holyoak, K.J., Gentner, D. & Kokinov, B.N. (2001) Introduction: The place of analogy in cognition, in Gentner, D., Holyoak, K.J. & Kokinov, B.N. (eds.) *The Analogical Mind: Perspectives from Cognitive Science*, pp. 1-19, Cambridge, MA: MIT Press.

Holyoak, K. J. & Hummel, J.E. (2008) No way to start a space program. Commentary on R. Leech, D. Mareschal and R.P. Cooper, Analogy as relational priming: A developmental and computational perspective on the origins of a complex skill, *Behavioral and Brain Sciences*, **31**, pp. 388-389.

Hristova, P. (2009a) Unconscious analogical mapping?, in Taatgen, N.A. & van Rijn, H. (eds.) *Proceedings of the 31th Annual Conference of the Cognitive Science Society*, Austin, TX: Cognitive Science Society.

Hristova, P. (2009b) Unintentional and unconscious analogies between superficially dissimilar but relationally similar simple structures, in Kokinov, B., Holyoak, K. & Gentner, D. (eds.) *New Frontiers in Analogy Research: Proceedings of the Second International Conference on Analogy*, Sofia: NBU press.

Hummel, J. & Holyoak, K. (1997) Distributed representations of structure: A theory of analogical access and mapping, *Psychological Review*, **104**, pp. 427-466. Reprinted in Polk, T.A. & Siefert, C.M. (eds.) (2002) *Cognitive Modeling*, Cambridge, MA: MIT Press.

Hummel, J.E. & Holyoak, K.J. (2003) A symbolic-connectionist theory of relational inference and generalization, *Psychological Review*, **110**, pp. 220-263.

Ji, L.J., Zhang, Z.Y. & Nisbett, R.E. (2004) Is it culture or is it language? Examination of language effects in cross-cultural research on categorization, *Journal of Personality and Social Psychology*, **87**, pp. 57-65.

Keane, M.T. (1990) Incremental analogising: Theory & model, in Gilhooly, K.J., Keane, M.T.G., Logie, R.H. & Erdos, G. (eds.) *Lines of Thinking*, Chichester: Wiley.

Kokinov, B. & French, R.M. (2003) Computational models of analogy-making, in Nadel, L. (ed.) *Encyclopedia of Cognitive Science*, vol. 1, London: Nature Publishing Group.

Kokinov, B. & Hristova, P. (2012) Conscious and unconscious processes in human thinking, in Larrszabal (ed.) *Cognition, Reasoning, Emotion and Action*, CogSc-12, Bilbao: University of the Basque Country Press.

Kokinov, B. & Petrov, A. (2001) Integration of memory and reasoning in analogy-making: The AMBR model, in Gentner, D., Holyoak, K. & Kokinov, B. (eds.) *The Analogical Mind: Perspectives from Cognitive Science*, Cambridge, MA: MIT Press.

Kokinov, B., Hristova, P. & Petkov, G. (2004) Does irrelevant information play a role in judgment?, in Forbus, K., Gentner, D. & Reiger, T. (eds.) *Proceedings of the 26th Annual Conference of the Cognitive Science Society*, Hillsdale, NJ: Erlbaum.

Krawczyk, D.C., McClelland, M.M., Donovan, C.M., Tillman, G.D. & Maguire, M.J. (2010) An fMRI investigation of cognitive stages in reasoning by analogy, *Brain Research*, **1342**, pp. 63–73.

Krott, A., Gagné, C.L. & Nicoladis, E. (2009) How the parts relate to the whole: Frequency effects on children's interpretations of novel compounds, *Journal of Child Language*, **36**, pp. 85–112.

Kruglanski, A.W. & Gigerenzer, G. (2011) Intuitive and deliberative judgements are based on common principles, *Psychological Review*, **118**, pp. 97–109.

Leech, R., Mareschal, D. & Cooper, R. (2008) Analogy as relational priming: A developmental and computational perspective on the origins of a complex cognitive skill, *Behavioral and Brain Sciences*, **31**, pp. 357–414.

Lin, E.L. & Murphy, G.L. (2001) Thematic relations in adults' concepts, *Journal of Experimental Psychology: General*, **130**, pp. 3–28.

Lin, Z. & Murray, S.O. (2014) Unconscious processing of an abstract concept, *Psychological Science*, **25** (1), pp. 296–298.

Livins, K.A., Doumas, L.A.A. & Spivey, M.J. (2016) Shaping relations: The effects of visuospatial priming on relational reasoning, *Journal of Experimental Psychology: Learning, Memory, and Cognition*, **42** (1), pp. 127–139.

Luo, Q., Perry, C., Peng, D., Jin, Z., Xu, D., Ding, G., et al. (2003) The neural substrate of analogical reasoning: An fMRI study, *Cognitive Brain Research*, **17**, pp. 527–534.

Maguire, M.J., McClelland, M.M., Donovan, C.E., Tillman, G.D. & Krawczyk, D.C. (2012) Tracking cognitive phases in analogical reasoning with event-related potentials, *Journal of Experimental Psychology: Learning, Memory, and Cognition*, **38**, pp. 273-281.

Maguire, P., Maguire, R. & Cater, A.W.S. (2010) The influence of interactional semantic patterns on the interpretation of noun-noun compounds, *Journal of Experimental Psychology: Learning, Memory, and Cognition*, **36**, pp. 288-297.

Markman, A.B. & Moreau, C.P. (2001) Analogy and analogical comparison in choice, in Gentner, D., Holyoak, K.J. & Kokinov, B. (eds.) *Analogy: Theoretical and Empirical Research*, Cambridge, MA: MIT Press.

Mather, E., Jones, L.L. & Estes, Z. (2014) Priming by relational integration in perceptual identification and Stroop colour naming, *Journal of Memory and Language*, **71**, pp. 57-70.

Medin, D.L., Goldstone, R.L. & Gentner, D. (1993) Respects for similarity, *Psychological Review*, **100** (2), pp. 254-278.

Mirman, D. & Graziano, K.M. (2012) Individual differences in strength of taxonomic versus thematic relations, *Journal of Experimental Psychology: General*, **141** (4), pp. 601-609.

Mitchell, M. (1993) *Analogy-Making as Perception: A Computer Model*, Cambridge, MA: MIT Press.

Novick, L.R. (1988) Analogical transfer, problem similarity, and expertise, *Journal of Experimental Psychology: Learning, Memory, Cognition*, **14**, pp. 510-520.

Overcast, T.D., Murphy, M.D., Smiley, S.S. & Brown, A.L. (1975) The effects of instructions on recall and recognition of categorized lists by the elderly, *Bulletin of the Psychonomic Society*, **5**, pp. 339-341.

Pavlova, M. & Kokinov, B. (2014) Analogy causes distorted memory by blending memory episodes, in Bello, P., Guarini, M., McShane, M. & Scassellati, B. (eds.) *Proceedings of the 36th Annual Conference of the Cognitive Science Society*, Austin, TX: Cognitive Science Society.

Perrott, D., Gentner, D. & Bodenhausen, G. (2005) Resistance is futile: The unwitting insertion of analogical inferences in memory, *Psychonomic Bulletin & Review*, **12**, pp. 696-702.

Petkov, G., Kiryazov, K., Grinberg, M. & Kokinov, B. (2007) Modeling top-down perception and analogical transfer with single anticipatory mechanism, *Proceedings of the European Conference on Cognitive Science*, Hillsdale, NJ: Erlbaum.

Popov, V. & Hristova, P. (2014) Automatic analogical reasoning underlies structural priming in comprehension of ambiguous sentences, in Bello, P., Guarini, M., McShane, M. & Scassellati, B. (eds.) *Proceed-*

ings of the 36th Annual Conference of the Cognitive Science Society, Austin, TX: Cognitive Science Society.

Popov, V. & Hristova, P. (2015) Unintentional and efficient relational priming, *Memory & Cognition*, **43** (6), pp. 866–878.

Qiu, J., Li, H., Chen, A. & Zhang, Q. (2008) The neural basis of analogical reasoning: An event-related potential study, *Neuropsychologia*, **46**, pp. 3006–3013.

Raffray, C.N., Pickering, M.J. & Branigan, H.P. (2007) Priming the interpretation of noun–noun combinations, *Journal of Memory and Language*, **57**, pp. 380–395.

Schunn, C.D. & Dunbar, K. (1996) Priming, analogy, & awareness in complex reasoning, *Memory and Cognition*, **24**, pp. 289–298.

Sharp, D., Cole, M. & Lave, C. (1979) Education and cognitive development: The evidence from experimental research, *Monographs of the Society for Research in Child Development*, **178**, pp. 1–109.

Simmons, S. & Estes, Z. (2008) Individual differences in the influence of thematic relations on similarity and difference, *Cognition*, **108**, pp. 781–795.

Smith, R., Keramatian, K., Smallwood, J., Schooler, J., Luus, B. & Christoff, K. (2006) Mind-wandering with and without awareness: An fMRI study of spontaneous thought processes, in Sun, R. (ed.) *Proceedings of the 28th Annual Conference of the Cognitive Science Society*, Vancouver: Erlbaum.

Speed, A. (2010) Abstract relational categories, graded persistence, and prefrontal cortical representation, *Cognitive Neuroscience*, 1 (2), pp. 126–137.

Spellman, B.A., Holyoak, K.J. & Morrison, R.G. (2001) Analogical priming via semantic relations, *Memory & Cognition*, **29**, pp. 383–393.

Storms, G. & Wisniewski, E.J. (2005) Does the order of head noun and modifier explain response times in conceptual combination?, *Memory & Cognition*, **33**, pp. 852–861.

Thibaut, J.P., French, R.M., Missault, A., Gérard, Y. & Glady, Y. (2011) In the eyes of the beholder: What eye-tracking reveals about analogy-making strategies in children and adults, in Carlson, L., Hölscher, C. & Shipley, T.F. (eds.) *Proceedings of the 33rd Annual Meeting of the Cognitive Science Society*, Austin, TX: Cognitive Science Society.

Vankov, I. & Kokinov, B. (2009) Grounding relations in action, in Kokinov, B., Holyoak, K. & Getner, D. (eds.) *New Frontiers in Analogy Research*, Sofia: NBU Press.

Vendetti, M.S., Wu, A. & Holyoak, K.J. (2014) Far-out thinking: generating solutions to distant analogies promotes relational thinking, *Psychological Science*, **25** (3), pp. 1–6.

Yan, J., Forbus, K. & Gentner, D. (2003) A theory of rerepresentation in analogical matching, *Proceedings of the 25th Annual Meeting of the Cognitive Science Society*, Boston, MA: Psychology Press.

Waxman, S.R. & Namy, L.L. (1997) Challenging the notion of a thematic preference in young children, *Developmental Psychology*, **33**, pp. 555–567.

Javier Bernacer and
Jose Ignacio Murillo

Habits and the Integration of Conscious and Non-Conscious Human Actions

1. Introduction

The interest that the theoretical and empirical sciences of the mind have shown in habits is discontinuous. According to Barandiaran and Di Paolo's (2014) genealogical map of the concept of habit, this expression of human behaviour was a critical point in the train of thought of several influential classical thinkers such as Aristotle or Aquinas, or of modern philosophers like Descartes and Spinoza. In the eighteenth and nineteenth centuries, however, many schools re-launched the interest in habit, splitting its interpretation into two main branches, namely associationist and organicist. The former includes British and American philosophers like David Hume, Thomas Reid, John Stuart Mill, or William James, as well as the most popular behaviourists (Thorndike, Watson, Pavlov, or Skinner, for instance). The latter covers a variety of philosophical schools, such as German idealism, French spiritualism, phenomenology, and Gestalt psychology, among others. Remarkably, the most common current understanding of habits in the cognitive sciences, including neuroscience, is 'much impoverished' according to Barandiaran and Di Paolo. It is directly inspired by cognitivism and computational-modular approaches to the mind, such as those of Minsky and Fodor. Consequently, 'the result of this development is the current convergence of machine learning and reinforcement learning with neuroscience (Sutton and Barto, 1998; Dezfouli and Balleine, 2012; Daw *et al.*, 2005) where habits have been subsumed under networks of conditional probabilities of expected rewards associated with a set of available actions under specific conditions, or simply reduced to stimulus-triggered responses reinforced only by repetition (Dickinson, 1985)' (Barandiaran and Di Paolo, 2014, p. 5). As the authors explain, the concept of habit is being substituted for the notion of representation in this trend of the cognitive sciences. On the other hand the present organicist view, represented by enactivism, embodied cognition, and dynamic systems interpretations, struggles to recover habits as an 'integral part of individual embodied intentionality'.

Our main goal is to carry out a critical review of the role of consciousness in the different interpretations of habits. Since an exhaustive historical approach is beyond the scope of this text, we will focus on the current understanding of habit and the main inspiration behind its present interpretation in regards to consciousness: Descartes. We will also summarize the contribution of Paul Ricoeur, a philosopher who, starting from a Cartesian position, proposed habits to overcome the conscious/unconscious dichotomy. After this historical summary of habits, we present our own proposal about the relationship between habits and consciousness: habits intrinsically *improve* consciousness. Although this hypothesis seems to be provocative, it is based on the classical contribution of Aristotle, the first philosopher that systematically characterized this important aspect of human behaviour. From this perspective, habits are intimately related with creative learning, and ultimately with behavioural freedom: they allow the agent to set further and higher goals, thanks to their role as integrators of conscious and non-conscious actions.

2. Habits and Consciousness in the Cognitive Sciences

Habits can be understood in many different ways. In fact, it is a common folk term used to explain our behaviour, which can either stress our agency in a particular action ('I have the habit of waking up at 6am to be more productive'), or be used to justify the inflexibility of our behaviour ('I can't help smoking a cigarette right after waking up; I have the habit of doing it'). Ann Graybiel, expert neuroscientist on the study of habits, expresses this idea as follows: 'a habit is a behavior that we do often, almost without thinking. Some habits we strive for, and work hard to make part of our general behavior. And still other habits are burdensome behaviors that we want to abolish but often cannot, so powerfully do they control our behavior' (Graybiel, 2008, p. 360). The role of habits in human behaviour is very wide indeed, and experimental approaches need a much narrower topic. For that reason, the empirical cognitive sciences—in particular experimental psychology—have restricted the study of habits to cover only some of their features. In this section, we will explain how habits are most commonly understood in the cognitive sciences, emphasizing the role of consciousness in their acquisition and performance.

According to Seger and Spiering (2011), the understanding of habit in the cognitive sciences goes hand in hand with the analytical progress in memory. In their critical review on the discovery through history of the neuroanatomical bases of habit learning, they explain that Richard Hirsh was the first to establish the relationship between memory and habits. After ablating the hippocampus of rats—a region consistently described as the brain correlate of memory (Scoville and Milner, 1957) —he observed that 'in the absence of the hippocampus, associative retrieval operates. Behavior is completely controlled by external stimuli and learning is a matter of habit formation. Readers familiar with learning theory will realize that… [behavior] of hippocampally ablated animals is held to be everything for which early S-R [stimulus-response] theorists could have wished' (Hirsh, 1974, p. 439). The investigation on habit learning and memory rapidly moved to the field of consciousness.

The concept of implicit memory was first experimentally assessed by Graf and Schacter (1985). In this research work about context-dependent word completion, the authors conclude that the task is mediated by implicit memory, which is independent of explicit recollection. How do they define implicit memory? 'For descriptive purposes, we use the terms implicit memory and explicit memory to distinguish between these forms of memory. Implicit memory is revealed when performance on a task is facilitated in the absence of conscious recollection; explicit memory is revealed when performance on a task requires conscious recollection of previous experiences' (*ibid.*, p. 501). To our knowledge, they do not explicitly relate implicit memory with habits, although they cite the work by other authors who do so (Mishkin *et al.*, 1984). Consciousness, therefore, is the critical criterion to differentiate explicit and implicit memory.

The next step in the evolution of memory in the cognitive sciences was the definition of declarative and non-declarative memory by Squire and Zola (1996). These authors clearly identify declarative with conscious, and non-declarative with non-conscious. According to them, 'the key distinction is between the capacity for conscious recollection of facts and events (declarative memory) and a heterogeneous collection of nonconscious learning capacities (non-declarative memory) that are expressed through performance and that do not afford access to any conscious memory content' (*ibid.*, p. 13515). Among the tasks involving non-declarative memory, they cite priming, non-associative learning, simple classical conditioning, and procedural learning, including skills and habits. Hence, the authors directly link habits with non-conscious processes. In a previous contribution, Squire et al. define habits as 'dispositions and tendencies that are specific to a set of stimuli and that guide behavior' (Squire et al., 1993, p. 471). Interestingly, the authors conclude their text with the following lines: 'In no small part, by virtue of the nonconscious status of these forms of memory, they create much of the mystery of human experience. Here arise the dispositions, habits, and preferences that are inaccessible to conscious recollection but that nevertheless are shaped by past events, influence our behavior, and are a part of who we are' (ibid., p. 486). In a recent review, Squire and Dede state that non-declarative memory is the source of habits and preferences: 'Here arise the habits and preferences that are inaccessible to conscious recollection, but they nevertheless are shaped by past events, they influence our current behavior and mental life, and they are a fundamental part of who we are' (Squire and Dede, 2015, p. 3). Therefore, habits are understood as one of the main unconscious components of our selves. In her extensive review on implicit learning, Carol Seger defines it as 'nonepisodic learning of complex information in an incidental manner, without awareness of what has been learned' (Seger, 1994, p. 163). Hence, the link between implicit/non-declarative memory, habits, and non-conscious processes was consolidated between the '80s and the '90s.

Apart from the connection between habit learning and memory in cognitive psychology, the most successful definition of habit within the neurosciences came from experimental psychology. In this case, habits are understood as a case of instrumental conditioning and henceforth opposed to goal-directed actions. In a set of experiments with rodents, Anthony Dickinson and his collaborators demonstrated that the behaviour of animals after extensive training was insensitive to outcome devaluation—a noxious liquid as 'reward' instead of food—or extinction—reward cessation (Adams and Dickinson, 1981; Dickinson, 1985). In other words, the rats did not direct their behaviour to achieve a goal—obtaining a reward—but the movement itself—lever pressing, for example—seemed to be rewarding itself. Habit learning was characterized as slow, since it was necessary to extensively train the animals in order to achieve this goal-independent stimulus-response pairing. These experimental findings were directly transposed to humans, and consequently the dichotomy between habits and goal-directed actions has been accepted in the cognitive sciences for humans as well. In this context, a habit is a rigid behaviour mostly identified with compulsions, addictions, and other psychiatric conditions (Gillan et al., 2011; Everitt and Robbins, 2005; Graybiel, 2008), which is automatically triggered in certain circumstances.

Even though the dichotomy between conscious and unconscious processes was not present in the first categorization by Dickinson, automaticity was rapidly identified with performance outside conscious awareness. Consequently, Graybiel asserts: 'Studying this process [habits as an end point of the valuation process] should help investigators identify the neural systems underlying the shift from deliberative behavior controls to the nearly automatic, scarcely conscious control that we associate with acting through habit. Tracking this process may help us to understand the conscious state itself' (Graybiel, 2008, p. 378). Current trends in the study of habits and goal-directed actions focus on neurocomputational models that account for behaviour, and the interplay between both in clinical and pre-clinical conditions (Gillan et al., 2016; Alvares et al., 2016; Dolan and Dayan, 2013; Dezfouli and Balleine, 2012). Although some authors discuss the radical distinction between habits and goal-directed behaviour from a computational perspective (Wood and Neal, 2007; Dezfouli and Balleine, 2013), the non-conscious and conscious nature of habitual behaviour and goal-directed actions, respectively, seems unquestionable (Everitt and Robbins, 2016).

In conclusion, we have shown in this section how the understanding of habit in the cognitive sciences is opposed to goal-directed actions. More importantly for the main thesis of our text, it has been exposed that habits are a kind of non-declarative or implicit memory, and therefore their performance — and in some cases, even their acquisition — is non-conscious. In the next section we will explain the implications of understanding habits outside of consciousness, and the philosophical inspiration underlying it: Descartes's *cogito ergo sum*.

3. The Cartesian Heritage:
Consciousness as the Defining Feature of the Mind

We have explained elsewhere the philosophical inspiration of the current view of habits in the cognitive sciences (Bernacer and Murillo, 2014). Consequently, we will just present a brief summary of this issue, especially focused on consciousness. The understanding of habits as automatic, non-conscious, and inflexible routines is indebted to William James, who in turn received a strong influence from associationism, utilitarianism, and spiritualism (Blanco, 2014). The short but rich article by Blanco about the principal sources of James's idea of habit is extremely relevant to understanding the role of consciousness in habits, at least in the way it is viewed in the cognitive sciences. According to Blanco, James accepts the hypotheses of William Benjamin Carpenter (1874) and Hermann von Helmholtz (1867) about the importance of unconscious processes: 'thought and perception would operate, to a large extent, without awareness, and we would remain unconscious about a substantial body of mental phenomena which we consider rooted in the deepest powers of consciousness' (Blanco, 2014, p. 1). By applying these principles to the human mind, James realized that the main characteristic of habits is to diminish conscious attention in well-known actions, an idea previously presented by the French spiritualist Maine de Biran.

The main goal of this section is to extend the line that we drew in the previous one: 1) habits are understood as an unconscious behaviour following the contributions of psychology on learning and memory; 2) the main philosophical inspiration for this view on habits is William James, who stressed the tight relationship between habits and the unconscious; now, we propose that 3) there is a common conception of consciousness in all these authors, initiated by Descartes, which pushes habits into the unconscious. We will explain this point here. In the next section, we will discuss the efforts of a Cartesian philosopher, Paul Ricoeur, to overcome this limitation.

After Descartes' philosophy, consciousness, which is understood for the first time as an independent substance, is the main defining characteristic of the mind and the distinctive feature of all the topics that are studied by psychology (Murillo, 2016). Before that, the Latin terms *synaisthesis*, *syneidesis*, or *conscientia*, or the German *Bewusstsein*, referred to a concomitant activity, instead of an individual *res*. The Cartesian turn is clearly viewed, for example, in the following text of his *Principles of Philosophy*: 'By the word "thought", I understand all those things which occur in us while we are conscious, insofar as the consciousness of them is in us. An so not only understanding, willing and imagining, but also sensing, are here the same as thinking' (Descartes, 1991, p. 5). The importance of consciousness as a defining feature of the human mind remains clear in this text. This trend was to be consolidated by philosophers such as Leibniz and Kant, and would find its highest splendour in German idealism, where self-consciousness was at the central position of knowledge.

The relationship between consciousness and the empirical sciences reflects this evolution in the interpretation of the human mind. From a broad perspective, the study of consciousness has been set aside from science, due to the difficulty in carrying out an objective analysis of its attributes. For that reason, it is considered an epiphenomenon with no causal role whatsoever. Nevertheless, this scientific approach to the study of consciousness has evolved only in the last century. According to the Aristotelian point of view, psychology is a special discipline within the philosophy of nature, whose main scope is living beings. Living beings are different from other natural beings because they can perform and are constituted through a certain kind of activities: imamnent actions, such as nutrition, growth, or reproduction. Note that *psyche*, the etymological root of psychology, was the Greek word for soul, understood as the principle of activity of material and organic living beings. Hence, classical psychology studied all living beings for the activities they were able to perform. However, since Descartes, modern psychology has shown a declining interest in the dynamic vital perspective of its object, and it became restricted to the human *psyche*. According to modern philosophy, the distinctive feature of the human soul is (self-)consciousness, and therefore this is what became the main object of modern psychology. Remarkably, the *res cogitans* is an inalterable and stable substance, whose dynamicity is hardly considered.

In any case, modern psychology has defined itself in relation to consciousness and has studied conscious experience in many different ways. Obviously, it is not restricted to the study of this kind of experience, although every phenomenon under its umbrella is considered or delimited from consciousness. To cite some examples, structuralist psychology studies psychic phenomena that is reliant on the conscious perception of the subject; adaptive psychology is mainly oriented towards behaviour, dealing in part with the stream of consciousness; psychoanalysis looks into the unconscious and attempts to *invoke* it to make it reach consciousness; and behaviourism emphatically excludes consciousness as a condition for the scientific study of behaviour.

In all these cases, consciousness is unarguably the point of reference, although it is considered hard or impossible to tackle from a scientific perspective. However, there is a current trend in the cognitive sciences to reintroduce its study in the discourse of science. The so-called *cognitive turn* is an attempt to understand what is happening inside the *black box* which, according to behaviourism, is the human mind. This new thrust is based on the cybernetic paradigm and the intention to build machines that simulate human mental operations (Dupuy, 2009). The interpretation of the brain as an information processing machine, similar to a computer, ended up in an increased interest in mental processes. Nowadays, neuroscience is understood as a multidisciplinary approach to the study of the brain and mind, including disciplines such as engineering, informatics, and physics, for instance. In our opinion, the study of consciousness has been able to join so many different disciplines because it is beyond any ordinary scientific approach. However, among all the qualities and characteristics of the mind, the scientific perspective just focuses on the most empirical of them: its condition as a phenomenon (Murillo, 2016).

Two fundamental problems arise when considering consciousness just as a phenomenon: 1) unifying all the diverse typologies of consciousness in the same undifferentiated phenomenon; and 2) obviating its relationship with the subject who exercises it. In our opinion, these problems are a consequence of the Cartesian philosophy. In this context, what is the position of habits with respect to the human mind? Let us remember that for Descartes the self is the conscious self: *cogito ergo sum*. In the course of modern philosophy, besides, habits are increasingly understood as psychological resources that transfer teleological actions into the unconscious, so they are performed automatically and without conscious supervision. The logical consequence is that habits are extraneous to the self: we are responsible for our goal-directed actions, but we are not in control of our habits. The Cartesian heritage forces us to decide which attributes of the human mind belong to the conscious, and which ones remain in the dark realms of the unconscious. In the next section we will reveal the attempt of the French philosopher Paul Ricoeur to release habit from this dichotomy. We advance that, in our opinion, the success of his endeavour was only partial.

4. Paul Ricoeur's Solution: Habits as the Unconscious Power of the Conscious Will

Since Descartes, the history of philosophy has reached a *status quo* with respect to consciousness that is manifested as a high tension between two opposite poles: what guides human behaviour? Is it the conscious or the unconscious? If it is the latter, are we free to decide? Does this 'we' include the unconscious? This tension has been naturally inherited by psychology, the cognitive sciences, and current neuroscience. As we have already mentioned, some trends within psychology, such as psychoanalysis, tried to expose the supremacy of the unconscious with respect to consciousness in the configuration of the self. It should be noted that this is not an attempt to integrate both realms, but a method of bringing the contents of the unconscious to the conscious foreground. Is there any way to link the two domains? Paul Ricoeur (1913-2005) devoted his life to philosophical anthropology, and especially to understanding the self as an agent responsible for its actions. In his early works, he utilized an existential and phenomenological approach to the philosophy of the will, although he abandoned the project to focus on hermeneutics. In our opinion, the contribution of the early Ricoeur regarding the unity of the voluntary and the involuntary is an interesting approach intended to overcome the dichotomy mentioned above. In fact, habits have a fundamental role in the search for this unity, since they are understood as the unconscious power of the conscious will. We will develop his idea in this section, and will also expose the limitations that, in our opinion, it entails.

In 'The Unity of the Voluntary and the Involuntary as a Limiting Idea', Ricoeur intends 'to show how the study of the reciprocity of the voluntary and the involuntary can revitalize the classical problem of the relations between "freedom" and "nature", by proposing between them a "practical mediation"' (Ricoeur, 1978; 1951, pp. 3–4). Hence, his main interest is to discuss the responsibility of human actions by confronting the usual dichotomies: voluntary and involuntary, freedom and nature, spontaneity and inertia, mind and body, and so on. Furthermore, the French philosopher believes that there are two human forms of involuntary action that illustrate the reciprocity between the voluntary and the involuntary: emotion and habit. According to his point of view, the former represents irregularity, the irruption of the moment; the latter, however, is previously acquired and the fruit of what has endured. Therefore, Ricoeur employs again a confronting dichotomy to study, in this case, human involuntary actions, and uses it as a means to explain the conscious will: 'In short, the body is a practical spontaneity, by turns emotive and habitual, which mediates our volitions… In both cases [the activation of knowledge and of bodily powers], the empty intention of the project finds in the naturalization of the will the instrument of its efficacy' (Ricoeur, 1978; 1951, p. 14). The 'naturalization of the will' is an important concept in Ricoeur's proposal, although it was previously introduced by Felix Ravaisson (Gaitan and Castresana, 2014; Ravaisson, 1997): the will—conscious, non-physical, expression of freedom—is open to being naturalized— and thus made unconscious, physical, and inertial. We will explain in this text Ricoeur's interpretation of habits, their impact on involuntary actions, and, ultimately, their influence on voluntary behaviour. This analysis is mostly based on research work by Morales (2016).

Following the Cartesian approach, emotion and habit are independent of the *cogito*: the former is something suffered by the self, and the latter is a movement unlinked to the agent. Ricoeur believes that both are not against the freedom of the agent, but they are involuntary expressions of the will. Involuntariness, in turn, cannot be understood independently of the voluntariness of actions. He also applies the dialectic method to habits themselves, arguing that they are voluntary in their acquisition, but also involuntary because they are spontaneous in their performance. Moreover, a habit is not just an automatism: it is a kind of behaviour deeply naturalized in the body, but with a radical relationship with the *cogito*—with the mind. Where Ravaisson defends that the habit is an inclination that substitutes the will, Ricoeur states that it is an empowerment of the will. How is this possible? Because habits help the conscious will to achieve its intended goals. Going back to the special interplay between emotions and habits, Ricoeur suggests that the former makes the agent move towards the object, whereas the latter organizes the agent's behaviour to find the best way to achieve it. Note that this is in sharp contrast with the associationist view of William James, and consequently with the understanding of habit in the cognitive sciences: habits are some sort of stimulus-response behaviour opposite to goal-directed actions. Following Ricoeur's philosophy of the will and dialectics, habits are between spontaneity and inertia, between the living being and the automaton. Spontaneity and inertia are dissociated naturally in old age and pathology, and artificially by psychology when a habit is pulled away from the *cogito* that performs it. Following Marcel (1949), Ricoeur believes that humans *have* habits, but also *are* habits, reinforcing the role of habits as second nature as it was already proposed in Ancient Greece. Ricoeur thus concludes that habits are the archetypical conjunction between mind and body. Finally, he defends a tight relationship between habits and freedom, inasmuch as habits are able to open new ways for the conscious will according to the behavioural history of the subject.

In summary, Paul Ricoeur struggles against the view of habits as automatisms, and assigns them a central role in the reciprocity between conscious and unconscious processes. They are in charge of guaranteeing the unity of the self across a changing life, even though they are acquired dispositions that do not originally belong to the self. They are engrained in the *res extensa*, that is, in the physical reality of human beings; however, they also serve the *cogito* as enhancers of the conscious will.

In our opinion, Ricoeur's proposal is extremely interesting because it relates habits with creativity and freedom. This is a fact that should not be overlooked in the philosophical, psychological, and neuroscientific analysis of habits (Bernacer *et al.*, 2014; Bernacer and Gimenez-Amaya, 2013; Bernacer and Murillo, 2014). As we usually exemplify, ignoring this issue denies the fact that expert musicians are able to improvise: starting from well-learned routines, they are able to create new pieces that are different from those routines. This is unquestionably due to the liberating dimension of habits (Güell, 2014; Güell and Nuñez, 2014). However, Ricoeur cannot escape from the main problem of Cartesian philosophy: dualism. He plays dialectics to reach the reciprocal synthesis in any dichotomy he encounters, while it could be more profitable to avoid some of these dichotomies from the outset. Furthermore, he recognizes the behavioural relationship between habits and creativity, although his proposal remains obscure due to the separation between the *cogito* and *nature*. Eventually, in some respects, it seems that habit is his particular pineal gland, trying to find a reciprocal relationship between two completely different substances.

More broadly, the mind–body dualism impedes a common standpoint between philosophy and the empirical sciences, since both of them deal with disparate realities. Furthermore, the Cartesian physical reality will always be linked to inertial necessity, so it is impossible to escape from hard determinism. Ricoeur tries to do it, once again, by using a dialectic approach where the synthesis seems logically impossible.

In the following section, we will present another alternative to explain the relationship between consciousness and habits. It is based on Aristotelian philosophy, and the main difference with respect to Ricoeur's is the avoidance of dualisms. The most striking consequence, as we shall see, is that habits are not just a link between the conscious and the unconscious, but a cognitive tool to *improve* consciousness itself.

5. A New Proposal Based on Aristotle: Habits and the Improvement of Consciousness

In his article about the history and *status quaestionis* of implicit memory, Schacter (1987) surprisingly states that Plato and Aristotle 'appear to have been concerned exclusively with explicit memory' (p. 502). However, as we discussed above, habit is one of the main examples of implicit memory in current psychology and neuroscience, and it is undeniable that Aristotle showed a special interest in the study of habits. In fact, his theory was the first scientific reflection about habits, and it has conditioned, in an explicit or implicit way, all later theories about it. In order to explain the relationship between the Aristotelian understanding of habits and consciousness, we will need first to outline his philosophy of action. It may seem strange to revisit this philosopher for such an interesting topic within the cognitive sciences, considering the scant knowledge that he had about the nervous system. However, it should be kept in mind that Darwin proposed his theory about natural selection with a similar limited knowledge about heritability and the dynamics of genetic material. Thus, we will explain the immanent character of human actions, the importance that goals have in them, and the increasing dynamicity of goals. In a sentence, habits allow the agent to search for further goals and to be more successful in achieving them.

The reason why this classical proposal may be interesting for present philosophers and scientists is that it is not conditioned by dichotomies of modern psychology and philosophy: conscious and unconscious, mind and body, freedom and determinism, etc. Aristotle coins the concept of habit to refer to a wide range of human capacities. Although they may seem diverse in nature, all of them can be considered as involving some kind of 'learning'. The empirical ground that

Aristotle provides for this theory is the capacity of human beings to reconfigure their behaviour according to their previous activity. A stone, as he says, can be thrown up to the sky once and again and it will continue falling to the earth in the same way; and, moreover, it would be necessary to use the same force to throw it up again (Aristotle, 2002). Conversely, living beings can adapt themselves to the activities they repeatedly perform: they operate a kind of disposition to better perform these repeated activities.

In the case of human beings, this feature is extremely important because they have '*logos*' as a defining trait of their lives. In other words, according to Aristotle, they have an almost inexhaustible capacity for learning new things and integrating them into their behaviour. This allows the invention of new tools and the setting of new goals, ending up in a global reconfiguration of their behaviour. How can this view contribute to the understanding of habits by the cognitive sciences? For a start, using Aristotle's view means the main focus to understand habits has changed: it is not so important to study the role of consciousness in acquisition or performance, but the fact that habit is a kind of learning. From this description it is obvious that Aristotle does not understand learning primarily as the acquisition of some cognitive contents that we can reproduce, but as a kind of activity that, by reaching its goal, can change our future behaviour and generate a new disposition to act. It is important to note that this disposition *cannot* be acquired in a passive way: habits are generated through previous *activity*, and their quality and strength depend on the characteristics of the activity that has generated them. As we will see later, if the activity that precedes the habit is cognitively enriched, the habit will result in a cognitive enrichment of behaviour. Conversely, if a habit is mechanistically acquired, it will routinize the agent's behaviour. Before developing this, we will explain in more detail how Aristotle understands learning, and its role in habit acquisition.

Aristotle starts his philosophy of action — which is inseparable from his ethics, that is, the global study of human behaviour — wondering about the goal that all things pursue (Aristotle, 2002). All things, including all human beings, organize their behaviour to reach some good. This should be understood as referring not only to the final aim of someone's existence, but also to every single action: each human action aims to reach some good. In order to understand what a 'good' is in this context, we should briefly discuss the Aristotelian philosophy of action. In general terms, Aristotle distinguishes between movement — *kinesis* — production — *poiesis* — and action — *praxis* (Aristotle, 2007; 1986). A better understanding of goal-directed human behaviour requires a more detailed explanation of productions and actions. The main difference between them is that, whereas in productions the effect of the behavioural performance falls on the product made or on the receiver of the act, the key characteristic of actions is that the effect remains within the agent itself. For example, painting is a production because the main effect is the portrait we have painted; but perceiving is an action because the main effect—the percept—remains in the subject. Actions are also known as *immanent* actions in the Latin Aristotelian tradition, stressing the permanency of the effect in the subject. Note that productions may also have an immanent character, since painting a portrait has a *by-product*, namely becoming a painter. In other words, as we have written elsewhere, 'Since human actions are driven and controlled by cognition, each new action leaves a footprint in the agent as a kind of learning' (Bernacer and Murillo, 2014, p. 4). This may be a different approach to implicit learning, but it is beyond any doubt that Aristotle is contributing to this topic.

How is this related to habits? The 'footprint left' in the agent by immanent actions can be understood in the same terms as the naturalization of the will proposed by Ravaisson and Ricoeur. Remarkably, William James also supports this perspective when he cites Léon Dumont as follows: 'Every one knows how a garment, after having been worn a certain time, clings to the shape of the body better than when it was new; there has been a change in the tissue, and this change is a new habit of cohesion' (James, 1890; Dumont, 1876, p. 1 (online)). Hence, human actions—or at least some of them—have an impact on the subject who performs them. Furthermore, habits are acquired through the repetition of actions—except in some cases, i.e. cognitive habits, where this is not strictly necessary—due to its immanent character: 'For the things we have to learn before we can do them, we learn by doing them, e.g. men become builders by building' (Aristotle, 2002, p. 21). As a matter of fact, there is common agreement in considering habits as second nature, from Aristotle (2002) to present neuroscience (Graybiel, 2008). However, there is a fundamental nuance in the Aristotelian proposal: this 'footprint' should not be understood just as physical, but it is rather a psychical feature. His anthropology is not dualist, although a deeper explanation of Aristotle's philosophy of nature is far beyond the scope of this chapter.

Thus, the first conclusion is that habits are a kind of learning. We will discuss next *what kind* of learning they are, in order to argue for their role in the improvement of consciousness. In previous sections of this chapter, we showed that habits are considered as implicit learning in the cognitive sciences. This means that, through action repetition, there is some learning unconsciously and automatically deposited in the subject, which is retrieved with the same two attributes. According to this view, habits are involuntarily learnt and, once they have been acquired, learning is over. Moreover, when habits are understood as a simple stimulus-response pairing, it is obvious that learning the pair is useful in order to improve performance (mainly decreasing reaction time), but it prevents the learning of new goal-directed actions (see, for example, De Wit *et al.*, 2012). Recently, this competition between the goal-directed and habitual system has been explained in neuro-anatomical terms: the circuit responsible for goal-directed actions is inhibited when behaviour is habitually performed (Gremel *et al.*, 2016). Let us go back to the Aristotelian proposal, in order to underpin the differences with this understanding of habitual learning: human actions are generally goal-directed, since the agent pursues some good; in addition, they usually have an immanent character, because they leave a 'footprint' in the agent's *psyche* that may affect its future actions.

These two elements of actions—goals and their immanent character—are intimately related: thanks to the 'learning trace' that is deposited in the subject after performing immanent actions, the agent can aim for further goals. Taken from a functional perspective, learning is defined as 'changes in the behavior of an organism that are the result of regularities in the environment of that organism' (De Houwer *et al.*, 2013; Balderas, 2014, p. 631). Therefore, according to Aristotle's philosophy, habit learning is *incremental* for two reasons: firstly, because acquiring a habit involves a change in the organism's behaviour that makes it apt for reaching for new goals; secondly, because habits are some kind of growing for the organism and they themselves can grow. Note that the interpretation of habit by the cognitive sciences does not allow one to understand it properly as learning, since habitual behaviour is rigid and ends up in the regularization of the animal's behaviour. Contrary to this view, Aristotle considers habit as a kind of growing. This is not casual: the most characteristic feature of living beings is not that they plainly maintain their existence, but rather that they can *grow*. Since a living being is an organism, growing has a limit: unlimited growth would be a monstrosity. However, life is by itself a sort of liberation from the limitations of materiality. This is clear in those living beings that can perform activities like cognition: even though they depend on matter, they are not confined to it. If Aristotle had known the nature of the nervous system, he would have understood its plasticity as the possibility for growing beyond an organic constitution.

Now that we have explained why habits can be considered as learning, we will discuss in the following paragraphs their relationship with consciousness. However, it is important to avoid two assumptions that would lead to a misinterpretation of our thesis. First, consciousness should not be identified with the self or the person: consciousness is an activity of the living being. Second, consciousness is not determined and immovable, but dynamic and open to growth.

Our hypothesis is that habits *improve* consciousness, which is intimately related to the understanding of habits as learning. This assertion needs to fulfil a condition to be true: a habit improves consciousness if that habit is a cognitive enrichment of behaviour. According to Aristotle, and as we can see in everyday life, there are good and bad habits. If habitual behaviour were a mere stimulus-response routine, this normative distinction would be impossible: what is the criterion for considering a stimulus-response pair as good or bad? Following the Aristotelian analysis, a good habit — i.e. a 'virtue' — is intimately related with the '*logos*' — i.e. cognition. By acquiring a good habit, we have enriched our behaviour with some kind of learning that opens up new possibilities for us, as we will discuss below. On the other hand, a bad habit — or 'vice' — routinizes the agent's behaviour, and constrains its search for new goals. The main example of a bad habit is a routine or automatized behaviour. Bad habits may have a strong influence on our behaviour, because once they are released it is difficult to prevent their completion. Hence, they are acquired without the intervention of the '*logos*', and consequently they impoverish behaviour from a cognitive point of view. Examples of good and bad habits are proper musical learning and addiction, respectively. Good habits, therefore, make the agent *more conscious* of its situation, thanks to the improved cognitive control of actions and the opening of higher goals. However, bad habits impoverish consciousness, since the agent remains captive to the routine's execution. In fact, Aristotle wrote that 'vice is unconscious of itself' (Aristotle, 2002). In the following paragraphs, we will explain the improvement of consciousness by habits in three aspects: 1) the release of consciousness from the motor aspects of actions to concentrate on the final goal; 2) the capacity to consciously monitor the motor aspects of actions to improve performance; 3) the improvement of the global experience of the action.

Interestingly, there is a common point between habits-as-learning and habits-as-routines: the release of consciousness from the motor aspects of the action. After training, the pianist can liberate her conscious resources from the hand movements to concentrate on the global performance of the piece. Ann Graybiel supports this view by asserting that 'fully acquired habits are performed almost automatically, virtually nonconsciously, allowing attention to be focused elsewhere' (Graybiel, 2008, p. 361). William James is also clear about this topic when he cites Henry Maudsley: 'If an act became no easier after being done several times, if the careful direction of consciousness were necessary to its accomplishment on each occasion, it is evident that the whole activity of a lifetime might be confined to one or two deeds — that no progress could take place in development' (Maudsley, 1889; James, 1890, p. 1 (online)). It is important to remark that the liberation of consciousness allows it to be focused on *something else*. Classically, cognitive neuropsychology has related this with dual tasking, i.e. the ability of the subject to perform a well-known and a novel task simultaneously (Hikosaka et al., 2013): thanks to habit acquisition, the agent can concentrate on a different task. This is true for some kinds of routine activities. However, humans are able to perform other habits-as-learning where the 'liberated consciousness' is placed on the final goal of the activity, such as in the case of the pianist. Therefore, the agent has an improved consciousness inasmuch as its attention and cognitive resources may be directed towards the final goal of the action.

Second, habit acquisition improves the capacity to consciously monitor the motor aspects of actions, in order to improve global performance. We have just explained how habits allow one to *overlook* the motor component of a well-known behaviour. However, when a habit has been correctly acquired — i.e. mediated by cognition — it also lets consciousness go *back* to the motor performance for error monitoring and improvement. This is understood in current neuroscience as *switching* between automatic and controlled behaviour (Hikosaka and Isoda, 2010), due to the Cartesian influence that we have explained above. However, there is a critical issue that may be helpful for refining this view: both the habitual and non-habitual aspects of behaviour are related to cognition, as we have just discussed. Thus, the best strategy is to consider both as aspects of the same cognitively-guided behaviour. In the case of musical performance, routinized movements help to improve the global *conscious* interpretation of the piece; at the same time, a proper mastering of the instrument allows the musician to modify motor routines to improve performance. In conclusion, consciousness is improved by habits because their correct acquisition enhances the interplay between the cognitive and motor aspects of behaviour.

Lastly, habits result in the improvement of the global experience of the action. We have argued in the previous paragraphs that 'good' habits liberate consciousness from motor routines in order to focus on the goal of the action, but also that they improve the ability to integrate the cognitive and motor aspects of behaviour. In this last point, we suggest that habits expand the quality of the goals that the agent can reach for. Let us conclude with the example of the musician — a piano player. An inexperienced pianist starts learning motor routines, continuously repeated until they are ingrained in her performance. As experimental neuroscience proposes, the pianist will become quicker and more accurate in those routines, at the expense of having a rigid interpretation and not being able to introduce new notes within the learnt sequences. However, the human musical experience shows that motor training is followed by an improved interpretation, which is manifested as a more fluent global performance, stressing the emotional component of the music, understanding better the dynamics of the piece, and ultimately leading to a more creative execution.

In our opinion, the understanding of habits as routines cannot account for this fact. However, the Aristotelian view of habits as learning is in line with this behavioural reality: the inexperienced pianist aims to achieve the simple goal of correctly repeating the motor routines. When this is achieved, she can reach for a further goal and correctly interpret complex pieces. Finally, she would be able to aim to the highest goal in musical interpretation, namely a creative performance. This incremental goal pursuit is the final aspect of the improvement of consciousness. This interpretation of habits can also be exemplified by a contemporary description of positive psychology, so-called optimal performance or flow. Flow is defined as 'a sense that one's skills are adequate to cope with the challenges at hand, in a goal-directed, rule-bound action system that provides clear clues as to how well one is performing' (Csikszentmihalyi, 1991, p. 71). It relies on the performance of a highly practised skill, although 'it should be stressed that the body does not produce flow merely by its movements. The mind is always involved as well' (*ibid.*, p. 95). Aside from the dualist view of the author, we believe that the Aristotelian characterization of habits that we have presented here and especially its relationship with consciousness are along the same lines since both link habitual performance with creativity (Dietrich, 2004).

6. Conclusions

The opposition between conscious and non-conscious behaviour has been a highly relevant topic for the last few decades of research in psychology, neuroscience, and philosophy. Habits are one example of human behaviour whose interpretation strictly depends on this opposition. We have shown in this chapter that the prevailing understanding of consciousness is built on the bricks made by Descartes, according to whom consciousness is identified with the self. Thus, habits and all kinds of activities somewhat related to the unconscious are considered extraneous to the self. This position was not satisfying for Paul Ricoeur who, in spite of having a strong Cartesian influence, proposed the role of habits as an unconscious power of the conscious will. His solution, however, led to a series of contradictions that were hard to reconcile. We have developed the initial proposal by Aristotle to suggest that habits, if properly acquired, can cognitively enrich human behaviour. This cognitive enrichment can be understood as an improvement of consciousness. In the development of our hypothesis, we defined consciousness as an activity of the living being that is open to being enriched — and also impoverished. We found that consciousness can be improved by habits in three different respects: 1) to liberate attention from the motor aspects of actions in order to concentrate on the final goal; 2) to consciously monitor the motor aspects of actions in order to improve performance; 3) to enhance the global experience of the action, which is revealed through an easier, more accurate, and more enjoyable expression of behaviour.

References

Adams, C. & Dickinson, A. (1981) Instrumental responding following reinforcer devaluation, *Quarterly Journal of Experimental Psychology: Section B*, **33**, pp. 109–121.

Alvares, G.A., Balleine, B.W., Whittle, L. & Guastella, A.J. (2016) Reduced goal-directed action control in autism spectrum disorder, *Autism Research*, in press, doi:10.1002/aur.1613.

Aristotle (2007) *Metaphysics*, Mineola, NY: Dover.

Aristotle (2002) *Nicomachean Ethics*, New York: Oxford University Press.

Aristotle (1986) *On the Soul*, New York: Penguin Books.

Balderas, G. (2014) Habits as learning enhancers, *Frontiers in Human Neuroscience*, **8**, pp. 10–12.

Barandiaran, X.E. & Di Paolo, E.A. (2014) A genealogical map of the concept of habit, *Frontiers in Human Neuroscience*, **8**, 522.

Bernacer, J. & Gimenez-Amaya, J. (2013) On habit learning in neuroscience and free will, in Suarez, A. & Adams, P. (eds.) *Is Science Compatible With Free Will?*, New York: Springer.

Bernacer, J., Balderas, G., Martinez-Valbuena, I., Pastor, M. a & Murillo, J.I. (2014) The problem of consciousness in habitual decision making, *Behavioral & Brain Sciences*, **37**, pp. 21–22.

Bernacer, J. & Murillo, J.I. (2014) The Aristotelian conception of habit and its contribution to human neuroscience, *Frontiers in Human Neuroscience*, **8**, 883.

Blanco, C.A. (2014). The principal sources of William James' idea of habit, *Frontiers in Human Neuroscience*, **8**, 1–2.

Carpenter, W. (1874) *Principles of Mental Physiology*, New York: Appleton.

Csikszentmihalyi, M. (1991) *Flow: The Psychology of Optimal Experience*, New York: Harper Collins Publishers.

Daw, N.D., Niv, Y. & Dayan, P. (2005) Uncertainty-based competition between prefrontal and dorsolateral striatal systems for behavioral control, *Nature Neuroscience*, **8**, pp. 1704–1711.

Descartes, R. (1991) *Principles of Philosophy*, Miller, V. & Miller, R. (eds.), Dordrecht: Kluwer Academic Publishers.

Dezfouli, A. & Balleine, B.W. (2012) Habits, action sequences and reinforcement learning, *European Journal of Neuroscience*, **35**, pp. 1036–1051.

Dezfouli, A. & Balleine, B.W. (2013) Actions, action sequences and habits: Evidence that goal-directed and habitual action control are hierarchically organized, *PLoS Computational Biolology*, **9**, e1003364.

Dickinson, A. (1985) Actions and habits: The development of behavioural autonomy, *Philosophical Transaction of the Royal Society B: Biological Sciences*, **308**, pp. 67–78.

Dietrich, A. (2004) Neurocognitive mechanisms underlying the experience of flow, *Consciousness and Cognition*, **13**, pp. 746–761.

Dolan, R.J. & Dayan, P. (2013) Goals and habits in the brain, *Neuron*, **80**, pp. 312–325.

Dumont, M. (1876) De l'habitude, *Revue Philosophique*, **1**, pp. 321–366.

Dupuy, J. (2009) *On the Origins of Cognitive Science*, Cambridge, MA: MIT Press.

Everitt, B.J. & Robbins, T.W. (2005) Neural systems of reinforcement for drug addiction: From actions to habits to compulsion, *Nature Neuroscience*, **8**, pp. 1481–1489.

Everitt, B.J. & Robbins, T.W. (2016) Drug addiction: Updating actions to habits to compulsions ten years on, *Annual Review of Psychology*, **67**, 150807174122003.

Gaitan, L.M. & Castresana, J.S. (2014) On habit and the mind–body problem: The view of Felix Ravaisson, *Frontiers in Human Neuroscience*, **8**, 1–3.

Gillan, C.M., Papmeyer, M., Morein-Zamir, S., Sahakian, B.J., Fineberg, N.A., Robbins, T.W. & de Wit, S. (2011) Disruption in the balance between goal-directed behavior and habit learning in obsessive-compulsive disorder, *American Journal of Psychiatry*, **168**, pp. 718–726.

Gillan, C.M., Kosinski, M., Whelan, R., Phelps, E.A. & Daw, N.D. (2016) Characterizing a psychiatric symptom dimension related to deficits in goal-directed control, *Elife*, **5**, pp. 1–24.

Graf, P. & Schacter, D.L. (1985) Implicit and explicit memory for new associations in normal and amnesic subjects, *Journal of Experimental Psychology: Learning, Memory, and Cognition*, **11**, pp. 501–518.

Graybiel, A.M. (2008) Habits, rituals, and the evaluative brain, *Annual Review of Neuroscience*, **31**, pp. 359–387.

Gremel, C.M., Chancey, J., Atwood, B., Luo, G., Neve, R., Ramakrishnan, C., Deisseroth, K., Lovinger, D. & Costa, R. (2016) Endocannabinoid modulation of orbitostriatal circuits gates habit formation, *Neuron*, in press. doi:http://dx.doi.org/10.1016/j.neuron.2016.04.043.

Güell, F. (2014) Pre-dispositional constitution and plastic disposition: Toward a more adequate descriptive framework for the notions of habits, learning and plasticity, *Frontiers in Human Neuroscience*, **8**, 2012–2015.

Güell, F. & Nuñez, L. (2014) The liberating dimension of human habit in addiction context, *Frontiers in Human Neuroscience*, **8**, 64.

Helmholtz, H. von (1867) *Handbuch Der Physiologischen Optik*, Leipzig: L. Voss.

Hikosaka, O. & Isoda, M. (2010) Switching from automatic to controlled behavior: Cortico-basal ganglia mechanisms, *Trends in Cognitive Sciences*, **14**, pp. 154–161.

Hikosaka, O., Yamamoto, S., Yasuda, M. & Kim, H.F. (2013) Why skill matters?, *Trends in Cognitive Sciences*, **17**, pp. 434–441.

Hirsh, R. (1974) The hippocampus and contextual retrieval of information from memory: A theory, *Behavioral Biology*, **12**, pp. 421–444.

De Houwer, J., Barnes-Holmes, D. & Moors, A. (2013) What is learning? On the nature and merits of a functional definition of learning, *Psychonomic Bulletin and Review*, **20**, pp. 631–642.

James, W. (1890) *Principles of Psychology, Toronto: An Internet Resource*, developed by C.D. Green, [Online], http://psychclassics.yorku.ca/James/Principles/prin4.htm [Accessed 1 June 2016].

Marcel, G. (1949) *Being and Having*, Westminster: Dacre Press.

Maudsley, H. (1889) *The Physiology of Mind*, New York: D. Appleton and Company.

Mishkin, M., Malamut, B. & Bachevalier, J. (1984) Memories and habits: Two neural systems, in McGaugh, J., Lynch, G. & Weinberger, N. (eds.) *Neurobiology of Learning and Memory*, New York: Guilford Press.

Morales, I. (2016) *Cuerpo y libertad. Las formas involuntarias del querer según Paul Ricoeur*, [Online], https://www.educacion.gob.es/teseo/mostrarRef.do?ref=1235226.

Murillo, J.I. (2016) Es posible una aproximación científica al problema de la conciencia? La ciencia, la filosofía y el problema de la conciencia, in *De simios, cyborgs y dioses. La naturalización del hombre a debate*, pp. 83–92, Madrid: Biblioteca Nueva.

Ravaisson, F. (1997) *De l'habitude*, Paris: Payot & Rivages.

Ricoeur, P. (1951) The unity of the voluntary and the involuntary as a limiting idea, *Bull. la Société française Philos.*, **45**, pp. 1–29.

Ricoeur, P. (1978) *The Philosophy of Paul Ricoeur: An Anthology of His Work*, Reagan, C. & Stewart, D. (eds.), Boston, MA: Beacon Press.

Schacter, D.L. (1987) Implicit memory: History and current status, *Journal of Experimental Psychology*, **13**, pp. 501–517.

Scoville, W.B. & Milner, B. (1957) Loss of recent memory after bilateral hippocampal lesions, *Journal of Neurology, Neurosurgery & Psychiatry*, **20**, pp. 11–21.

Seger, C.A. (1994) Implicit learning, *Psychological Bulletin*, **115**, pp. 163–196.

Seger, C.A. & Spiering, B.J. (2011) A critical review of habit learning and the basal ganglia, *Frontiers in System Neuroscience*, **5**, 66.

Squire, L.R., Knowlton, B.J. & Musen, G. (1993) The structure and organization of memory, *Annual Reviews of Psychology*, **44**, pp. 453–495.

Squire, L.R. & Zola, S.M. (1996) Structure and function of declarative and nondeclarative memory systems, *Proceedings of the National Academy of Sciences*, **93**, pp. 13515–13522.

Squire, L.R. & Dede, A.J.O. (2015) Conscious and unconscious memory systems, *Cold Spring Harbor Perspectives in Biology*, **7**, a021667.

Sutton, R. & Barto, A. (1998) *Reinforcement Learning*, Cambridge, MA: MIT Press.

De Wit, S., Standing, H.R., Devito, E.E., Robinson, O.J., Ridderinkhof, K.R., Robbins, T.W. & Sahakian, B.J. (2012) Reliance on habits at the expense of goal-directed control following dopamine precursor depletion, *Psychopharmacology (Berlin)*, **219**, pp. 621–631.

Wood, W. & Neal, D.T. (2007) A new look at habits and the habit-goal interface, *Psychological Reviews*, **114**, pp. 843–863.

Massimiliano Cappuccio

Flow, Choke, Skill
The Role of the Non-Conscious in Sport Performance

1. Embodiment and Background: Philosophy and Psychology on the Sporting Field

Do expert athletes rely preferably on conscious or non-conscious[1] mental processes in order to peak their sport performances? Today, this question is passionately debated by sport psychologists and cognitive scientists. Given the elusive nature of the notion of consciousness, it could hardly be answered without simultaneously addressing some interrogatives that are exquisitely normative in nature or that importantly involve philosophical clarification: what truly is consciousness, and what does make an experience conscious? What is the relationship between conscious and non-conscious behaviours? Providing a definitive answer to these interrogatives is well beyond the scope of this chapter, which only aspires to concretely illustrate how the question of peak performances and the question of the nature of consciousness are intertwined and need to be addressed together. That is why this chapter will not simply assume that the philosophical clarification of the notion of consciousness is a prerequisite to understand peak

---

[1] In describing non-conscious psychological processes, I will assume a classical tripartition. All of the cognitive processes and information that are accessed by the system at the subpersonal level, without them explicitly informing phenomenal experience (as they are outside the field of one's conscious awareness/attention), are called *non-conscious*. The domain of the non-conscious comprises two subsets: *subconscious* dynamics (cognitive processes and bodies of information that currently are not, but that in principle can be and could have been, consciously accessed at the personal level, if the cognitive agent directed its attention to them); and *unconscious* dynamics (cognitive processes and bodies of information that are not, and that in principle cannot be, accessed by consciousness, as structural limitations or intrinsic incompatibilities in the cognitive architecture prevent the cognitive agent from consciously accessing them at the personal level).

performance; it will also show that, in turn, an empirical understanding of the psychological conditions that enable peak performance can significantly enrich our philosophical understanding of consciousness. In order to do this, it will examine the cognitive processes and psychological states that allow professional sportsmen and sportswomen to excel on the field.

Some philosophers might be tempted to ask what is authentically philosophical, if anything, in an investigation on the cognitive architecture underlying the conscious experience of expert athletes. Philosophical inquiry has often tended to investigate consciousness exclusively as a metaphysical category, a fundamental property or principle identified by the conditions of its actual or counterfactual existence within the logic of possible worlds, conditions that in turn are to be inferred through *a priori* reasoning (for example: Lewis, 1980; Searle, 1995; Chalmers, 1996). However, another genuinely philosophical approach is possible, one that investigates the experiential nature of consciousness in real-life scenarios, exploring how situated cognitive agents interact with their world-environments, and how the patterns of this interaction concretely shape the modes of their awareness (Zahavi, 2005; Gallagher and Zahavi, 2008; Marbach, 2010; Thompson, 2014). The reflection offered by this chapter is inspired by the general terms of such an approach: instead of *assuming* consciousness as the mysterious fulcrum of transcendent forces or as an illusion to be dispelled, the goal is to *discover* consciousness as a mundane phenomenon coherent with scientific accounts of the mind, with the naturalistic view that the human psyche is immanent and incarnated, and with the evidence that mental activity is intrinsically intentional, relational, and social.

This approach presupposes a deep entanglement between one's cognitive performances and the normative dimension of one's conscious experience, and searches for a systematic correspondence between the phenomenological and the functional dimension of the mind. This correspondence appears with particularly compelling evidence when we observe the negative modulations of one's consciousness, i.e. the distinctive experiences of unawareness or distraction that, in a specific range of situations, predict a decrease (or, more surprisingly, as we shall see, an increase) of one's cognitive abilities. Real-life cases (in particular, those related to the challenges faced by professional athletes) show that manipulating one's consciousness state modifies her cognitive performances too. This suggests that between phenomenal consciousness and mental function there is actually deep causal codependence, and not only extrinsic correspondence.

Why is the psychology of sportsmen and sportswomen particularly significant for a general inquiry into consciousness? In so far as our cognitive systems are most heavily tested by challenging sport competitions and demanding psychophysical training, sport and performance psychology—the scientific discipline that studies the cognitions, the emotions, and the motivations of athletes—can tell us a few important things about the fundamental architecture of the mind, and especially about the power of its non-conscious components. Sport became a strikingly fecund domain of inquiry for both philosophical psychologists and cognitive scientists: a privileged perspective to study the limits and the potential of the mind 'in action' is offered by the empirical examination of the feats of ability accomplished by sportsmen and sportswomen. During these performances, humans are required to make the most out of cognitive faculties like perception, motor control, action planning, and decision in complex, possibly fast changing, real-life scenarios. Sport performance allows us to observe these cognitive faculties under a magnifying lens because it exposes them when they are both extraordinarily developed, nearing the edge of human perfection, and extraordinarily under stress, nearing the edge of exhaustion and collapse. Cognitive science can use the lens provided by sport psychology to infer the nature of the mental resources needed by well-trained humans to accomplish their superior feats of ability, but its goal is to generalize these observations in order to model the fundamental normative dimensions that define the skilful abilities of all humans.

It is remarkable that the psychologists and the cognitive scientists who work on sport seem naturally inclined towards embodied cognition theory and tend to embrace at least its most general and fundamental tenet (Varela, Thompson and Rosch, 1991; and Wilson and Foglia, 2011): that the processes involving body and environment importantly contribute to shape and scaffold higher forms of cognition. More specifically, this means that perception and action, considered as dynamic forms of coupling with the environment, participate in judgment and decision no less than the information internally stored by or produced in isolation within the central nervous system (see Beilock, 2008; Holt and Beilock, 2006; Beilock and Holt, 2007; Gray, 2014). These tenets resonate deeply with the ecological doctrines that distinctively contributed to the theoretical foundation of sport psychology (Gibson, 1979).

In fact, the performances of expert athletes, in revealing the deep integration and mutual presupposition of perceptual experience, action, and adaptive intelligence, illustrate in the most compelling way the embodied and situated dimension of the cognitive mechanisms that govern skill learning and ability. With regard to the theme of this chapter, referring to the embodied view of cognition is somehow inevitable for an inquiry into the powers of the non-conscious mind, not only by reason of the huge influence that embodied theory exerted on philosophy of mind and cognitive science during the past thirty years, but also because this perspective has inspired the most influential philosophical approaches to skill and performance, some of which are characterized by a frankly anti-intellectualist (Fridland, 2014) and anti-representationist stance (Dreyfus, 2002; Hutto and Sanchez-Garcia, 2007).

The characterization of skill and expertise provided by the first theorists of embodiment (Dreyfus and Dreyfus, 1980) was in substantial accord with the most classical works in sport psychology (Fitts and Posner 1967): experts paradigmatically deliver their skilful performances in the form of unreflective action (i.e. through automatic routines), and expertise itself can be developed only when complex action patterns are automatized (i.e. when they become less demanding in terms of cognitive control). It is only when conscious control and explicit decision are replaced by well-trained habitual responses that the most effective and fastest results become possible. It is in these circumstances that, according to the aforementioned theorists of embodiment, the body shows its own intelligence, a skilfulness that is more primordial than intellectually and verbally educated intelligence: it is when a pre-reflective and prenoetic intelligence takes over that the body shows its inherent capability to effectively attune and respond to worldly contingencies without relying on rules of thumb, conceptual models, rational inferences, explicit instructions, or intellectual representations. Importantly, sport psychologists have ascertained that automated behaviours, despite being produced without explicit control or reflection, are nonetheless adaptive, i.e. richly flexible, dynamic, and not stereotyped (e.g. Beilock et al., 2002a; 2004; 2008; Beilock and Gonso, 2008). Correspondingly, various philosophers — interested in the notions of 'habit' or 'pre-reflective intelligence' — have characterized skill as an intelligent form of unaware or semi-aware responsiveness to the material and perceptual contingencies in which the body is embedded (Dreyfus, 2002; Kelly, 2002; Rietveld, 2008; Brownstein, 2014).

According to these theorists, what makes the live action of an expert fluid and quickly adaptive is the fine-grained arrangement of its dynamic elements and its transient dynamics into a highly coherent and temporally unitary system (Silva et al., forthcoming; Araújo, Davids and Hristovski, 2006). The principles that guide and shape skilful action are complex, holistically interconnected, and truly intelligent because they are situated against a background of tacit know-how that the experts can *use* skilfully but can hardly be explicitly *aware of*

(Radman, 2012). According to this view, the cognitive preconditions of skilful action sit on this irrepresentable background and are infinite: a bottomless abyss of implicit knowledge that is ingrained within the very structure of the living body but that exceeds human verbalization and conceptualization (Wheeler, 2008; Dreyfus, 2012; Cappuccio and Wheeler, 2012). Is this background the real source of the power of the non-conscious? In order to understand if this view is correct, it is useful to discuss the theoretical principles of embodied cognition and their applications to the practical cases addressed by sport psychology.

2. Flow, Choke, and the Phenomenology of Trying Too Hard

I have already mentioned that one of the questions most often debated by sport psychologists is whether the best sport performances can be achieved through conscious or (at least partly) non-conscious modes of perception and action (e.g. Breivik, 2013). In other words, the question is whether sport skills are facilitated or rather disrupted by explicit attention and aware control. This question seems to pose a dilemma: on the one hand, many sport practitioners find it obvious that acting while simultaneously reflecting on their ongoing performance is detrimental and disrupts both action and reflection, probably because they require two different and, to some extent, competing or even mutually exclusive forms of engagement; on the other hand, certain sport practices also suggest that an athlete has to be completely present to her actions, remaining focused and avoiding distracting thoughts, if she wants to perform at the best of her capabilities. Is it the *excess*, then, or rather the *lack* of awareness that damages our expert abilities the most? Is it convenient for an athlete to try to dedicate more or less attentional resources to the execution of her actions in order to succeed? From the point of view of the aprioristic definitions of consciousness often favoured by metaphysicians, these two options are antithetical and mutually exclusive, and that is why philosophers have often tended to oppose them categorically. In truth, a careful examination of the actual relation between conscious awareness and performance, as it emerges in the concrete experience of the athletes, allows for other, more nuanced, solutions.

Optimal performance is usually associated with a particular psychological condition called 'flow' or, colloquially, 'the zone' (Jackson and Csikszentmihalyi, 1999). Flow is described by practitioners as a sustained state of absorption into their skilful performance, a condition of moderate arousal and enjoyment in which challenging tasks are accomplished without apparent effort (Swann, 2016). Some studies suggest that the experience of flow actually correlates with an objective increase of the performance level (Swann *et al.*, 2016). Although not universally endorsed, the conviction that flow is more easily achieved under conditions of implicit or semi-conscious attention is a very old and influential one (often associated with east-Asian doctrines, see Slingerland, 2014). In fact, the belief that the 'effortless' phenomenology of flow is typically accompanied by non-conscious cognitive processing is linked to two somewhat common-sensical notions: that conscious awareness is involved whenever we are 'trying hard' to achieve optimality, and that 'trying too hard' tends to reduce one's capability to act skilfully.

For a long time, the notion of 'trying too hard' has been character-ized as 'overthinking', and it was often associated with the folk psychological assumption that 'thinking' and 'acting' are reciprocally exclusive types of mental acts (Dreyfus and Kelly, 2001). Recently, this contrastive dualism has been criticized for being too vague to help scientific inquiry (see Cappuccio, 2015; in particular Christensen, Sutton and McIlwain 2015). On the other hand, 'trying hard' can be understood in a very different way, as 'over scrutinizing one's own performances', i.e. an excessive focus on movement execution (Carr, 2015). This connotation resonates with a different view that is well represented in sport psychology, and that has received support from a large set of bodies of evidence, including both anecdotal testimonies (Swann *et al.*, 2012), experimental data (Beilock and Carr, 2001), and comprehensive theoretical models (Wulf, 2013).

The current debate on skill acquisition and disruption is informed by the urge to combine the applied know-how of the practitioners with the scientific knowledge of cognitive systems, and this is why today, more than ever before, the study of skill and expertise is a multidisci-plinary endeavour, involving competing coalitions of sport psychol-ogists, philosophical psychologists, and various types of cognitive scientists. It doesn't focus on the opposition of reflection (thought) and performance (action), but invests in more sophisticated distinctions to critically evaluate the role played by non-conscious processing in peak performance. The progress in this field of research inevitably led to various diverging interpretations of this role.

The competition between different interpretative options becomes fierce when they try to explain the debilitative phenomenon commonly called 'choking under pressure', or 'choking effect', or just 'choke' (Beilock, 2011), which in many respects represents the antithesis of flow, and a key element for understanding flow itself. Choke, feared by many sportsmen and sportswomen as one of the most dangerous threats to their professional activity, is usually construed as the expert athletes' incapability to successfully perform actions that would otherwise be routine due to psychological factors such as stress and tension. According to Baumeister (1984), it is associated with a paradox many practitioners are familiar with: the situations that overemphasize the expectation to produce excellent results are those that most likely bring about below-standard performances. This phenomenon is related to a multifaceted complex of cognitive, motivational, and emotional dynamics that have been studied widely, but ultimately without consensus, by sport psychologists (Hill *et al.*, 2010; Mesagno, 2013). Even though the scientific definition of choking may vary significantly across different schools (reviewed by Christensen, Sutton and McIlwain, 2015), and there is no universal agreement about its nature and causes, researchers have tended to focus on the allocation of attentional resources and on the role played by conscious and non-conscious processing (Jackson and Beilock, 2008).

While other explanatory narratives certainly exist (e.g. Carr, 2015; Sutton *et al.*, 2015; Jordet, 2010; Mesagno, Harvey and Janelle, 2012; Mesagno and Hill, 2013), three competing models of choking are particularly relevant to our discussion on consciousness: *distraction* (Wine, 1971), *overload* (often conflated with distraction), and *self-monitoring* (Beilock and Carr, 2001) theories. They impute the negative effects of choking to the intrusion of external competing attentional stimuli, the overflowing of internally available computational resources, or the disruption of fluid habitual motor routines, respectively. Note that, while the causes indicated by these models may tend to overlap, at least in part, it is indispensable to distinguish them conceptually: in fact, distraction is not necessarily caused by overload, whose origin and phenomenology are, in turn, different from self-monitoring.

While certain qualitative studies have been influential (Sutton, 2007; Hill and Shaw, 2013), most of the scientific debate of the last fifteen years orbits around the lines of experimental research that Sian Beilock, Tom Carr, Gabriele Wulf, and their fellow colleagues have conducted on the habitual skills that choking disrupts. The empirical results obtained by in-lab experimentation have provided some compelling arguments in favour of the third model of choking, a broad constellation that comprises all the self-monitoring theories, including preeminently 'self-focus', 'execution focus', and 'explicit control' theories. These empirical results have exposed the limits of the distraction and overload models (without entirely ruling them out), and have corroborated the claim that poor performances in experts are associated with conscious processing, or at least the particular form of consciousness involved in the explicit control of one's own component movements.

The most important arguments favouring self-monitoring theories of choking over the competitors are fourfold.

First of all, various authors (following Beilock and Carr, 2001) maintain that distraction theories are inconsistent with the fact that choking typically occurs in conditions of focused attention (when the athletes try to maintain total concentration on their task), while it normally doesn't affect experts in conditions of split attention (as in concurrent dual tasks, such as performing a golf putt while being asked to simultaneously recognize a given tone).

Secondly, they argue that cognitive overload theories can't account for the fact that choking doesn't normally affect experts who perform in fast action conditions (under speed instructions of the kind 'putt as quickly as possible'), while self-paced action conditions (under attention instructions such as 'putt only when you think you are ready') typically increase their chances of choking (Beilock et al., 2004). In turn, self-monitoring theories account for poor performances by experts in dual-task conditions in which the secondary task demands attentive analysis of the primary motor tasks (e.g. 'tell me when the swing movement is finished'), or under attention instructions that stress the component processes of the performance (e.g. 'keep the knees bent').

Thirdly, self-monitoring theories can better explain the skill-level specificity of choking: split attention and fast action conditions have a negative effect only on expert athletes that perform well-practised sensorimotor skills, while novices typically exhibit an inverted pattern of skill disruption, as they experience a drop in their performance under exactly the conditions that don't damage the experts (i.e. dual tasks that involve neither self-monitoring nor speed instructions). On the contrary, distraction and heavy cognitive loads affect novices as much (or possibly more than) the experts.

Last but not least, theories based on distraction and overload struggle to explain why distractors and multiple tasks are consistently and successfully used in therapeutic interventions on athletes that are particularly prone to choke. Despite the higher cognitive burden, the performance of athletes is improved by any expedient that forces them to stop overscrutinizing the details of their own movements, redirecting their attention to external foci (Jackson et al., 2006; Wulf and Su, 2007).

Taken together, these correlations can be used to argue that the disruption of the fluid course of familiar automatized routines is due to self-monitoring (or, more precisely, to 'execution focus', as elucidated by Carr, 2015), which is the reason why choking is often defined also as 'paralysis by awareness/analysis' (Jackson and Beilock, 2008). The main assumption underlying their theory is that performance anxiety typically solicits self-monitoring in the athletes during competition; self-monitoring, in turn, is likely to disrupt the continuous flow of automatic action by breaking its holistic and highly organic structure into an unnaturally segmented, mechanic array of inflexibly juxtaposed, discrete kinematic components.

However, this explanation is not universally accepted. Various objections have been raised against self-monitoring theories of choking: to begin with, not everybody sees a necessary causal link connecting anxiety to self-monitoring and self-monitoring to a consistent decrease in performance (e.g. Montero, 2014; 2016; in press). Second, it has been pointed out that the small decrease in performance artificially produced in the laboratory has little or nothing to do with the catastrophic effects engendered by choking during real-life competitions, and that the data collected by the experimenters lack ecological validity, as the laboratorial setting inevitably fails to reproduce the pressure-filled environment in which choking typically occurs (Jordet, 2010; Christensen, Sutton and McIlwain, 2014). Third, it has been noted that explicit control is not only possible but indispensable to successfully completing many sport tasks, and that attention to one's own movements is necessary to effectively strategize and take appropriate decisions during complex interactive sport activities (Sutton, 2007).

These objections, in particular the third one, suggest that the self-monitoring theory of choking assumes a simplistic notion of expertise, one that emphasizes the velocity and fluidity of routine skilful actions executed in familiar situations but doesn't recognize that expert skills are also characterized by superior capabilities of decision making, problem solving, and creative improvisation in problematic or unfamiliar situations (Carr, 2015; Cappuccio, 2015). In order to assess the applicability of these objections, it is useful to identify and circumscribe the specific contribution offered by non-conscious processing to embodied cognitive performances.

3. The Role of the Non-Conscious

Whatever opinion one has about flow or choke, it should be quite uncontroversial that a significant component of the cognitive activity necessary for expert athletes to carry out their sport tasks is done at a non-conscious level. The stimuli subconsciously processed by the athletes on the field play a very important role for their sport tasks. There are cases in which subconscious stimuli affect performance level more than if they were explicitly perceived.

For example in soccer, during penalty kicks, the position of the goalkeeper on the goal line affects the penalty taker's choice of the direction in which he will send the ball (penalty takers typically choose to send the ball in the direction of the goal area that is least reachable by, i.e. furthest from, the goalkeeper). Masters, van der Kamp and Jackson (2007) have shown that the preference for one side or the other is stronger when the penalty taker is not consciously aware that the goalkeeper's position is displaced on one side of the goal line, i.e. when the length of displacement from the midpoint is far enough to be subconsciously processed by the penalty taker as significant for his task, but not sufficiently far enough to trigger awareness. The length of the displacement must be small and must be perceived subliminally in order to affect the decision of the penalty taker. One reason could be that, if the displacement of the goalkeeper on the goal line were so salient that the penalty taker became aware of it, then the penalty taker would have reason to think that the goalkeeper is intentionally assuming that position in order to influence the penalty taker's direction choice, strategically soliciting him to send the ball in a predictable direction. This kind of conscious consideration may trigger in the penalty taker explicit inferences capable of neutralizing the otherwise subliminal effect of the solicitations provided by the cue.

But what happens when subliminal cues compete with more supraliminal ones? To clarify whether implicit information exerts an intrinsic influence on the decisions of experts, a follow-up study (B. Noël, van der Kamp and Memmert, 2015) ascertained that subconsciously accessed perceptual information about the asymmetric disposition of the goalkeeper on the goal line can affect the penalty takers' decision of the goal side in a 'rather pervasive' way also when 'many supraliminal sources of information' are available. In these cases, the key variable is whether the expert is acting automatically or deliberately monitoring his own actions. In this experiment, 'implicit influences of the goalkeeper's position on goal side selection were consistently overridden by the (conscious) perception of the direction of the goalkeeper's dive… only if the penalty takers deliberately monitored the goalkeeper's action and the goalkeeper committed early enough for penalty takers to respond.' This means that the actions of experts can be flexible and adjust to context when decisions about the environment are taken nonconsciously, as experts have a remarkable capability to extract rich and useful information from implicit cues when they are performing without self-monitoring.

This result is confirmed by another study, by Bocanegra and Hommel (2014), which infers that explicit monitoring is useful only within a circumscribed scope: cognitive control is beneficial for challenging tasks or in unpredictable situations, but is overall detrimental for routine activity, which tends to be more flexible and adaptive if executed automatically. The authors maintain that 'top-down control can impair and interfere with the otherwise automatic' actions, making behaviour less efficient. This is especially true of contextual scenarios that are appropriate to the execution of routine actions, i.e. environmental circumstances that provide the agent with sufficient familiar information to easily predict the consequences of his actions and hence 'behave on autopilot based on automatic processes alone'.

To test this hypothesis, the subjects of this study were asked to perform a decision task responding to various visual stimuli that varied in a number of features, including shape, size, and colour. Unbeknownst to the volunteers, the colour feature always predicted the correct response. The task was replicated in easy (familiar) conditions in which decisions tended to occur automatically, without explicit reflection, as well as in challenging/unfamiliar scenarios that prompted explicit attention. The statistical distribution of successful and unsuccessful trials showed that the subjects subconsciously picked the useful information from the colour and unconsciously used it to solve tasks, but only when the subjects were not dedicating too much attention to the decision, namely in the 'easy/familiar' condition of the task. On the contrary, when the subjects were 'exerting mental control in a predictable situation, when automatic response was enough', their performances were impaired, suggesting that 'the unneeded mental effort appeared to interfere with what is a perfectly adequate automatic performance.'

This result seems to support the belief that automaticity, the capacity to perform effectively and take decisions without explicit control or reflection, not only statistically correlates with expert performance, it actually causally contributes to the mechanisms of expert behaviour, enabling more accurate and efficacious performances. But this leaves open a question of a normative kind (stressed for example by Gottlieb, 2015), concerning what behaviour can legitimately be considered 'skilful': is adaptive automatic behaviour just a prerogative and a distinctive mark of skilful expertise (i.e. is it only a sufficient condition to be considered an expert, a feature precluded to inexperienced agents); or is it the case that adaptive automatic behaviour is also a necessary and constitutive condition of skilful expertise, in the sense that the capability to subliminally manipulate implicit information offered by predictable task environments is one of the prerequisites of expert action? In order to answer this question I will examine the notion of consciousness involved in skilful performances, first introducing and then following the interpretative grid proposed by Birch *et al.* (in press).

To counter the intellectualist view that sporting skills are not cognitive, Birch *et al.* (in press) defend the claim that sensorimotor activities involving a deep dynamic integration of neuronal and bodily processes, like playing basketball, are not less cognitive and intelligent than a purely contemplative and static task such as solving an equation. Two traditional dogmas must be rejected in order to rediscover the unjustly underestimated embodied dimension of cognition: 'the first is the belief that sporting skills are not knowledge; the other is that sporting skills go on (or should go on) without consciousness' (*ibid.*). Against these dogmas, Birch and colleagues argue that many embodied activities are properly cognitive and intelligent in so far as they involve *knowledge* and *consciousness*. Knowledge itself, according to Birch and colleagues, can be characterized as a defining characteristic, or even a particular form, of consciousness. As such, it pervasively informs the performances of the experts.

4. Defining Consciousness: Nine Dimensions

Birch *et al.* (in press) consider the view that sees skilful activity as necessarily non-conscious misleading and they reject it: while some self-paced closed disciplines put a higher stress on the perfect repetition of standardized sensorimotor schemata, without involving high levels of active control and decision, in all sport disciplines consciousness is a key element for a perfectly executed performance. Building importantly on the philosophical reflection advanced by Breivik (2007; 2013), Birch *et al.* defend this claim analytically, arguing that all the defining features of consciousness are actually preserved in skilled action and even during repetitive routine performance. Skilful action involves both *access* (or 'psychological') *consciousness* ('a mental state which plays a causal/functional role in the production of behavior') and *phenomenal consciousness* ('how it feels' and 'what is it like' to have a certain experience), in accord with the famous distinction introduced by Block (1995). Phenomenal consciousness is the distinctive characteristic of the subjective qualitative mental states experienced by all beings having a body and inhabiting a world (as opposed to 'robots or zombies').

The key functional features of psychological consciousness, on the other hand, are captured by eight dimensions: *awakeness* (which demarcates conscious states from unconscious ones like sleep and coma); *introspection* (the capability to monitor and interrogate both one's own internal states, such as emotions, and physical states, like body posture and motion, including proprioception and interoception); *reportability* (representing in declarative terms events and features of one's experience); *self-consciousness* (the sense of one's personal identity, as distinct from others'); *attention* (the active and selective allocation of cognitive resources for processing online stimuli; it can be wider or narrower, focused or split, etc.); *awareness* (i.e. the sensibility and responsiveness to trigger stimuli; unlike attention, it is a passive and general, i.e. non-selective, form of promptness); *voluntary control* (one's sense of agency, involving both a sense of initiation and a sense of control); and finally *knowledge* (which includes the capabilities of adaptation, explanation, and prediction that allow the agent to efficaciously deal with familiar and unfamiliar situations based on stored information and previous experience).

Each of these nine dimensions is represented to some degree in the typical skilful performance. Most, if not all, are indispensable to sporting activities.

1) Athletes certainly are *phenomenally conscious* when they perform. Their first-hand experience involves qualia, i.e. the private mental states that characterize as intrinsically subjective our percepts, feelings, sensations, and emotions during intense physical training or performance exactly like during any other moment of our lives. In particular conditions, athletes are not fully aware of how the environment affects them (e.g. subliminal cues), or they are not capable of fully reporting on tasks they were immersed in; and yet, this doesn't prove they are not phenomenally conscious, as their performance was always subjectively oriented and coloured by a distinctive ego-centred experience (e.g. certain moods or emotions such as arousal, fear, discomfort, fun, or a complex blend of them). The performance certainly 'felt' in some particular way, qualitatively irreducible to its verbal description.

The study of skilful performances offers good arguments in favour of the real (as opposed to epiphenomenal) status of qualia: not only do qualia colour the athlete's personal experience in a particular manner; they also modify the subpersonal cognitive processes in the athlete's mind. As reminded by Birch *et al.* (in press), this view is well represented in cognitive neuroscience: Edelman (1992, pp. 135–6) doesn't just claim that phenomenal consciousness accompanies automated action; it also contributes to its success, exerting an effective causal impact on 'discriminating sensory perceptions' and subsequently on 'action and behavior, whether we are able to report it or not'.

The intrinsically qualitative dimension of the athlete's experience is not a supplementary but a constitutive and causally relevant component of the cognitive processes that allow optimal attunement of an agent with her world. This explains why all of the eight functional dimensions of psychological consciousness are involved during skilful performances. Their systemic dynamic integration would be harder to explain if we didn't factor in phenomenal consciousness, in that the functioning of the athlete's mind as a highly coherent unitary system is unintelligible without the unifying factor represented by his personal experience.

2) Revisiting the features of psychological consciousness previously mentioned, Birch et al. (in press) note first of all that athletes are certainly *awake* when they perform, as they don't seem asleep or in a trance.

3) Also, athletes often need to *introspect*, as they must be capable of monitoring their body, for example comparing their posture with a pre-given model of their movements (or a choreography) to correct the position and the kinematic of the body parts; also, they may need to control their internal states and attend to their own psychophysical condition in order to detect when emotions, feelings, and somatic sensations deviate from an optimal standard, enacting appropriate compensatory or regulatory responses (for example when they are reaching the dangerous edge of physical exhaustion, sensorial overload, or emotional meltdown).

4) Definitely, athletes can *verbally report* about their performances (for example, football players interviewed after a match), providing elaborate descriptions of specific aspects of their behaviours, decisions, and the circumstances that motivated them.

5) They appear *self-conscious*, as they interact with inanimate objects, the other athletes, and the people in the audience in ways that suggest the stability of their personal identity and the structural persistence of their self.

6) As for *attention*, Birch et al. admit that it can be modulated in very different ways through different sporting disciplines ('focused, selective, spotlight, integrating, divided, alternating, etc.'): the type, the amount, and the robustness of the needed attentional resources may vary significantly between tasks; however, the authors exclude that any sporting discipline could ever be successfully practised by players affected by severe lack of attention, probably because this would prevent concentration.

7) *Awareness* is also inalienable from the experience of an expert player because, regardless of the depth of her absorption in the task at hand, and her capability to isolate herself from external contingencies, she is still sensitive and responsive to a broad perceptual horizon contextually dependent on her action possibilities. Such tight relation to the context is key to staying tuned into the continuous changes in the experiential flow while remaining in 'the zone'.

8) Regarding the *voluntariness* of the actions executed by sport people, the authors remark that they 'are certainly not random or lucky'; rather, they are the result of well-formed intentions. Even when

an agent is not entirely aware of the action she is going to perform, its movements spawn from an act of volition (as opposed to being merely reflexive).

9) Last but not least, Birch et al. argue that sporting skills embody a rich form of knowledge characterized in procedural and not declarative terms, i.e. an operative know-how that informs the body's flexible adaptivity to the environment. Unlike traditional cognitivist approaches, the embodied view of cognition maintains that this knowledge is less accurately characterized in propositional terms than in terms of *motor intentionality*: this concept, developed at first by phenomenologist Maurice Merleau-Ponty, refers to a non-representational form of directedness, or 'aboutness', of the mind that makes it possible for a situated agent to directly perceive the world as a rich landscape of meaningful possibilities of interaction (Merleau-Ponty, 1945; see Dreyfus, 1996, for a discussion of its contemporary relevance). Unlike other forms of cognition, the knowledge incarnated by motor intentionality is not normed by conditions of satisfaction, in that this knowledge is not about possessing a more or less accurate representation of the external world, but conditions of improvement, which define how our actions deviate from an optimal standard, i.e. a certain implicit balance with the world that is only tacitly sensed, never fully represented (Dreyfus, 2002).

While this kind of embodied knowledge is usually associated with automaticity, Birch et al. (in press) find that this association needs to be clarified: the know-how enacted by the body through its motor intentional attunement with the world relies on subpersonal cognitive mechanisms (like those localized in our premotor cortex) that work in an automatic way, i.e. without explicit decisions being possible or necessary; but that doesn't mean the agent is compelled to act in a certain way when she is guided by motor intentionality, as motor intentionality (unlike the reflexes that are pre-programmed by genes or hardwired in the system through associative mechanisms) is a relational and dynamic knowledge involving all the aforementioned elements of consciousness. This is fundamentally different from an automatic predisposition to deliver fixed responses.

5. The Other Side of the Story: Non-Conscious Modulations of Conscious Awareness

I believe this analysis is essentially accurate in revealing how the main components or features of consciousness, conceptually identified by philosophers, are present in the experience of athletes. But the conclusion presented by Birch et al. invites an additional interpretative effort to recognize the specific role played by legitimately non-conscious dynamics during skilful activities.

Birch *et al.* seem to reduce subconscious or semi-conscious experiences, *qua* personal experiences, to the positive realm of consciousness: the point maintained by the authors is that, even when we drive home on 'automatic pilot' (to use the famous example by Armstrong, 1981), we are *not entirely* (and certainly *not reflectively*) *aware*, but we clearly live the qualitative circumstances of driving as a subjective experience. We are still—at some level—aware of the situation, and more or less ready to respond to sudden changes; consciousness of the road conditions may be in its minimal form, but it never completely abandoned us.

While I agree with this positive account, I wonder whether it would be satisfactory if we didn't complement it with an empirically and phenomenologically accurate account of the negative modulations that specifically characterize consciousness during automatic skilful action. Consciousness is present in both reflective and autopilot driving, but is it equally present in both of them? Conscious engagement in automatic skilful action involves a different, negative or privative, dimension. It is a particular form of absence and disengagement that makes it possible to characterize the distinctive phenomenology of this experience, which is both qualitatively (it feels differently) and functionally (it works differently) irreducible to the experience of positive presence associated with fully aware and reflective driving.

The expert's experience and the novice's are paradigmatically different in this respect, as the novice's certainly includes the fully aware and reflective dimension of action, but hardly involves the automated and reflective one. Could we ever capture the specificity of the minimally articulated form of consciousness associated with the absorbed performances of expert athletes, if we excluded negative attributes and privative notions such as diminishment, absence, or disengagement? As I believe the answer to this question is negative, my proposal is to stop treating automatic action as a zombie-like mechanical procedure, considering it instead as a phenomenological *chiaroscuro* in which explicit and implicit intentional acts are intertwined, forming a complex patchwork of foreground and background knowledge that can be fully, only partially, or not at all accessible at the personal level, in different degrees and at a different times (refer to the dynamic model of background knowledge first proposed in Cappuccio and Wheeler, 2012). This involves a particularly nuanced pattern of features that we usually associate either with full awareness or with unconsciousness, respectively. Their dynamic interplay can be recognized by examining once again the nine characteristics of consciousness indicated by Birch *et al.* (in press).

To begin with, athletes absorbed in their skilful performances are certainly *awake* (1) and vigilant, but a more interesting question is whether their state of wakefulness is qualitatively and functionally the same when they are not absorbed in their performance. Without suggesting that their condition were in any significant sense comparable to a kind of 'trance', or a blind repetition of mechanical routines, we cannot underestimate the specificity of their particular way of being awake either. This specificity is attested to by the fluctuations in the performance levels that an expert athlete experiences when the modes and the targets of her attention vary.

For example, we know that the condition of flow importantly correlates with three factors: skill level (Catley and Duda, 1997; Engeser and Rheinberg, 2008; Jackson, 1996), performance, and experience, meaning that expert athletes are more likely than novices to achieve a condition of deep absorption and maximize the benefits of this condition during challenging tasks. The phenomenological specificity of this experience is the reason why many experts believe that flow consists in a peculiar modulation of consciousness: flow increases the success rates in expert sport performance because the cognitive functions recruited for the task at hand are contingent upon the sub- or semi-conscious processing typically favoured by flow. Hence, from a cognitive point of view, being awake while 'in the zone' is very different from being just generically awake.

6. Introspection and the Attentional Focus Effect

The specificity of this condition is reflected by the particular limitations typically suffered by the athletes' capability of *introspection* (2) during peak performance. Importantly, these limitations seem *an enabling condition* to achieve peak performance: obviously, athletes can introspect during their performances, in the sense that they never lose their sheer capability to monitor their postures, movements, and their internal states; but athletes don't exert their full introspective potential when they produce top performances, or when they simply want to be efficacious in their sport activity. The empirical evidence in favour of this claim is abundant.

I am referring to the 'attentional focus effect', largely documented by the experimental activity by Gabriele Wulf and her colleagues (Wulf, Shea and Park, 2001; Wulf and Su, 2007; Wulf, Lauterbach and Toole, 2009; Wulf, 2013), and replicated by several other research groups. This comprehensive body of empirical data corroborates a theory of attention that distinguishes between internal and external foci. The attentional focus effect model goes hand in hand with the theory of choking as 'paralysis by awareness/analysis' or 'explicit monitoring' and seems to back it, although there are differences. The model essentially claims that attending to the constitutive elements of one's performance (paradigmatically observing the components of one's own actions) usually brings about substandard outcomes. Attention to 'internal foci' (typically one's own body parts, or aspects of one's movements) damages the precision and the coordination of one's actions even when they are just familiar routines. On the contrary, paying attention to 'external foci', like the immediate transformational effects of one's action, or their physical targets, doesn't reduce the quality of one's performance and in many cases can actually help improve it, offering a dynamic indicator that co-varies with one's actions without being a component of these actions.

This is useful to strategically regulate one's own action or stabilize one's posture and movements in an indirect manner, without needing to directly attend to internal elements of one's performance. The paradox captured by this theory is that the best way to adjust or correct one's action is to avoid attention to the action itself, exploiting features, processes, or events that are produced by or predicted from one's action. They provide all the information necessary to update one's movements without directly attending to them. This theory and its variants discourages coaches and trainers from using instructions that explicitly describe (or even mention!) body parts, movements, and postures; rather, it suggests focusing on the expected outcomes and the immediate effects of one performance, or on indirect instructions intended to facilitate its achievement.

When it is indispensable to refer to internal components, the internal focus theory recommends the use of metaphorical descriptions or other means of analogical representation (in accord with implicit learning theory: Master, 1992; Komar *et al.*, 2014). If the athlete's attention is recurrently — possibly unintentionally and compulsively — being drawn by internal foci, then an intervention based on distractors and secondary tasks, to direct the attention away from the internal foci, may be necessary. Such forms of intervention prove highly beneficial to mitigate the choking effect (Wulf *et al.*, 2002). According to Wulf (2015; 2016), the model can be generalized to all kinds of expert activities because of the robustness of the attentional foci effect and its very high replicability across very different sport disciplines and typologies of tasks. Moreover, Wulf maintains that this effect is independent of the individual characteristics of the agents (e.g. age, gender, disability) and of their skill level (unlike the phenomenon of choking by awareness/

analysis/self-monitoring, which clearly correlates to expertise and automatization).

While the large amount of empirical data in support of this model cannot be ignored, their universality has been questioned by psychologists and philosophers who defend a different interpretation of attention in skilful performance. Toner, Montero and Moran (2014) and Montero, Toner and Moran (in press) have questioned the internal/external focus distinction, claiming that it is ill-defined. In particular, the apparent vagueness and extreme variability of the notion of external focus would confound the domain of applicability of the theory: external focus seems to indicate alternatively the immediate effects of an action (scoring a goal), an external object or perceptual cue, a quality ('speed, tempo, going fast, swimming hard') of the action, or its metaphorical representation; however, particular features or metaphorical descriptions of an action count as internal, if they refer to a particular isolated body part or movement, as opposed to the movement of the whole body: for example, 'focusing on the angle of one's wrist and focusing on one's wrist and hand taking the shape of a candy cane, for example, would both count as internal foci since they identify isolated bodily parts' (Montero, Toner and Moran, in press). Also, Montero and colleagues impute 'contradictory' claims to Wulf (2016), such as the claim that 'adopting an external focus does not mean that the performer is not aware of her or his body movements' (p. 337), apparently contradicting the claim that an external focus is 'related to the *planning* of the movement, but *has nothing to do with the processing of intrinsic feedback or bodily awareness*'. Montero, Toner and Moran (in press) deem these claims contradictory in that they indicate that 'an external focus both allows and excludes an internal focus (bodily awareness).'

This contradiction might be only apparent. If interpreted within the framework of control theory, the two claims by Wulf seem perfectly compatible with each other and with the attentional focus theory: it is perfectly possible for a forward system to control the online action simply by trying to reduce the difference between the effects produced by the action and the prediction of these effects generated by the system according to an internal model of the action. The system doesn't use detailed information about the execution of the action itself, because this information could impair the online capability of adjustment by abnormally over-intensifying the sensitivity of the prediction-matching system, which in turn would generate a cascade of false-errors capable of inhibiting the correct execution of the planned movements. According to the model presented by Clark (2015, pp. 218–20), predictive processing theory is consistent with the attentional focus effect.

Like the metaphors used in implicit learning, the predictions generated by the system to anticipate the effects of its own actions vicariate the role of an internal focus by replacing it with an external focus. A confirmation of this intuition is offered by the fact that neuro-bio-feedback systems are effective in accelerating training and peaking performance in tasks that require high levels of concentration (as in marksmanship; see Berka *et al.*, 2012). The reason is exactly that such a system allows the agent to adjust her actions without directly attending to them: the system, through portable dynamic EEG technology, receives information from the brainwaves of the agent and, as soon as it registers high levels of relaxation and deep concentration (as in flow), uses tactile feedback to make the agent aware that it is the best time to shoot. The shots executed in these circumstances are in fact statistically more successful. The agent is trained to try to reach and maintain the optimal mental state that the EEG system is programmed to detect; the subject modulates the responses of the system with their sheer will until the conditions for an optimal performance arise. The agent corrects his execution without introspecting (without focusing on his current internal state), simply relying on an indirect representation of the internal state calculated by the artificial system. Attention has an internal focus only in an indirect, mediated, and delayed sense: the human agent alone (unlike the EEG system that is specifically designed for this purpose) wouldn't be naturally capable of directly monitoring his own brain state. The prediction of the optimal brain states at the same time informs and is produced by the execution of his actions, and no internal focus is ever directly targeted in this closed feedback action loop.

A different objection against the attentional focus effect theory comes from Sutton (2007). The accurate analysis of the reports by cricket batsmen provided by this study highlights that self-monitoring and attention to component movements are not just possible and likely to happen on the field without apparent decrease; they are also indispensable to allow the professional players to correctly adjust their posture and maximize the effectiveness of their actions. This qualitative study contributed to undermining the myth that expert athletes don't (*and can't, qua experts*) consciously control the component processes of their actions: it provided theoretical grounds to critically discuss the maximalist interpretations of the self-monitoring theory of choking (i.e. those interpretations that see self-monitoring and explicit control as necessarily and ultimately detrimental to all kind of skilful activities, as suggested by Dreyfus, 2002).

At a first glance, this seems sufficient to question the universal validity of the attentional focus effect theory: certain sport tasks (such as swimming; see Swann, 2016), because of their closed nature, benefit from the explicit awareness of one's posture and the kinematic components, especially for those actions that, in order to be optimized, need very precise trajectories and fine-grained mechanical adjustments that can neither be automatized, regardless of the amount of experience acquired, nor controlled through indirect strategies based on external foci or metaphorical representations. While this is true, it might not be enough to exclude the existence of a trade-off (perhaps just a minor one) between self-monitoring and quality of the execution, as per attentional focus theory. It is possible, in fact, that the execution of certain types of tasks receives both a benefit and a drawback from attending to internal foci, and that the benefits happen to be higher or more visible than the costs. If that is the case, then a general disruptive effect due to the attentional focus might still be present, although it would be largely exceeded — and therefore masked — by the concurrent benefits produced by the finer control, involving a more precise coordination brought about by explicit monitoring.

The existence of a general disruptive effect on performance due to the explicit control of one's own actions cannot be excluded *a priori* if we consider only tasks that cannot be carried out without such explicit control, as there would be no valid term of comparison (control condition) to measure the corresponding performance decrease. On the contrary, various experiments show that the negative effect engendered by internal foci is perspicuous regardless of whether the tasks were carried out with or without self-monitoring (i.e. automatable tasks that don't strictly rely on body awareness; actions that don't require fine adjustments in the posture or calibration of the movements against a pre-given model). This is the kind of setting that allows us to systematically measure the decrease in performances (see Wulf, 2015; 2016).

In this kind of setting, introspective consciousness (at least the kind of introspection that targets internal foci) is still systematically correlated with substantive decreases in performance. For the purposes of our analyses, this means that the sense of body awareness of an expert athlete who performs successfully by relying on his automated embodied routines is modulated in a specific negative way — one that naturally tends (or has been exercised) to avoid or reduce introspective attention.

7. Reportability, Attention, Awareness

Limited capabilities of introspection imply limited capabilities to *verbally report* (3) on one's actions. In fact, the reports provided by experts, in comparison to novices, suggest a less accurate and a coarser-grained capability of recollecting information about events and features experienced by experts absorbed in the execution of their tasks (even if at least some of this information was obviously accessed and processed, probably subconsciously, during the task). This effect has been documented systematically over different disciplines, is particularly evident in closed self-paced tasks, and its intensity seems positively correlated to skill level.

Beilock, Wierenga and Carr (2002) use the expression 'expertise-induced amnesia' to indicate the impoverished episodic memory experienced by skilled athletes: they display a generally reduced 'declarative accessibility', i.e. a lower 'openness to introspection and report'. This phenomenon justifies the inference that experts don't primarily rely on declarative forms of cognition to execute their routine tasks, especially when the sporting task doesn't require the use of working memory, i.e. the kind of memory that specifically manipulates declarative representations: various experiments show that the cognitive system is not overloaded by secondary tasks that use working memory, if the primary task is a sport task that doesn't involve working memory either (e.g. Beilock and Carr 2005; Beilock et al., 2006). This separation is accounted for by 'dual-process' theories that postulate the fundamental autonomy of the different cognitive subsystems that, respectively, use and that don't use working memory (Furley, Schweizer and Bertrams, 2015).

Expertise-induced amnesia doesn't contradict the well-known fact that experts can typically produce copious, rich, and articulated commentaries about their own performances (or others', as done by TV sport commentators): skilled players, rather than relying on the precision of their episodic memory, exploit their superior familiarity with the discipline (procedural memory) to imaginatively reconstruct the circumstances of their actions by elaborating on minimal recollections. This way, experts find it easier to compensate for the lack of precise factual information about the occurrence of particular events during the task or about the presence/absence of particular features associated with their performance.

How are *attention* (4) and *awareness* (5) specifically modulated in the experience of expert athletes? The answer is not simple, as it largely depends on the kind of task and the training done to maximize one's performance. Attention can be more or less narrowly funnelled during the sport task—depending on the nature of the sport discipline itself, and whether the task is closed or interactive, self-contained or continuous, self-paced or synchronized to other agents or event. In any case, the attentional field of the athlete is shaped by the way he concretely inhabits the scenario of his performances, including first of all his expectations about the perceptual environment and the prospects opened by his previous performative knowledge. It is necessary to point out that such know-how, in turn, depends importantly on his direct familiarity with the typical circumstances of the performance, on the motoric and perceptual habits that he has developed through his training, and ultimately on the way his embodied adaptive dispositions (including a primordial background of purely reflexive reactions, emotions, and visceromotor sensations like pain) have been moulded by the repetition of previous trials.

Each way of engaging in interaction with the world brings about its own specific manners of attending to it and perceiving it. Different practical stances and goals literally change how athletes subjectively see, hear, and feel things: for example, the size of a baseball seen by a batter depends on how well he hits (Witt and Proffit, 2005), and the diameter of a golf hole is seen as larger by golf players when they putt successfully (Witt et al., 2008). Similarly, the perception of the width of the gap between two goal posts varies with the facility that the football has been kicked between them (Witt and Dorsch, 2009). Durgin et al. (2012) and Firestone (2013) have questioned the interpretation of these results, claiming that these effects are not produced by an actual modification of one's perceptual patterns, but on retrospective biases. However, Witt (2015) and Foerster, Gray and Cañal-Bruland (2015) have convincingly argued that perception of the action itself, not the judgment about the performance's outcome, is specifically affected by the agent's performative expertise. These experiments manipulate perceptual feedback to dissociate the perceived outcome of one's performance from the objective quality and precision of the performance: this way, they have proven that a subject's tendency to perceive larger-than-real-size targets correlates only to the higher quality and precision of his performance (which in turn depends on the depth of his absorption into the task and, of course, on his skill level), not on the perceived outcome of his performance.

Expert athletes don't see the world like a general agent would see it; their senses are fine-tuned to pick only the information that is relevant to their tasks, information that is intrinsically characterized in an interactional, practical, and motoric way. Both Gibson's theory of affordances (1979) and Merleau-Ponty's theory of motor intentionality (1945), mentioned by Birch, Breivik, and colleagues, efficaciously account for the cognitive mechanisms underlying these empirical results: what makes the experts expert is exactly that they see opportunities for action and practical occasions that invite intervention or careful response, where average observers can only see pragmatically neutral, objective, uncharacterized environments. Neuroscientific investigation confirms both the phenomenological intuition and the ecological principles behind this concept. For example, the discovery of canonical neurons explains why the perception of affordances in the environment (e.g. graspable objects) builds on the capability to execute the corresponding actions (precision grasp): the perceptual system recruits the competences stored in the repertoire of goal-oriented motor actions that is encoded in the premotor cortex of humans' and monkeys' brains (Fadiga et al., 2000; Rizzolatti and Umiltà, 2013).

This specific calibration of the perceptual system, regulated by the motoric proficiency in particular classes of goal-oriented skills, implies that the attentional resources available to the system are recruited selectively, in accord with the ongoing modulation of one's state of conscious awareness, which in turn depends on the task at hand and the subject's practical familiarity with it (see premotor theory of attention; Craighero et al., 1999; Rizzolatti and Craighero, 2010). This selectivity suggests that consciousness in expert athletic performance is characterized by a negative or privative dynamics, as it is modulated through a task-specific or goal-specific type of obstruction or redirection of the available global attentional resources. The creation of areas of disengagement and unawareness is a by-product of this filtering process: while the senses of the athletes are receptive to external, even distant, unanticipated stimuli, and ready to non-consciously process them, their zone of awareness is curved to filter out the percepts that are irrelevant to the success of their goal-oriented motor activity.

8. Self-consciousness, Volition, and Knowledge

If motoric expertise and skill level modulate perception, then they are likely to modulate self-perception too. This is documented by the diminished sense of self-consciousness (i.e. decreased awareness of self and of social evaluation) reported during flow by approximately 30% of athletes, most often combined with other perceptual modifications such as transformation of time (Swann et al., 2012, p. 13; see also Jackson and Marsh, 1996). Why do athletes suppress self-consciousness? One of the reasons might be that, in order to achieve a flow state, athletes have to replace negative thoughts with positive feelings of fulfilment, and fear of the judgment of others can be one of the most disruptive thoughts (Mesagno, Harvey and Janelle, 2012).

In fact, it is well-known that the kind of awareness of oneself solicited by social evaluation (intended as explicit judgment about oneself, or reflection on the possible opinions or beliefs of others) is often detrimental to one's performance and undermines flow: 'stereotype threat', in particular, can significantly affect performances by prompting negative thoughts that undermine confidence and weaken motivation. Even just mentioning negative stereotypes can disrupt performance (Beilock et al., 2006; Beilock, 2011). This is either because recurrent thoughts overload the cognitive system (saturating the working memory) or because they prompt reflective tendencies that increase the chances of compulsive self-monitoring. In either case, successful athletes have reasons to avoid social judgment in order to prevent negative self-thoughts.

Self-consciousness is related to performance also in other ways: i.e. not just thinking of the judgments of others can be detrimental; also observing their poor executions can be too. In various sports, especially team-based ones, expert players are prone to contagion effects that can be explained—for example—in terms of emotional or motoric resonance between players. One player performance is affected negatively or positively by the unsuccessful or successful performances of the other players. Different models try to explain the cause of contagion: for example, it may depend on the (de)motivation arising from the general mood of the affiliated players (Moll, Jordet and Pepping, 2010); or on the fact that the observer's motor system is primed in specific ways when it is recruited to internally simulate the observed actions of others (Ikegami and Ganesh, 2014); or on motivational mechanisms related to the natural tendency to preserve consistency between one athlete's beliefs about himself, about the other players, and about the quality or value of their respective skills (as in Fritz Heider's Balance Theory; see Cartwright and Harary, 1956). In any of these cases, the self-conscious dimension of one athlete is directly modulated by interactions between team players. A psychological link between the actions of the subjects involved in a match exists, and produces remarkable effects during intense competitions, when attention to one's own and others' performances is dramatically increased by the psychological pressure. The boundaries of one's consciousness are not autonomously defined, but depend by the presence of others.

This idea is consistent with the 'shared manifold hypothesis' in social neuroscience (Gallese, 2007): regardless of what we are doing, our brain constantly simulates the behaviour of surrounding agents. Specific brain areas (motor, emotional) become selectively active not only when an agent executes an action or experiences a certain emotion, but also when the agent perceives other agents executing the same action or experiencing the same emotion. This strengthens the idea that team players reciprocally coordinate through an intersubjective network of resonating brain activities. When they successfully attune to one another, they create a supra-personal system of cognitive interactions.

This hypothesis is not enough to suggest that the boundaries of one's self-consciousness could be transcended by intersubjective resonating brain processes: one's subjective experience remains fundamentally, irremediably ego-centred (Zahavi, 2005). It remains, in some sense, not entirely communicable. But this hypothesis certainly invites us to abandon a solipsistic way of understanding consciousness as the isolated prison of personal identity: identity is not the effect of a self-contained system, but a relational phenomenon structurally open to the experience of others, and emerging from a network of conscious interactions. This dynamic is even more marked during challenging skilful activities, when the emotional and cognitive attunement between agents is key, for the better or for the worse, in determining the outcome of their individual actions in a collective context.

Sense of agency, another key feature in the phenomenology of self-consciousness, is also peculiarly modified in skilful activities. Sense of agency is crucial because our capability to experience our own actions as voluntary or involuntary (8) depends on it: according to control theory, the feeling of being an agent, i.e. the one who intentionally initiated a certain movement, derives from a comparison of the sensory predictions generated when a motor command is sent to initiate an action with the sensorial feedback actually produced by the execution of that action. This is how the system distinguishes between endogenous and exogenous movements, informing the agent at once about the actual origin of the ongoing changes in the perceptual flow and about the boundaries between internal and external world (e.g. it tells me whether the room seems to be spinning because my body is rotating or because the walls are actually moving around me).

Now, there is evidence that sense of agency is qualitatively modified during situations of flow involving deep absorption. While athletes' capabilities of identifying the actions that they have voluntarily started is not substantially modified (unlike in schizophrenic patients), the subjective experience associated with this function is modulated in a characteristic way. Expert athletes—especially in disciplines that emphasize fast reactivity and appropriate responsiveness to an opponent's actions, like martial arts—often describe their movements as springing spontaneously, seemingly without explicit intention or active volition, without thought or decision being necessary or possible: a situation in which the difference between activity and passivity, moving or being moved, seems to dissolve.

The essence of this phenomenon is paradigmatically epitomized by martial artist Bruce Lee's famous statement, 'I do not hit, it hits all by itself' (whose speculative meaning, clearly influenced by traditional Taoist doctrines, is developed in Lee's sport and philosophical treatises;

see Thomas, 2005). But a similar subjective characterization also emerges from self-reports produced by expert athletes when asked to describe the phenomenological specificity of their engagement in situations of flow (Swann, 2016). One of the most common phenomena, reported by almost 75% of athletes is 'action-awareness merging' (i.e. total absorption, or 'feeling completely at one with the activity' during the flow state).

A recent study with elite golf players (Swann et al., 2016), documents that flow corresponds both to a feeling of calm/relaxation and a sense of total control, often described in terms of semi-automaticity of movement, effortlessness, and enjoyment. The athlete often describes her attitude in this condition in terms of 'letting it happen', as opposed to more effortful modes of engagement defined as 'making it happen', an expression that indicates an increased sense of alertness and effortful control. The complete awareness of the intense challenge faced by the athlete during flow, combined with the high level of performance typically achieved in this condition, results in a positive sense of exceeding one's expectations and 'surfing' problematic contingencies with ease. Crucially, this condition is experienced as a feeling of relief, a mitigated sense of effort, and a greater capability to rely on automatic subconscious routines that don't require voluntary decision and reflection.

Sense of agency and voluntary action don't disappear in this condition (the subject is still aware of being in charge), but their experiential valence is modified: the subject does not experience his involvement as a way to transform the situation, but as a way to respond to it, compensating the contingencies to find a balance with them. Volition (or intention) is not an antagonistic force emanating from the core of the self to oppose the world and alter it according to a plan; rather, volition is encountered during flow as an effect, a surface tension that vibrates through the dialectic of internal and external forces, emerging from the encounter of trained predispositions and challenging contingencies.

Last but not least, also knowledge (9) is distinctively modulated during flow: motor intentionality certainly constitutes a kind of knowledge, as pointed out by Birch et al. (1996); however, Merleau-Ponty (1945) also points out that it differs at root from conceptual knowledge. The know-how embodied by motor intentionality is never entirely accessible by aware consciousness, not even through careful introspection (Kelly, 2000; 2002). For example, Merleau-Ponty famously noted that experienced stenographers perfectly know where to find each key on a keyboard while they are typing, but wouldn't be able to indicate the overall positions of the keys on a diagram of the keyboard if asked to do so. This dissociation between a 'spatiality of situation' (mapping the environment in terms of opportunities for action) and a 'spatiality of position' (mapping the environment in terms of objective metric relations) is confirmed by clinical neuropsychology, which has documented how brain damage that affects the capability to execute particular goal-oriented actions doesn't necessarily affect the capability to conceptually decompose those actions into neutral geometrical elements, and vice versa. This dissociation has been variously interpreted, for example through the two visual pathways theory (Milner and Goodale, 2008). The fact that two different systems superintend these two different types of knowledge can explain why we intuitively know how to manipulate objects and how to recognize the environmental circumstances that afford their manipulation, even if our capability to report on our motor-intentional dispositions is disturbed. In fact, explicit judgment through a declarative stance involves an *a posteriori* reconstruction that can and often does rely on motor-intentional competences, but that is not motor-intentional in itself.

This explains why athletes are not always capable of telling us 'how they did it', especially when their motor-intentional abilities (silently stored as bodily readiness to act) made them capable of improvising new actions and unexpected strategies that are not reducible to verbal instructions or models received during training. If these two types of knowledge are not always equally accessible (because motor-intentional schemata are not available to analysis through conscious control and introspection), this is not only because the two systems were contingently separated; it may well be that there is a radical incommensurability between the two kinds of information encoded as motor intentionality and as cognitive intentionality, respectively. While know-that is based on detached, explicit representation of facts, know-how is intrinsically interactional and non-contentful in nature, i.e. devoid of propositional or figurative content; the former's normative structure is defined by conditions of satisfaction, the later by conditions of improvement (Dreyfus, 2002).

Due to its fundamentally non-representational nature, motor intentionality cannot be reduced to a collection of facts or bits of information expressed in a declarative format; motor intentionality builds on a direct practical and interactive acquaintance with the world, and emerges on an holistic, unquantifiable background of bodily and material preconditions that are essentially 'transcendent', i.e. always exceeding our possibilities of verbalization and conceptualization. If this picture is correct, then, although motor intentionality scaffolds the higher levels of our cognitive engagement with the world (verbalization, conceptualization), and provides the embodied root of our intellectual categories, there would be no assurance that motor-intentional knowledge could ever be entirely accessible through conscious awareness, reflection, or introspection (and, even if it were accessible, that would happen only reconstructively, *a posteriori*). That is why, even if deploying intellectual faculties can certainly help during sport performance, higher forms of cognition and explicit thought are never enough to accomplish excellent results and cannot replace embodied expertise.

9. Conclusions

I think the analysis of the components of consciousness proposed by Birch *et al.* (in press) efficaciously proves that athletes are not zombies without qualitative experiences or automata deprived of decision, reflection, or control. Both the phenomenology of consciousness (at the personal level) and its functional components (at the subpersonal level) play a crucial role in the mental life of athletes when they engage with challenging tasks. We can consider this role as legitimately 'cognitive', in that it importantly involves embodied forms of cognition and emotional and practical intelligence.

At the same time, a careful analysis of performance under pressure and skilful activity in a flow state reveals that consciousness and its functional components are modulated in highly specific ways, with a complex interplay of background and foreground elements. This specificity reveals some important features of the basic architecture of our minds, including how we acquire and deploy complex sensorimotor skills governed by embodied forms of intelligence (e.g. motor intentionality). These basic forms of cognition are intelligent because they allow rich and flexible forms of adaptive behaviour and very fine-grained capabilities of perceptual discrimination, practical categorization, and responsiveness to context; at the same time, our analysis confirms that these forms of basic cognition are largely or fundamentally automatic and pre-reflective. Not only in the sense that they operate at a subpersonal level through the work of mechanisms situated well below the level of our consciousness, as correctly maintained by Birch and colleagues; but also in the sense that, whenever consciousness is involved in skilful performances, the fundamental features of conscious experience are specifically modified or partly suppressed. When the best performances have to be achieved, some particular forms of our awareness are adumbrated or selectively filtered to minimize attention to oneself and maximize the allocation of resources to the goals or expected consequences of the agent's actions, in order to increase her readiness and boost her reactiveness to the circumstances.

I hope I have been able to persuasively argue that, studying the cognitive processes underlying the performances of the athletes, it is necessary to take seriously both the fact that athletes are aware and conscious and the fact that their conscious awareness during peak performance is characterized by the selective suppression or attenuation of particular patterns of perception and attention.

Acknowledgments

This research has been sponsored by a UPAR grant by UAE National Research Foundation 'Sport and Brain Science: Technological Applications for Peaking Performances' (grant code: 31H087-UPAR (3) 2014). The author is thankful to Shaun Gallagher, Jesse Prinz, and Dan Zahavi for commenting on an early version of this paper when it was presented for the first time at the Inter-University Center in Dubrovnik, in April 2013, and to Giacomo Bertacchi, Jens Birch, Geir Jordet, and Gunjan Khera for kindly reviewing its latest version.

References

Araújo, D., Davids K. & Hristovski, R. (2006) The ecological dynamics of decision making in sport, *Psychology of Sport and Exercise*, **7**, pp. 653–676.

Armstrong, D. (1981) What is consciousness?, in *The Nature of Mind, and Other Essays*, pp. 55–67, Ithaca, NY: Cornell University Press.

Baumeister, R.F. (1984) Choking under pressure: Self-consciousness and paradoxical effects of incentives on skillful performance, *Journal of Personality and Social Psychology*, **46** (3), pp. 610–620.

Beilock, S.L. (2008) Beyond the playing field: Sport psychology meets embodied cognition, *International Review of Sport and Exercise Psychology*, **1** (1), pp. 19–30.

Beilock, S.L. (2011) *Choke*, Boston, MA: Little, Brown Book Group.

Beilock, S.L. (2015) *How the Body Knows Its Mind: The Surprising Power of the Physical Environment to Influence How You Think and Feel*, New York: Atria Books.

Beilock, S.L. & Carr, T.H. (2001) On the fragility of skilled performance: What governs choking under pressure?, *Journal of Experimental Psychology: General*, **130**, pp. 701–725.

Beilock, S.L., Wierenga, S.A. & Carr, T.H. (2002) Expertise, attention, and memory in sensorimotor skill execution: Impact of novel task constraints on dual-task performance and episodic memory, *The Quarterly Journal of Experimental Psychology*, **55A** (4), pp. 1211–1240.

Beilock, S.L., Bertenthal, B.I., McCoy, A.M. & Carr, T.H. (2004) Haste does not always make waste: Expertise, direction of attention, and

speed versus accuracy in performing sensorimotor skills, *Psychonomic Bulletin & Review*, **11** (2), pp. 373–379.
Beilock, S.L. & Carr, T.H. (2004) From novice to expert performance: Memory, attention and the control of complex sensorimotor skills, in Williams, A.M., Hodges, N.J, Scott, M.A. & Court, M.L.J. (eds.) *Skill Acquisition in Sport: Research, Theory and Practice*, pp. 309–328, London: Routledge.
Beilock, S.L. & Carr, T.H. (2005) When high-powered people fail: Working memory and 'choking under pressure' in math, *Psychological Science*, **16** (2), pp. 101–105.
Beilock, S.L. & Holt, L.E. (2007) Embodied preference judgments: Can likeability be driven by the motor system?, *Psychological Science*, **18** (1), pp. 51–57.
Beilock, S.L., Bertenthal, B.I., Hoeger, M. & Carr, T.H. (2008) When does haste make waste? Speed-accuracy tradeoff, skill level, and the tools of the trade, *Journal of Experimental Psychology*, **14** (4), pp. 340–352.
Beilock, S.L. & Gonso, S. (2008) Putting in the mind versus putting on the green: Expertise, performance time, and the linking of imagery and action, *The Quarterly Journal of Experimental Psychology*, pp. 1–13.
Birch, J.E., Breivik, G. & Fusche Moe, V. (in press) Knowledge, consciousness and sporting skills, in Cappuccio, M. (ed.) *MIT Press Handbook of Embodied Cognition and Sport Psychology*, Cambridge MA: MIT Press.
Block, N. (1995) On a confusion about a function of consciousness, *The Behavioral and Brain Sciences*, **18**, pp. 227–287.
Bocanegra, B.R. & Hommel, B. (2014) When cognitive control is not adaptive, *Psychological Science*, **25** (6), pp. 1249–1255.
Breivik, G. (2007) Skillful coping in everyday life and in sport; a critical examination of the views of Heidegger and Dreyfus, *Journal of Philosophy of Sport*, **34** (2), pp. 116–134.
Breivik, G. (2013) Zombie-like or superconscious?: A phenomenological and conceptual analysis of consciousness in elite sport, *Journal of the Philosophy of Sport*, **40**, pp. 85–106.
Brownstein, M. (2014) Rationalizing flow: Agency in skilled unreflective action, *Philosophical Studies*, **168** (2), pp. 545–568.
Cappuccio, M. (ed.) (2015) Unreflective action and the choking effect, *Phenomenology and the Cognitive Sciences*, **14** (2).
Cappuccio, M. (ed.) (in press) *MIT Press Handbook of Embodied Cognition and Sport Psychology*, Cambridge, MA: MIT Press.
Cappuccio, M. & Wheeler, M. (2012) Ground-level intelligence: Inter-context frame problem and dynamics of the background, in Radman, Z. (ed.) *Knowing without Thinking: Mind, Action, Cognition*

and the Phenomenon of the Background, Besingstoke: Palgrave Macmillan.

Cartwright, D. & Harary, F. (1956) Structural balance: A generalization of Heider's theory, *Psychological Review*, **63** (5), pp. 277–293.

Catley, D. & Duda, J.L. (1997) Psychological antecedents of the frequency and intensity of flow in golfers, *International Journal of Sport Psychology*, **28** (4), pp. 309–322.

Chalmers, D.J. (1996) *The Conscious Mind: In Search of a Fundamental Theory*, New York: Oxford University Press.

Christensen, W., Sutton, J. & McIlwain, D.J.F. (2014) Putting pressure on theories of choking: Towards an expanded perspective on breakdown in skilled performance, *Phenomenology and the Cognitive Sciences*, **14** (2), pp. 253–293.

Christensen, W., Sutton, J. & McIlwain, D.J.F. (2016) Cognition in skilled action: Meshed control and the varieties of skill experience, *Mind and Language*, **31** (1), pp. 37–66.

Clark, A. (2015) *Surfing Uncertainty: Prediction, Action, and the Embodied Mind*, Oxford: Oxford University Press.

Craighero, L., Fadiga, L., Rizzolatti, G. & Umiltà, C. (1999) Action for perception: A motor-visual attentional effect, *Journal of Experimental Psychology: Human Perception and Performance*, **25** (6), pp. 1673–1692.

Davids, K., Araújo, D., Hristovski, R., Passos, P. & Chow, J.Y. (2012) Ecological dynamics and motor learning design in sport, in Williams, M. & Hodges, N. (eds.) *Skill Acquisition in Sport: Research, Theory & Practice*, 2nd ed., pp. 112–130, London: Routledge.

Dreyfus, H.L. (1996) The current relevance of Merleau-Ponty's theory of embodiment, *The Electronic Journal of Analytic Philosophy*, **4**.

Dreyfus, H.L. (2002) Intelligence without representation—Merleau-Ponty's critique of mental representation, *Phenomenology and the Cognitive Sciences*, **1**, pp. 367–383.

Dreyfus, H.L. (2008) Why Heideggerian AI failed and how fixing it would require making it more Heideggerian, in Husbands, P., Holland, O. & Wheeler, M. (eds.) *The Mechanical Mind in History*, pp. 331–371, Cambridge, MA: MIT Press.

Dreyfus, H.L. (2012) The mystery of the background qua background, in Radman, Z. (ed.) *Knowing without Thinking: The Background in Philosophy of Mind*, pp. 1–10, Basingstoke: Palgrave Macmillan.

Dreyfus, S.E. & Dreyfus, H.L. (1980) *A Five-Stage Model of the Mental Activities Involved in Directed Skill Acquisition*, Washington, DC: Storming Media.

Durgin, F.H., Klein, B., Spiegel, A., Strawser, C.J. & Williams, M. (2012) The social psychology of perception experiments: Hills, backpacks, glucose, and the problem of generalizability, *Journal of Experimental Psychology: Human Perception and Performance*, **38**, p. 1582.

Edelman, G. (1992) *Bright Air, Brilliant Fire: On the Matter of the Mind*, New York: Basic Books.

Engeser, S. & Rheinberg, F. (2008) Flow, performance and moderators of challenge-skill balance, *Motivation and Emotion*, **32**, p. 158.

Fadiga, L., Fogassi, L., Gallese, V. & Rizzolatti, G. (2000) Visuomotor neurons: Ambiguity of the discharge or 'motor' perception?, *International Journal of Psychophysiology*, **35** (2–3), pp. 165–177.

Firestone, C. (2013) How 'paternalistic' is spatial perception? Why wearing a heavy backpack doesn't and couldn't make hills look steeper, *Perspectives on Psychological Science*, **8**, pp. 455–473.

Fitts, P.M. & Posner, M.I. (1967) *Human Performance*, Belmont, CA: Brooks/Cole.

Foerster, A., Gray, R. & Cañal-Bruland, R. (2015) Size estimates remain stable in the face of differences in performance outcome variability in an aiming task, *Consciousness and Cognition*, **33**, pp. 47–52.

Freeman, W.J. (1991) The physiology of perception, *Scientific American*, **264**, pp. 78–85.

Fridland, E. (2014) They've lost control: Reflections on skill, *Synthese*, **191**, art. 2729.

Furley, P., Schweizer, G. & Bertrams, A. (2015) The two modes of an athlete: Dual-process theories in the field of sport, *International Review of Sport and Exercise Psychology*, **8** (1), pp. 106–124.

Gallagher, S. & Zahavi, D. (2008) Phenomenological approaches to self-consciousness, in *Stanford Encyclopedia of Philosophy*, https://plato.stanford.edu/entries/self-consciousness-phenomenological/.

Gallese, V. (2007) The shared manifold hypothesis: Embodied simulation and its role in empathy and social cognition, in Farrow, T.F.D. & Woodruff, P.W.R. (eds.) *Empathy in Mental Illness*, pp. 448–472, Cambridge: Cambridge University Press.

Gibson, J.J. (1979) *The Ecological Approach to Visual Perception*, Boston, MA: Houghton Mifflin.

Gottlieb, G. (2015) Know-how, procedural knowledge, and choking under pressure, *Phenomenology and the Cognitive Sciences*, **14** (2), pp. 361–378.

Gray, R. (2014) Embodied perception in sport, *International Review of Sport and Exercise Psychology*, **7** (1), pp. 72–86.

Hill, D.M. & Shaw, G. (2013) A qualitative examination of choking under pressure in team sport, *Psychology of Sport and Exercise*, **14** (1), pp. 103–110.

Hill, D.M., Hanton, S., Matthews, N. & Fleming, S. (2010) Choking in sport: A review, *International Review of Sport and Exercise Psychology*, **3** (1), pp. 24–39.

Holt, L.E. & Beilock, S.L. (2006) Expertise and its embodiment: Examining the impact of sensorimotor skill expertise on the representation of action-related text, *Psychonomic Bulletin & Review*, **13** (4), pp. 694–701.

Hutto, D.D. & Sánchez-García, R. (2015) Choking REctified: Embodied expertise beyond Dreyfus, *Phenomenology and the Cognitive Sciences*, **14** (2), pp. 309–331.

Ikegami, T. & Ganesh, G. (2014) Watching novice action degrades expert motor performance: Causation between action production and outcome prediction of observed actions by humans, *Scientific Reports*, **4**, 6989.

Jackson, R.C. & Beilock, S. (2008) Attention and performance, in Farrow, D., Baker, J. & MacMahon, C. (eds.) *Developing Elite Sports Performers: Lessons from Theory and Practice*, New York: Routledge.

Jackson, R.C., Ashford, K.J. & Norsworthy, G. (2006) Attention focus, dispositional reinvestment, and skillful motor performance under pressure, *Journal of Sport and Exercise Psychology*, **28**, pp. 49–68.

Jackson, S.A. (1996) Toward a conceptual understanding of the flow experience in elite athletes, *Research Quarterly for Exercise and Sport*, **67** (1), pp. 76–90.

Jackson, S.A. & Marsh, H.W. (1996) Development and validation of a scale to measure optimal experience: The Flow State Scale, *Journal of Sport and Exercise*, [Online], http://fitnessforlife.org/AcuCustom/Sitename/Documents/DocumentItem/8983.pdf

Jackson, S.A. & Csikszentmihalyi, M. (1999) *Flow in Sports*, Champaign, IL: Human Kinetics.

Jordet, G. & Hartman, E. (2008) Avoidance motivation and choking under pressure in soccer penalty shootouts, *Journal of Sport & Exercise Psychology*, **30**, pp. 450–457.

Kelly, S.D. (2000) Grasping at straws: Motor intentionality and the cognitive science of skillful action, in *Essays in Honor of Hubert Dreyfus*, vol. II, Cambridge, MA: MIT Press.

Kelly, S.D. (2002) Merleau-Ponty on the body, *Ratio*, **15** (4), pp. 376–391.

Komar, J., Chow, J.-Y., Chollet, D. & Seifert, L. (2014) Effect of analogy instructions with an internal focus on learning a complex motor skill, *Journal of Applied Sport Psychology*, **26** (1), pp. 17–32.

Lewis, D. (1980) Mad pain and Martian pain, in Block, N. (ed.) *Readings in the Philosophy of Psychology*, vol. I, pp. 216–222, Cambridge, MA: Harvard University Press.

Marbach, E. (2010) Is there a metaphysics of consciousness without a phenomenology of consciousness? Some thoughts derived from Husserl's philosophical phenomenology, *Royal Institute of Philosophy Supplement*, **67**, pp. 141–154.

Masters, R.S.W. (1992) Knowledge, knerves and know-how: The role of explicit versus implicit knowledge in the breakdown of a complex motor skill under pressure, *British Journal of Psychology*, **83** (3), pp. 343–358.

Masters, R.S.W., van der Kamp, J. & Jackson, R.C. (2007) Imperceptibly off-center: Goalkeepers influence penalty-kick direction in soccer, *Psychological Science*, **18** (3), pp. 222–223.

Merleau-Ponty, M. (1945) *Phénoménologie de la perception*, Paris: Gallimard.

Mesagno, C. (ed.) (2013) Performance under pressure, *International Journal of Sport Psychology*, **44** (4).

Mesagno, C., Harvey, J.T. & Janelle, C.M. (2012) Choking under pressure: The role of fear of negative evaluation, *Psychology of Sport and Exercise*, **13** (1), pp. 60–68.

Mesagno, C. & Hill, D.M. (2013a) Choking under pressure debate: Is there chaos in the brickyard?, *International Journal of Sport Psychology*, **44** (4), pp. 288–293.

Mesagno, C. & Hill, D.M. (2013b) Definition of choking in sport: Reconceptualization and debate, *International Journal of Sport Psychology*, **44** (4), pp. 267–277.

Milner, A.D. & Goodale, M.A. (2008) Two visual systems re-viewed, *Neuropsychologia*, **46** (3), pp. 774–785.

Moll, T., Jordet, G. & Pepping, G.-J. (2010) Emotional contagion in soccer penalty shootouts: Celebration of individual success is associated with ultimate team success, *Journal of Sports Sciences*, **28** (9), pp. 983–992.

Montero, B.G. (2014) Is monitoring one's actions causally relevant to choking under pressure?, *Phenomenology and the Cognitive Sciences*, **14** (2), pp. 379–395.

Montero, B.G. (2016) *Thought in Action*, Oxford: Oxford University Press.

Montero, B.G., Toner, J. & Moran, A. (in press) Questioning the breadth of the attentional focus effect, in Cappuccio, M. (ed.) *MIT Press Handbook of Embodied Cognition and Sport Psychology*, Cambridge, MA: MIT Press.

Noël, B., van der Kamp, J. & Memmert, D. (2015) Implicit goalkeeper influences on goal side selection in representative penalty kicking tasks, *PLoS ONE*, **10** (8), e0135423.

Radman, Z. (ed.) (2012) *Knowing without Thinking: Mind, Action, Cognition and the Phenomenon of the Background*, Basingstoke: Palgrave Macmillan.

Rietveld, E. (2008) Situated normativity: The normative aspect of embodied cognition in unreflective action, *Mind*, **117** (468), pp. 973-1001.

Rizzolatti, G., Fogassi, L. & Gallese, V. (2001) Neurophysiological mechanisms underlying understanding and imitation of action, *Nature Reviews Neuroscience*, **2**, pp. 661-670.

Rizzolatti, G. & Craighero, L. (2010) Premotor theory of attention, *Scholarpedia Journal*, [Online], http://scholarpedia.org/article/Premotor_theory_of_attention

Rizzolatti, G. & Umiltà, M.A. (2013) Canonical neurons, in Runehov, A.L.C. & Oviedo, L. (eds.) *Encyclopedia of Sciences and Religions*, pp. 305-305, Dordrecht: Springer.

Royal Institute of Philosophy Supplement, **85** (67), pp. 141-154.

Searle, J. (1995) *The Mystery of Consciousness*, New York: New York Review of Books.

Silva, P., Kiefer, A., Riley, M. & Chemero, A. (in press) Trading perception and action for complex cognition: Application of ecological theory to the design of interventions for skill learning, in Cappuccio, M. (ed.) *MIT Press Handbook of Embodied Cognition and Sport Psychology*, Cambridge, MA: MIT Press.

Slingerland, E. (2014) *Trying Not to Try: The Art and Science of Spontaneity*, New York: Crown.

Sutton, J. (2007) Batting, habit and memory: The embodied mind and the nature of skill, *Sport in Society*, **10** (5), pp. 763-786.

Swann, C. (2016) Flow in sport, in Harmat, L., Andersen, F. Ø. , Ullén, F., Wright, J. & Sadlo, G. (eds.) *Flow Experience: Empirical Research and Applications*, pp. 51-64, Berlin: Springer.

Swann, C., Keegan, R.J., Piggott, D. & Crust, L. (2012) A systematic review of the experience, occurrence, and controllability of flow states in elite sport, *Psychology of Sport and Exercise*, **13** (6), pp. 807-819.

Swann, C., Keegan, R., Crust, L. & Piggott, D. (2016) Psychological states underlying excellent performance in professional golfers: 'Letting it happen' vs. 'making it happen', *Psychology of Sport and Exercise*, **23**, pp. 101-113.

Thomas, B. (2005) *Bruce Lee: Fighting Words*, Madison, WI: Frog Books.

Thompson, E. (2007) *Mind in Life: Biology, Phenomenology, and the Sciences of Mind*, Cambridge, MA: Harvard University Press.

Thompson, E. (2014) Dreamless sleep, the embodied mind, and consciousness, *Open MIND*, Johannes Gutenberg Universität Mainz.

Toner, J., Montero, B.G. & Moran, A. (2014) Considering the role of cognitive control in expert performance, *Phenomenology and the Cognitive Sciences*, **14** (4), pp. 1127–1144.

Toner, J., Montero, B.G. & Moran, A. (2015) The perils of automaticity, *Review of General Psychology: Journal of Division 1, of the American Psychological Association*, **19** (4), pp. 431–442.

Toner, J., Montero, B.G. & Moran, A. (2016) Reflective and prereflective bodily awareness in skilled action, *Psychology of Consciousness: Theory, Research, and Practice*, [Online], http://dx.doi.org/10.1037/cns0000090

Varela, F., Thompson, E. & Rosch, E. (1991) *The Embodied Mind: Cognitive Science and Human Experience*, Cambridge, MA: MIT Press.

Wheeler, M. (2008) Cognition in context: Phenomenology, situated robotics and the frame problem, *International Journal of Philosophical Studies*, **16** (3), pp. 323–349.

Wilson, R.A. & Foglia, L. (2011) Embodied cognition, in *Stanford Encyclopedia of Philosophy*, https://plato.stanford.edu/entries/emobodied-cognition/.

Wine, J. (1971) Test anxiety and direction of attention, *Psychological Bulletin*, **76**, pp. 92–104.

Witt, J.K. (2015) Awareness is not a necessary characteristic of a perceptual effect: Commentary on Firestone (2013), *Perspectives on Psychological Science*, **10**, pp. 865–872.

Witt, J.K. & Proffitt, D.R. (2005) See the ball, hit the ball: Apparent ball size is correlated with batting average, *Psychological Science*, **16**, pp. 937–938.

Witt, J.K., Linkenauger, S., Bakdash, J. & Proffitt, D. (2007) Golf performance makes the hole look as big as a bucket or as small as a dime, *Journal of Vision*, **7** (9), p. 291.

Witt, J.K. & Dorsch, T.E. (2009) Kicking to bigger uprights: Field goal kicking performance influences perceived size, *Perception*, **38** (9), pp. 1328–1340.

Wulf, G. (2013) Attentional focus and motor learning: A review of 15 years, *International Review of Sport and Exercise Psychology*, **6** (1), pp. 77–104.

Wulf, G. (2015) An external focus of attention is a conditio sine qua non for athletes: A response to Carson and Collins (2015), *Journal of Sports Sciences*, **34** (13), pp. 1293–1295.

Wulf, G. (2016) Why did Tiger Woods shoot 82? A commentary on Toner and Moran (2015), *Psychology of Sport and Exercise*, **22**, pp. 337–338.

Wulf, G., Lauterbach, B. & Toole, T. (1999) Learning advantages of an external focus of attention in golf, *Research Quarterly for Exercise and Sport*, **70**, pp. 120–126.

Wulf, G., Shea, C.H. & Park, J.-H. (2001) Attention in motor learning: Preferences for and advantages of an external focus, *Research Quarterly for Exercise and Sport*, **72**, pp. 335–344.

Wulf, G., McConnel, N., Gärtner, M. & Schwarz, A. (2002) Enhancing the learning of sport skills through external-focus feedback, *Journal of Motor Behavior*, **34** (2), pp. 171–182.

Wulf, G. & Su, J. (2007) An external focus of attention enhances golf shot accuracy in beginners and experts, *Research Quarterly for Exercise and Sport*, **78**, pp. 384–389.

Zahavi, D. (2005) *Subjectivity and Selfhood: Investigating the First-Person Perspective*, Cambridge, MA: MIT Press.

Zahavi, D. (2012) Mindedness, mindlessness and first-person authority, in Shear, J. (ed.) *Mind, Reason and Being-in-the-World: The Mcdowell-Dreyfus Debate*, London: Routledge.

Sam Wilkinson
and Charles Fernyhough

Auditory Verbal Hallucinations and Inner Speech
A Predictive Processing Perspective

1. Introduction

Inner speech is a pervasive feature of our conscious lives.[1] But what is inner speech, and what happens in unconscious processing that makes it the conscious experience that it is? A clue to answering this can be found in cases where the mechanisms that produce inner speaking behave unusually. In this paper, we suggest an account of a specific instance of this, namely, a particular subtype of auditory verbal hallucination (AVH), and draw some lessons about the processes that underlie normal inner speech.

An AVH involves, roughly, the experience of hearing a voice in the absence of anyone actually speaking. As a phenomenon, it varies enormously in a number of ways: in how it presents itself phenomenologically, in terms of the context in which it occurs, and arguably in what causes it. This has lead some theorists (Jones, 2010; Wilkinson, 2014; Smailes *et al.*, 2015) to claim that there are subtypes of AVHs, and that these amount to fundamentally different phenomena, underpinned by different mechanisms and different aetiologies. Three identified subtypes are memory-based, inner speech-based and hypervigilance hallucinations. As the names suggest, the 'raw materials' for memory and inner speech-based hallucinations are episodic memories and episodes of inner speech respectively. In contrast, hypervigilance hallucinations involve the moulding and boosting of ambiguous environmental stimuli into voices (as such, they are strictly speaking not so much

---

[1] At least for most of us; for individual differences see Hurlburt *et al.* (2013).

hallucinations as illusions). Our focus in this paper is on inner speech-based AVHs, and what they tell us about inner speech more generally.[2]

It is worth mentioning that the order of explanatory primacy is normally the reverse of what we are doing here. Theorists tend to use inner speech (which they take to be relatively un-mysterious) to make sense of AVHs (which they take to be relatively mysterious) and not vice versa. However, it seems to us that, despite its prevalence and familiarity, the nature of inner speech is far from self-evident. Given this, it makes sense to start, for the sake of inquiry, with the hypothesis that at least some AVHs are instances of pathological inner speech, and then to ask: what kind of thing must inner speech be in order for it to play this role in the generation of AVHs?

Before moving on, it is important for us to get clear on what kind of thing we are referring to by 'inner speech'. By that term one can be referring either to a particular experience, with a particular phenomenology, or to a particular feature of human cognition, which makes use of particular mechanisms, say, and which sometimes gives rise to that phenomenology, but which needn't always (for example, when it is disrupted in certain ways). In the former sense, the subtype of AVH that interests us is not an instance of inner speech, even though it may be generated by the processes that usually generate inner speech. In the latter sense, that subtype is, or is partly constituted by, an instance of inner speech. We will use the term 'inner speech' in the latter sense, although nothing of substance hangs on this terminological decision, and we acknowledge that both are valid senses of the term 'inner speech'.

2. A Predictive Processing Account of Auditory Verbal Hallucinations

In this section, we present an account of AVH that is built within the predictive processing framework (PPF). Since this account arose in part as a reaction to self-monitoring accounts of AVHs, we begin by presenting these accounts, and then move on to the predictive processing accounts. Then, in Section 3, we will explore the potential for predictive processing accounts of inner speech.

---

[2] Some theorists don't buy into subtypes, but if they adopt inner-speech based accounts across the board, then what we say will be of relevance to them. It is only those who think either that AVHs are homogeneous and nothing to do with inner speech, or who think that there are subtypes, but none of those subtypes are inner speech-based who will take issue with our starting point.

2.1. Self-monitoring accounts of AVH

Self-monitoring accounts are often viewed as unifying accounts of the positive symptoms of schizophrenia.³ Among these symptoms are delusions of control, AVHs, and thought insertion. What these symptoms all have in common is that they are instances of 'self-monitoring' having gone awry, which roughly means that a 'self-produced' or endogenous phenomenon fails to be recognized as such by the nervous system. These symptoms differ in so far as the phenomenon that is failing to be self-monitored differs. In AVHs and thought insertion, it is often taken to be inner speech. In delusions of control (and experiences of passivity) it is overt bodily action.

So what is this 'self-monitoring'? Perhaps the first theorist to make use of the idea of self-monitoring was Helmholtz (1866). He was not concerned with pathological cognition, but with healthy visual perception. In particular, he wondered, when an image moves across the retina, how does our brain know whether it is the world moving across our eyes or our eyes moving across the world? He suggested that when our eyes move there is a motor command, and that a copy of that motor command, later called an 'efference copy', is used by the brain to calculate a prediction of the sensory consequences of the upcoming eye movement. If this prediction is accurate and the predicted and actual sensory consequences match, then the brain 'infers' that the change was self-generated and the conscious percept is interpreted accordingly as a case of the eye moving across the world. We can experience for ourselves what happens when there is no motor command, and hence no adjustment, when we move our eye directly with our finger: the world itself seems to move, namely, the brain 'thinks' it is the world moving across the eye rather than vice versa.

These ideas were, much later, applied to psychosis (Feinberg, 1978). Although Feinberg's initial paper was on 'thought' (which he took to involve motor mechanisms) and thought insertion, we introduce the self-monitoring account with delusions of control, since it is clear that, if anything involves motor commands, it is overt bodily actions. In delusions of control, a subject claims that somebody else is controlling her actions. Frith and Done (1989) claimed that here there is a mismatch between the predicted and actual sensory consequences of the bodily movement, with the result that (as with Helmholtz's ocular example) the movement is attributed to an external source. In Helmholtz's

---

3 Needless to say, in reporting these accounts we are remaining silent on the validity of the concept of schizophrenia. For the record, we have doubts that all of those who standardly get the diagnosis of schizophrenia suffer from the same unified condition.

example, the recognition by the nervous system that a certain stimulus is self-produced, due to this matching between the predicted and sensory consequences of movement, causes a correction of the conscious percept. In contrast, in more typical bodily motor control, this matching results in 'sensory attenuation', namely a decrease in the intensity of the sensation. In effect, when there is sensory attenuation, your nervous system is telling you: 'You don't need to pay attention to this: it's only you.'

One striking datum that seems to support the hypothesis that something has gone wrong with this kind of self-monitoring in schizophrenia is the reported finding that subjects with diagnoses of schizophrenia can tickle themselves. The postulated explanation for this is that there is a mismatch between expected and actual sensory consequences and the sensory consequences are not attenuated: the tickling sensation is like being tickled by somebody else (Frith, Blakemore and Wolpert, 2000).

Several theorists (Feinberg, 1978; Frith, 1992; Jones and Fernyhough, 2007; Seal *et al.*, 2004) have attempted to explain AVHs in terms of these same self-monitoring abnormalities operating on inner speech. On these accounts, AVHs are instances of badly monitored, and hence unattenuated and externally attributed, inner speech.

2.2. Problems for the self-monitoring account of AVH

What is wrong with the self-monitoring account of AVH? As Wilkinson (2014) points out, there are potentially problems in accounting for (i) the phenomenology of AVH, and (ii) their variety. The first of these is effectively the issue of how we explain the transformation, in phenomenology, from inner speech to AVHs. The second of these concerns the issue of accounting for the wide varieties in AVHs with one model. This second worry can be overcome simply by saying that only *some* AVHs are misattributed episodes of inner speech arising from self-monitoring abnormalities. This kind of strategy seems like a sensible move regardless of what explanatory model you are trying to promote: AVHs are varied in how they present, in their contexts of occurrence, and in their apparent causes. Whether the first worry can be overcome is still a matter of debate (see, e.g. Cho and Wu's, 2013, attack on inner speech-based approaches and Moseley and Wilkinson's, 2014, defence), but it seems that acknowledging, on the one hand, the complexity and variety of inner speech phenomenology (McCarthy-Jones and Fernyhough, 2011), and the effect of the postulated lack of 'attenuation' resulting from failed self-monitoring, may go some way towards answering this worry.

Another worry may not come from whether the phenomenon to be explained (AVH) seems to fit the account, but rather from the viability of the very idea of monitoring inner speech (regardless of what phenomenon a deficit of such monitoring might generate). First of all, it is not obvious that inner speech involves motoric elements, and, so, where is the motor command that self-monitoring is supposed to exploit? This concern can be addressed, however. Motoric involvement in some forms of inner speech has been empirically supported by electromyographical (EMG) studies, some of which date as far back as the early 1930s (e.g. Jacobsen, 1931). Furthermore, later experiments made the connection between inner speech and AVH, showing that similar muscular activation is involved in healthy inner speech and AVH (Gould, 1948; McGuigan, 1966).

However, demonstrating motoric involvement in both inner speech and (at least some) AVHs doesn't let the self-monitoring theorist off the hook. It is not just motor commands that are important for self-monitoring, it is also the predicted and actual sensory consequence of the monitored phenomenon, and the match or discrepancy between the two. But what *are* the *sensory* consequences of inner speech? Is inner speech sensory at all? If so, where is the stimulus? Furthermore, since it doesn't occur in three-dimensional space, does it even *need* monitoring? These questions point towards a more fundamental worry, namely that the self-monitoring mechanism is not actually very well understood at the neural level. In a related manner, the postulated self-monitoring mechanism seems to be little more than a re-description of the computational task that any active system would need to do in order to distinguish what it does from what is done to it. In contrast, predictive processing accounts start from a general theory of what the brain does, and how this is implemented at the neural level (Friston, 2005). It then turns out that the self-monitoring task that needs to be achieved falls naturally out of this (along with many other tasks besides, e.g. making sense of noisy and ambiguous perceptual inputs).

Indeed, perhaps the main problem with the self-monitoring account actually has less to do with the account itself, and more to do with the overall *framework* within which the account operates, namely, how *cognition generally* is taken to work, how the brain processes information from the outside world and how that relates to conscious experience. Self-monitoring accounts try to explain, within a standard framework for understanding cognition, why someone is having an experience that usually occurs with a particular environmental stimulus (i.e. a speech sound), in the absence of that stimulus. The answer that the self-monitoring account gives is that there *is* a stimulus of sorts, it just hasn't been recognized by the nervous system (it may be so recognized by the *person*, as when a voice-hearer says 'I know it's just my brain') as a self-produced stimulus. That stimulus is inner speech. But what if this approach is doubly wrong? What if cognition generally, and healthy perceptual cognition, isn't really about the external stimulus in this way? And what if inner speech, more specifically, isn't about, and couldn't be counted as, a stimulus either? We present a general framework, and an account of AVH within it, that pursues precisely this line of questioning.

2.3. From self-monitoring to predictive processing

Some theorists (some of whom were, earlier, the main proponents of the self-monitoring account; compare Frith, 1992, with Fletcher and Frith, 2009) have proposed that the self-monitoring that is taken to go awry in AVH falls naturally out of a basic principle of brain function, namely, *prediction error minimization*. On this account, self-monitoring is not some *additional* aspect of cognition, but is a fundamental part of it (see Pickering and Clark, 2014). One upshot of this is that all of the varieties of AVHs can be accounted for (see Wilkinson, 2014), including those that may not involve motor commands from which forward models could be derived. For example, they can account for the 'hypervigilance' hallucinations (Dodgson and Gordon, 2009) we briefly mentioned in the introduction, in which environmental stimuli are boosted and shaped. This framework for thinking about cognition, and within which self-monitoring emerges from the basic functioning of cognition, is called the predictive processing framework (PPF).[4]

According to the PPF, the brain's main task is to 'infer' from incoming signals what the causes of those signals are. However, the incoming signal underdetermines distal causes: since inputs are noisy and ambiguous, the same stimulation can be brought about by two different distal causes (and different stimulation in different circumstances can be caused by the same distal cause). Given that more than one hypothesis is compatible with the incoming signal, the brain needs to take two things into account: first, the fit of the input with the hypothesis, and, second, how statistically likely that hypothesis is (the 'prior probability'). A hypothesis could fit the input extremely well, but its prior probability could be so low that it isn't even considered. Conversely, an hypothesis could have such a high prior probability, that, even though it doesn't fit the input well, it is settled upon.[5]

---

[4] For a fuller presentation of the PPF, see Clark (2013).
[5] A nice example of this is the Hollow Mask Illusion. When you are presented with the concave back of a mask, your brain 'corrects' the concave stimulus into a convex stimulus. This is due to the fact that the hypothesis 'convex face' (i.e.

What the selection of an hypothesis does is that it determines a set of predictions about subsequent inputs, namely, inputs that are compatible with the hypothesis. If the hypothesis does a good job of predicting inputs, it is kept. If it does a bad job, it is tweaked or abandoned altogether in favour of another hypothesis that does a better job. In other words, one hypothesis is selected rather than another if it better *minimizes prediction error*.

This picture has interesting consequences for how we are to view the role of input on sensory receptors and its impact on higher cortical regions, and also on conscious experience. According to the PPF, the only information that gets passed on up the cortical hierarchy is *prediction error*. This stands in sharp contrast to the standard view of perception and cognition according to which inputs come in, are processed, and passed on. According to the PPF, what determines your perceptual experience is what your brain has already predicted, your brain's best hypothesis.

This prediction error minimization is not only taken to account for perception and cognition, but for action as well (see, e.g. Adams *et al.*, 2013). Instead of there being motor commands, as on the standard picture, what you have are predictions, which are then fulfilled by the subsequent bodily movement, thereby also being a case of prediction error minimization. This is often called 'active inference', which Pickering and Clark (2014) helpfully gloss as follows: 'the combined mechanism by which perceptual and motor systems conspire to reduce prediction error using the twin strategies of altering predictions to fit the world and altering the world (including the body) to fit the predictions' (p. 1).

Another extremely important aspect of the PPF is that the hypotheses are hierarchically organized, with the hypotheses of one level providing the inputs for the next. 'Hypotheses' can also be talked about in a very 'zoomed out' way, to talk about the overall hypothesis, or in a very 'zoomed in' way, to talk about 'hypotheses' in early stages of perceptual processing. 'Higher' parts of the hierarchy are, roughly, those parts that are further away from the sensory stimulus. These tend to be at longer temporal timescales, and a higher level of abstraction. They might correspond, for example, to the belief that lions are dangerous. 'Lower' parts of the hierarchy are closer to the sensory stimulus. These tend to be operating at shorter timescales, and at low

---

normal face) has a very strong prior probability and that overrides the better fit that the 'concave face' hypothesis has with the incoming signal. This prior probability is generated by the expectation that the faces you will encounter will always be convex.

levels of abstraction. These, for example, might correspond to early stages of perceptual processing: your brain's early statistically-driven attempts to make sense of noisy inputs (see, for example, Gangepain *et al.*, 2012, for strong evidence for predictive processing in auditory word recognition). Of course, in order to express these neurally encoded predictions we need to use rough-and-ready descriptions in natural language (in this case English), but there is nothing linguistic about the priors/hypotheses ('light comes from above'/'This is a face') themselves.

Let's take an example (adapted from Pezzulo, 2014) to illustrate the predictive hierarchy. Suppose that, on the basis of a noise, which you take to be a squeaking window, two hypotheses present themselves about what is going on: either the wind blew the window, or a thief is clambering into your house. At the stage where those two hypotheses are competing, a great deal of ambiguity has already been resolved at lower levels of the hierarchy. For example, in early stages of auditory processing, the qualities of the sound will have been settled upon, giving rise to the conscious experience being a certain way, qualitatively speaking. Higher up the hierarchy, that sound gets interpreted as a creaking window, as opposed to something else. The direction of causation is from the (events represented in the) lower regions of the hierarchy to the (events represented in the) higher regions of the hierarchy. However, the direction of the inference is from the effects to the causes.

One final way in which the framework is made a bit more complex is that, in order to accurately form predictions in a world where the degree of noise varies from context to context, the brain needs also to predict the extent to which it can rely on its predictions. In other words, it needs to form second-order predictions, or estimate the precision of its predictions. In the predictive processing literature, this is called 'precision-weighting', and it amounts to the extent to which prediction error, once generated, is given weight. In contexts of high noise (e.g. in a dark room), the precision-weighting on bottom-up sensory prediction error will be low, and more influence will be placed on top-down influences. That is why at dusk you are more likely to see a tree trunk as a lurking aggressor.

2.4. The PPF and hallucinations

The PPF changes how one thinks of perceptual experience, and, by extension, radically changes one's explanatory focus in trying to account for hallucinations. On a standard framework, where front-line sensory stimuli get gradually processed and passed on up the hierarchy, hallucinations make one wonder, 'Where does this erroneous sensory stimulus come from?' As mentioned, we can see self-monitoring accounts as making attempts to answer this within a standard framework. Their answer is: they come from the quasi-sensory stimulation of inner speech, which is then misattributed. However, when instead we adopt the PPF, incoming stimuli play a much smaller role in determining the conscious percept, even where veridical perception is concerned. Given that a conscious percept is constituted by the hypothesis that best minimizes prediction error, we don't ask 'Where does the input come from?', since the input alone doesn't (and can't) determine the percept. Rather we ask, 'Why does this hypothesis minimize prediction error?' This general approach makes hallucinations both less perplexing, and less different from veridical perception.

Wilkinson (2014) has suggested at least three different ways in which the hypothesis corresponding to an AVH experience may be selected. These correspond to the three phenomenologically and aetiologically identifiable subtypes mentioned at the outset: inner speech-based, memory-based, and hypervigilance hallucinations. Both inner speech- and memory-based hallucinations are taken to be the result of aberrant weighting on prediction error. In other words, the self-generated hypotheses corresponding to inner speech and episodic memory turn out to generate unexpected levels of prediction error, which results in perception-like hypotheses being selected in an attempt to minimize this. This leads to a 'perceptualization' of the usual experiences of inner speech and episodic memory. In contrast, hypervigilance hallucinations are explained in terms of interoceptive predictive processing (Seth, 2013), where the hypothesis is selected not only based on how well it explains the incoming signal, but on how well it explains both the incoming signal and the subject's interoceptive emotional state. Thus the hypothesis that someone is insulting me explains not just a vague environmental stimulus, but also my state of anxiety and hypervigilance (see Wilkinson, 2014, for more details here).

In Wilkinson (2014), the more original contribution was taken to be the interoceptive account of hypervigilance hallucinations, with existing inner speech (Jones and Fernyhough, 2007) and memory-based (Badcock *et al.*, 2005) accounts merely requiring a slight reframing, from self-monitoring to predictive processing. However, such a reframing is not obviously achieved for either the inner speech or memory subtype, largely because it is not obvious how the PPF accounts for inner speech and episodic memory. In this chapter, we focus on inner speech, although an important area of future theorizing would involve an explanation of how the PPF accounts for episodic recollections.

3. A Predictive Processing Account of Inner Speech

It's all very well saying that AVHs are the result of changes to predictive processing, and that a subtype of AVH involves the mechanisms that are involved in inner speech. But what are the mechanisms involved in standard inner speech? In other words, what does a predictive processing account of inner speech look like?

3.1. Inner speech as 'internalized' outer speech

Before asking ourselves what the PPF makes of a given phenomenon, we need to be clear that we have successfully identified the phenomenon in question. So, what is inner speech, how does it develop, and what purpose does it serve? One very attractive theory, attributed to Lev Vygotsky, which carries both evolutionary and developmental plausibility, is that inner speech starts off as speech, namely, 'overt speech'. That is to say, whatever function inner speech plays, once it has developed, is played by overt speech in children who have not yet developed the capacity to engage in inner speech. This capacity to engage in inner speech is usually seen to involve, at least in part, the capacity to inhibit the overt production of speech (see Alderson-Day and Fernyhough, 2015, for a comprehensive review on the psychology and neuroscience of inner speech).

According to this story, inner speech is the end product of a developmental trajectory that begins with social speech, between an infant and primary caregiver, and then becomes overt private speech, before finally becoming inner speech. 'Private speech' refers to speech that is not produced for the benefit of anyone other than the speaker. Thus, although there is an important sense in which inner speech is always *de facto* private speech, pragmatics dictates that 'private speech' tends to refer to overt private speech, rather than inner speech (since inner speech is obviously private). Young children will first, under the guidance of a caregiver, learn to reason verbally, but out loud, for the benefit of guiding their thinking and attention. Over time, they learn to 'internalize' this speech, or, to phrase it in somewhat less misleading terms, to inhibit its overt production. However, as with many cases of motoric inhibition, vestiges of the motor processes often remain (as clearly seen in Jacobsen, 1931). Furthermore, the reason why an auditory phenomenology is often reported is quite simply because, as with any aborted overt action (motor imagery), the predictions of the sensory consequences of the action come into play, activating sensory (and somatosensory) cortices (this is central to feedback, which is crucial for all successful motoric activity).

3.2. Inner speaking and auditory imagery

What is going on when someone is engaged in inner speech? What constitutes inner speech? It is tempting to think of inner speech in terms of auditory imagery. Engaging in inner speech, on such a view, consists in imagining the sound of you speaking (or imagining hearing yourself speak). There is little doubt that one can imagine the sound of oneself speaking. It is like imagining hearing someone else speak, except that it has the properties of your voice. This, however, is not what inner speech, the phenomenon of primary interest to us, is. As we've seen, inner speech involves not just an auditory/imagistic component, but an articulatory/motoric component, too. Inner speech is agentive and more or less intentional (Jones and Fernyhough, 2007). To the extent that it is correct to speak of inner speech in terms of imagination at all, it does not consist in imagining hearing one's voice: it is the phenomenon of imagining oneself speaking (see Hurlburt, Heavey and Kelsey, 2013, for a phenomenological distinction between 'inner speaking' and 'inner hearing'). In any case, it seems misleading to speak of inner speech in terms of imagination, and here is why.

It is crucially important to differentiate imagination from imagery. Imagination is a personal-level phenomenon: people are engaged in acts of imagination. These acts of imagination enable them to appreciate, in potentially many different ways, non-actual scenarios, and, when they are engaged in such acts, they may be motivated to do so by a number of different things. They may be trying to remember the colour of someone's hair, judge whether they could have jumped over that river, reason about a social situation, or simply engage in imagination for the pleasure of it. These acts of imagination often will recruit or make use of imagery in many modalities, but there will also be aspects to the imaginative experience that aren't imagistic.

Imagery, in contrast, is not a personal-level event. Whereas people imagine things, people don't do imagery. When people imagine things, imagery may be involved. Imagery is also involved in personal-level events that aren't imaginings. For example, imagery may be involved in inner speech, indeed it may even be similar (or even the same imagery) to the imagery involved in imagined speech, but that doesn't make the personal-level act of inner speaking an act of imagining speaking. For a start, with inner speaking, you are not appreciating something non-actual: it is actual. You are speaking.

In short, it is important to understand the relationship between auditory imagery and inner speech, and, in a related manner, to understand that inner speech is not, in virtue of its recruitment of auditory imagery, simply a kind of imagined speech. Two things underpin this;

one is more sophisticated than the other. On the one hand, inner speech involves not just (and sometimes perhaps not even) auditory imagery, but motoric/articulatory imagery as well. In principle, however, there could be imagined speaking that made use of both motoric and auditory imagery, and this leads us on to the second more sophisticated reason why inner speech isn't imagined speech. Inner speech involves making a speech act, involves speaking your mind directly. The fact that someone is engaging in inner speech entails that they are speaking. The fact that someone in engaged in imagining themselves speak not only fails to entail that they are speaking, it actively entails that they are *not* speaking, since they are merely imagining it!

If we are to provide an account of inner speech, we need not only to account for the sensory and motoric imagery that are standardly part of acts of inner speech, but which can also potentially be part of other acts too, but also to account for what distinguishes inner speech from those other acts that make use of similar imagery.

3.3. Motoric and sensory imagery within the PPF

The PPF can very nicely accommodate the aspects of sensory and motoric imagery that are standardly part of inner speech. According to the PPF, all the brain ever does is minimize prediction error. As we've seen, this is taken to account for both perception and action. Whereas in perception hypotheses are selected to generate accurate predictions about the world, thereby minimizing prediction error, in action, predictions are generated which are then to be fulfilled by the action, thereby also minimizing prediction error. As a result, the notion of motor commands, at least as a type of neural activity in their own right, is dispensed with (we could, of course, still call the predictions that bring about actions motor commands—they do, after all, serve precisely that function).

Now, this presents us with an account of 'imagery' for both motoric and sensory domains. Although they are very different, they operate on exactly the same principles, namely, inhibition at a neural level, which within the PPF amounts to down-modulating the weighting/precision of prediction error. This turning down of the gain allows for a decoupling of the brain from the world. It is a way of minimizing prediction error without having to actually match the world (a relatively costly and difficult way, which is why it takes a while for children to master it, and why it is interfered with under conditions of cognitive load). And that is partly, and by definition, what imagery (as opposed to perception) is: something that represents something non-actual. It is a percept or action that isn't actually happening.

How does this relate to self-monitoring accounts? Imagery, both motoric and sensory (of which inner speech is composed), is not a self-produced *stimulus* in need of monitoring. There are no predicted and actual sensory consequences of imagery, where the latter can diverge from the former (as is the case with overt bodily action). Rather, the imagery, like any part of conscious experience, *is the prediction itself*, or, more specifically, a decoupled hypothesis that entails a bunch of deliberately unfulfilled (but prediction-error-minimized, through down-modulation) predictions.

A point of clarification is needed at this point. In this subsection, we have said nothing about inner speech *per se*. We have simply explained how sensory and motoric imagery, both of which seem to be involved in inner speech (as well as many other events besides), are to be viewed within the PPF. What does PPF have to say about inner speech more specifically?

3.4. A predictive processing account of inner speech

As we've said, an episode of inner speech (or, perhaps better, an act of inner speaking) is not an imaginative act. It is not imagining yourself speaking, indeed, it is not imagining anything: it is speaking. But just like overt speaking involves moving your mouth, throat, etc. as well as hearing yourself speak, so does inner speaking, at least often, involve the decoupled versions of these. This amounts to saying that inner speaking makes use of auditory and motoric imagery. So much we've already said. But what makes something an act of inner speaking as opposed to an act of imagined speech is that part of my experience is not only the low-level decoupled hypotheses that determine my imagery (both sensory and motoric), but high-level hypotheses about myself as an agent. Indeed there is a similar distinction when someone else is speaking to me, in a normal overt case, between my low-level hypotheses about sounds (or slightly higher up, phonemes, or higher up still, words, etc.) and my high-level hypotheses about the agent, their intentions, whether these speech sounds constitute a sincere speech act, and, if so, what kind of speech act, and what is the precise communicated content, etc. In inner speech we have all of these hypotheses about ourselves, as we are engaged in inner speech, and, what's more, they are almost always accurate. Verbally reprimanding myself in inner speech involves (i) me actually verbally reprimanding myself and (ii) in so far as I experience that reprimand, my brain having an hypothesis, not just about the words (or phonemes, etc.) used in the reprimand, but about the fact that I am reprimanding myself (which is clearly accurate in this case).

4. Consequences of a Predictive Processing Account of Inner Speech

A predictive processing account of inner speech has a number of interesting consequences. Some of these consequences are shared by compatible but higher-level or developmental accounts, such as Vygotskian accounts. Others are specific to the PPF and the hierarchical arrangement of hypotheses.

4.1. Epistemic consequences

The predictive processing account of inner speech fits nicely with the Vygotskian developmental story, at least in part because they make inner speech and outer speech very similar phenomena. However, this proximity between the inner and outer phenomenon raises an interesting epistemological issue, and it is as follows.

We very often talk to others in order to inform them, either directly or indirectly, of certain things, including states of the world, and our own states of mind. Granted, not all speech acts are informational: they can be imperative, expressive, etc. and the same applies to inner speech. However, in the cases where inner speech is informational, what motivates such speech acts? Why would I bother talking to myself if I already know what I'm going to say? One obvious way out of this problem is to insist that, contrary to our intuitions, we don't really know what we are going to say. Hence our utterances, in inner or outer speech, do not presuppose self-knowledge: they often *generate* it. This conclusion, though arrived at through logical argument, fits extremely well with the PPF. Within the PPF, there is no in-built provision for an introspective mechanism; there is simply the experience of certain percepts, actions, and emotions which all have the potential to feed into higher-level hypotheses I might have about myself.

A related upshot of this is that thinking is in some sense always dialogic. According to the PPF, simple, world-directed cognition involves coming up with accurate hypotheses about the world in the service of the organism's goals. 'Thinking' (among this woolly notion we include reasoning, supposing, wondering) emerges when the organism can decouple itself from the world in the service of goals represented *in absentia*. This involves the generation of things that stand as proxies for the absent (because they are future or distant or abstract) aspects of the world. Speech is a particularly helpful phenomenon that helps us do this (there are likely others), either overtly ('thinking aloud', a phenomenon that literally happens) or in inner speech. In these situations, we are both producers and recipients, and, as such, we are in a constant and inescapable dialogue with ourselves. This, coupled with the earlier anti-introspectionist epistemological consequence, may even suggest that this dialogicality (although it may not always use the medium of *speech*) is central to the robust self-awareness that humans are capable of.

4.2. The self and other in inner speech

At some relevant level of abstraction, it is clear that we represent other agents, when, among other things, we see them, think about them, hear them talk, etc. Within the PPF, this would correspond to hypotheses at a relevant level in the predictive processing hierarchy (generative models). When someone talks to us, we represent them (we retrieve a previous representation if it's someone we know already, or we use a generic model if it's someone new). It also seems that, in inner speech, we at some level represent ourselves at least implicitly. Now, there is one feature of AVHs, inserted thoughts, and cases of delusions of control that seems somewhat problematic for self-monitoring accounts, and it is that the subject doesn't simply claim to be *passive* in the face of these thoughts, utterances, and actions, but gets a sense that they are the responsibility of another, often quite richly represented, agent (see Wilkinson and Bell, 2015, for a focus on the representation of specific agents in AVHs). This is problematic for the self-monitoring account, because this account only tells us that the subjects ought to experience this as 'not me'. However, they do not explain the move from 'not me' to 'someone else'. Of course, these theorists could say that this is merely an abductive inference based on the feeling of 'not me' ('My actions or inner speech don't feel like me. How do I explain that? It must be someone else').

We note that this inferential step is an under-acknowledged feature of inner-speech models of AVHs. Two retorts to this inferential tactic are that, first, it doesn't explain why that is the explanatory inference so often resorted to (one would expect others), and, second, it wouldn't seem like a very good hypothesis to adopt. The hypotheses that other people can control your actions or insert thoughts into your head, or talk to you in their physical absence, ought to be assigned a pretty low probability compared to, say, the explanatory hypothesis that something is wrong with your nervous system.

What if there is a more straightforward story to be told about how the move is made from 'not me' to 'someone else'? The PPF may have the resources to do just that. From birth we learn about the world as our nervous systems become sensitive to statistical regularities, and this is manifested in hierarchically arranged hypotheses and expectations. Two very different kinds of stimuli, about which a (even moderately) developed human being's nervous system will have a host of different kinds of expectations, are inanimate objects, and animate objects (namely, agents). The expectations our nervous systems have about inanimate objects embody our *naïve physics*, so to speak, the expectation about agents, our *naïve psychology* (e.g. Spelke, 2000). Now, if a phenomenon exhibits basic statistical characteristics that activate our naïve psychology, this will be experienced as the work of an agent, and in a way that the subject may find very hard to override.

This may account for why agency, in a generic sense, is attributed, but as for why it is often *specific* agents, the answer may be as follows. Our nervous systems have expectations about types of thing. It, however, also makes sense that it should have expectations about—representations of—specific individuals (including oneself). Some of these individuals may be particularly salient as a result of the subject's past, or may be constructed and attached to a particular statistical pattern that is recurrently present in the experience. This idea amounts to a sort of merging of theories of agent tracking (see Bullot, 2009) and predictive processing. We see no reason to think that these two theories aren't compatible; indeed, the agent representation that is used in tracking could be viewed within the PPF as a generative model for that specific individual.[6]

Such agent-specific generative models won't only have utility in interacting with others (verbally, visually, etc.) but also in live interactions with oneself, where a generative model of oneself will be active and liable to being updated or diverged from. This would occur in inner speech, among other contexts (and may contribute to fleshing out just what is meant by a misattribution of inner speech, and the different ways in which it can be incurred). Furthermore, such generative models needn't be restricted to live interactions, but would come into play in simulated interactions with others (and indeed oneself). This would also encompass cases of dialogic inner speech where other individuals are represented (see McCarthy-Jones and Fernyhough, 2011).

4.3. Soundless voices

One important feature of the PPF is that hypotheses are hierarchically arranged in terms of how concrete and fine-grained they are. Thus, when we perceive things visually or auditorily, our brains are adopting

---

[6] Future avenues for research could tie the disruptions of these agent-specific generative models to delusional misidentification, where people claim that the misidentified person looks the same, but is somehow different. This ineffable difference may be due to changes to expectations about that person, which the person is no longer fulfilling (and hence there is a generation of prediction error).

hypotheses about specific colours and sounds, as well as higher-level hypotheses about, say, tables and chairs (in the visual case) and, say, melodies (in the auditory case). It is plausible to think that there are special intermediary hypotheses involved in linguistic cognition that correspond to specialized areas of linguistic expertise, from phonology, lexicon, grammar, all the way up to the literal and intended meanings of whole utterances. In perception, the low-level hypotheses tend to ground the higher-level ones: you experience a particular sentence because you experience particular words, which in turn you experienced because you experienced particular phonemes, and particular phonemes because particular sounds. Of course, the extraction of these is driven by top-down expectations, as the PPF would suggest. However, the high-level hypotheses tend not to be active in the absence of lower level ones: you don't auditorily perceive words in the absence of perceiving sounds. Things are somewhat different in imagination and in 'thought'. The higher-level hypotheses are activated with degraded or absent sensory hypotheses. That's arguably what more or less 'abstract thinking' is.

But what if something has the externality of a perceptual experience, but has the informational quality of one of these more 'abstract' episodes? That is precisely what we seem to get in the (not especially rare) cases of 'hearing soundless voices'. Higher-level hypotheses are activated, with an unusual perception-like vivacity, in the absence of lower-level ones. This would yield an unfamiliar sort of perception-like experience, in the absence of sensory qualities. Here is a self-report of such an experience:

> It's hard to describe how I could 'hear' a voice that wasn't auditory; but the words the voices used and the emotions they contained (hatred and disgust) were completely clear, distinct, and unmistakable, maybe even more so than if I had heard them aurally. (Woods *et al.*, 2015, p. 326)

This idea of higher-level hypotheses being active in the absence of those lower level ones that usually accompany them is in keeping with work examining the idea that the experience of communication may be at the heart of AVHs. In particular the idea is that sometimes what is experienced is the communicative intention, e.g. the intention to insult, which may or may not bring about an accompanying sensory auditory phenomenology (Deamer and Wilkinson, 2014). In principle, the PPF allows for the separation of the levels of the hierarchy, since the precision can be turned down at any point in the hierarchy, leading to one level no longer being answerable to the other (which is the same principle as decoupling from the world, but occurs within the nervous system itself).

5. Recap and Conclusion

What have we learnt about inner speech? Well, what inner speech fundamentally *is*, when viewed through the lens of the PPF, is the generation by my brain of a decoupled hypothesis that I am speaking (which I am doing for my own cognitive benefit). When I am speaking out loud, there are motoric and proprioceptive elements, and there are also auditory elements. Similarly, when I am engaged in inner speech there is both auditory and motoric imagery (predictions which are united under the same hypothesis—namely, that I am speaking to myself, or at least for my own benefit). How accurate is the hypothesis? Well, the hypothesis is *multi-layered*: there are low-level, decoupled predictions about auditory and proprioceptive stimulation, which, in a sense, are inaccurate, but unproblematically so, since they are deliberately decoupled. They are cases of imagery, not perception. There are also high-level predictions about my own agency and communicative intentions, and *these* are in an important sense *not* decoupled. But in a similar vein, they are also, at least usually, *accurate*: I *am* speaking, performing speech acts, when I experience my healthy, ecologically valid, inner speech. This combination, within a unified hypothesis, of coupled and decoupled predictions, this hybrid of imagination and self-perception, means that inner speech involves a delicate balance. The high-level hypothesis that this is *me*, and that I am saying *this*, is liable to be discarded in favour of another hypothesis (this is someone else, and they are saying something else), if there are disruptions to either aspects of the lower-level sensory and proprioceptive decoupled predictions, or to aspects of more high-level predictions. In particular, these predictions may remain *de facto* decoupled, but not recognized as such by the experiencing subject. This involves a perceptualization of imagery: a percept with perception-like vivacity, but which isn't answerable to what's happening in the world.

References

Adams, R., Shipp, S. & Friston, K. (2013) Predictions not commands: Active inference in the motor system, *Brain Structure and Function*, **218**, pp. 611–643.

Alderson-Day, B. & Fernyhough, C. (2015) Inner speech: Development, cognitive functions, phenomenology, and neurobiology, *Psychological Bulletin*, **141** (5), pp. 931–965.

Badcock, J.C., Waters, F.A.V., Maybery, M.T. & Michie, P.T. (2005) Auditory hallucinations: Failure to inhibit irrelevant memories, *Cognitive Neuropsychiatry*, **10** (2), pp. 125–136.

Bullot N. (2009) Toward a theory of the empirical tracking of individuals: Cognitive flexibility and the functions of attention in integrated tracking, *Philosophical Psychology*, **22** (3), pp. 353–387.

Cho, R. & Wu, W. (2013) Mechanisms of auditory verbal hallucination in schizophrenia, *Frontiers in Schizophrenia*, **4**, pp. 1–8.

Dodgson, G. & Gordon, S. (2009) Avoiding false negatives: Are some auditory hallucinations an evolved design flaw?, *Behavioural and Cognitive Psychotherapy*, **37** (3), pp. 325–334.

Feinberg, I. (1978) Efference copy and corollary discharge: Implications for thinking and its disorders, *Schizophrenia Bulletin*, **4**, pp. 636–640.

Friston, K. (2005) A theory of cortical responses, *Philosophical Transactions of the Royal Society of London B: Biological Sciences*, **360** (1456), pp. 815–836.

Frith, C. (1992) *The Cognitive Neuropsychology of Schizophrenia*, Hove: Lawrence Erlbaum.

Frith, C. & Done, D. (1989) Experiences of alien control in schizophrenia reflect a disorder in the central monitoring of action, *Psychological Medicine*, **19**, pp. 359–363.

Frith, C., Blakemore, S.-J. & Wolpert, D.M. (2000) Explaining the symptoms of schizophrenia: Abnormalities in the awareness of action, *Brain Research Reviews*, **31** (2–3), pp. 357–363.

Gagnepain, P., Henson, R.N. & Davis, M.H. (2012) Temporal predictive codes for spoken words in human auditory cortex, *Current Biology*, **22** (7), pp. 615–622.

Gould, L.N. (1948) Verbal hallucinations and activation of vocal musculature, *American Journal of Psychiatry*, **105**, pp. 367–372.

Helmholtz, H. von (1866) Concerning the perceptions in general, in *Treatise on Physiological Optics*, Southall, J.P.C. (trans.), 1925, Opt. Soc. Am. Section 26, reprinted, New York: Dover, 1962 (vol. III, 3rd ed.).

Jacobsen, E. (1931) Electrical measurements of neuromuscular states during mental activities, VII: Imagination, recollection, and abstract thinking involving the speech musculature, *American Journal of Physiology*, **97**, pp. 200–209.

Jones, S.R. (2010) Do we need multiple models of auditory verbal hallucinations? Examining the phenomenological fit of cognitive and neurological models, *Schizophrenia Bulletin*, **36** (3), pp. 566–575.

Jones, S.R. & Fernyhough, C. (2007) Thought as action: Inner speech, self-monitoring, and auditory verbal hallucinations, *Consciousness and Cognition*, **16** (2), pp. 391–399.

McCarthy-Jones, S.R. & Fernyhough, C. (2011) The varieties of inner speech: Links between quality of inner speech and psychopathological variables in a sample of young adults, *Consciousness and Cognition*, **20** (4), pp. 1586–1593.

McGuigan, F. (1966) Covert oral behaviour and auditory hallucinations, *Psychophysiology*, **3**, pp. 73–80.

Moseley, P. & Wilkinson, S. (2014) Inner speech is not so simple: A commentary on Cho and Wu (2013), *Frontiers in Psychiatry*, **5**, art. 42.

Pezzulo, G. (2014) Why do you fear the Bogeyman? An embodied predictive coding model of perceptual inference, *Cognitive, Affective, and Behavioral Neuroscience*, **14** (3), pp. 902–911.

Pickering, M. & Clark, A. (2014) Getting ahead: Forward models and their place in cognitive architecture, *Trends in Cognitive Sciences*, **18** (9), pp. 451–456.

Seal, M.L., Aleman, A. & McGuire, P.K. (2004) Compelling imagery, unanticipated speech and deceptive memory: Neurocognitive models of auditory verbal hallucinations in schizophrenia, *Cognitive Neuropsychiatry*, **9** (1–2), pp. 43–72.

Smailes, D., Alderson-Day, B., Fernyhough, C., McCarthy-Jones, S. & Dodgson, G. (2015) Tailoring cognitive behavioural therapy to subtypes of voice-hearing, *Frontiers in Psychology: Psychopathology*, **6**, art. 1933.

Spelke, E. (2000) Core knowledge, *American Psychologist*, **55**, pp. 1233–1243.

Wilkinson, S. (2014) Accounting for the phenomenology and varieties of auditory verbal hallucination within a predictive processing framework, *Consciousness and Cognition*, **30**, pp. 142–155.

Wilkinson, S. & Bell, V. (2016) the representation of agents in auditory verbal hallucinations, *Mind & Language*, **31** (1), pp. 104–126.

Woods, A., Jones, N., Alderson-Day, B., Callard, F. & Fernyhough, C. (2015) Experiences of hearing voices: Analysis of a novel phenomenological survey, *The Lancet Psychiatry*, **2** (4), pp. 323–331.

Jonathan Cole

'Efforts of the Cultivated Mind'
Neurological Impairment, Intention, and Attention to the Body

Despite what we presume to be our supremely evolved self-awareness, much of our movement and sensory information, whether in vision, hearing, or touch, lies beyond awareness and attention. If our richly developed consciousness defines our species, its limited reach into our own neural processes is profound. There are a number of ways in which the boundary between perception and the body, in action and sensory information, have been approached. In relation to movement, the early twentieth-century neurologist, Sir Henry Head, described the phenomenon thus,

> Every recognisable change enters consciousness already charged with its relation to something that has gone before, just as on a taximeter the distance is presented to us already transformed into shillings and pence.
> Recognition of posture and movement is obviously a conscious process. But the activities on which depend the existence and normal character of the schemata lie forever outside consciousness; physiological processes with no direct psychical equivalent. (Head, 1920, p. 605)

This has not prevented neuroscience researchers from exploring these boundaries too. Work on the preparation to move has teased out the timings related to action preparation and consciousness, the brain areas involved, and a large literature on the interpretation of the empirical findings (see Libet *et al.*, 1983; and, for instance, Haggard, 2008). Others have used functional imaging and other techniques to reveal how much the brain does at levels before, below, and in parallel with awareness, apparently automatically and autonomously (e.g. Decety and Cacioppo, 2012). Another approach, by Fernyhough and Wilkinson in this volume, is to discuss how and why unbidden thoughts and hallucinations arise in the brain and are presented to consciousness.

Underpinning all these are central unanswered questions about what consciousness is for.

My own approach in this chapter, neither empirical nor philosophical, but rather clinical and narrative, is to ask how neurological impairments which disturb the normal automaticity of action and perception present themselves to patients. The nineteenth-century physiologist, anatomist, and artist Charles Bell wrote:

> We use our limbs without being conscious, or at least, without any conception of the thousand parts which must conform to a single act... by an effort of the cultivated mind we must rouse ourselves to observe things and actions of which sense has been lost by long familiarity. (Bell, 1833, p. 14)

Can people bring to consciousness levels of action and intention, and changes in sensory perception which are unusual, or are they doomed to remain outside description, as Head suggested? Perhaps some aspects might lie beyond language, but the effects of such changes might be amenable to description. What is it like to be bereft of automatic movement, or sensation, or facial expression, and what effects might these have on the ways a person perceives their body and their self? Though patients do not discuss their situation in such terms, these are not simply questions of altered perception or agency, but challenges to their sense of self and their ontology.

When discussing this, movement may be divided into simple categories: instrumental and locomotor (how we dress, eat, walk, etc.); expressive (gesture, posture, and facial expression which express our mood and emotion); and autonomic (heart rate, breathing, sweating, bladder and bowel control, etc.). Intimately involved in these is sensation from skin, and the movement and other receptors in the body. One might make a further distinction between congenital problems, when no other state is known, and acquired conditions where a subject can compare their states before and after impairment, though such distinctions are not straightforward. I will consider four neurological conditions, cerebral palsy, spinal cord injury, Möbius syndrome (congenital absence of facial expression), and sensory neuropathy, and how their losses of movement and sensation are experienced and described.

Arguably, the most severe and catastrophic deficit, which begins suddenly and unexpectedly, is spinal cord injury at the neck. A person can become paralysed, insentient, and incontinent in a fraction of a second, forever. How do people in this situation describe it? One man who has written about this is Robert Murphy, who became tetraplegic (quadriplegia is the term in the US) late in life, following a spinal tumour rather than an accident, and he wrote:

> [A] quadriplegic's body can no longer speak a 'silent language'... the thinking activity can no longer be dissolved into motion, and the mind can no longer be lost in an internal dialogue with physical movement... My thoughts and sense of being alive have been driven back into my brain... (Murphy, 1987, p. 76)

The loss of movement and communion with the body in action was described by a man, 20 years after his paralysis, who had not adapted to his new situation:

> You just cannot substitute for the experience of being able to use this wonderful piece of equipment, the body, be it running, riding or shagging. I enjoyed contact with a whole body experience; the sheer total involvement and the physical contact. (Cole, 2004, p. 42)

Others in the same position had a very different perspective right from the beginning:

> I was concerned with the practicalities, I suppose, of getting on with life. I didn't lie there thinking all the time 'Oh My God what have I done, what's this going to mean?' I never burst into tears because, from the early stages of living with the injury, I have seen the whole thing as a challenge... (ibid., p. 66)

Generalizing people's responses to similar losses can be problematic. Soon after he became paralysed Murphy described how,

> for a while I tried to will the legs to move, but each futile attempt was psychologically devastating... I was saved from the edge of breakdown because the slow process of paralysis of my limbs was paralleled by a progressive atrophy of the need and impulse for physical activity. I was losing the will to move. (Murphy, 1987, p. 87)

For others, though, the need for action and agency continues to simmer over years and decades. Some with tetraplegia told me of their joy when given the means to move, either in a car or even a cycle; Graham, 30 years after his spinal injury:

> I enjoy cycling (via a tracker, a tricycle moved with arm cranks), because it is a raw physical release. I did not realise how badly I needed it. (Cole, 2004, p. 57)

Another man described a similar experience:

> Two years ago I bought a tracker, and I must admit that I have had more pleasure out of it than from anything else since I have been disabled. I am cycling, going from one place to another under my own control. I am amazed at how much pleasure it is giving me. (ibid., p. 32)

Though consciousness is clearly not essential for control of movement, and can indeed get in the way, we have a deep need for independence in action and for the simple, life affirming creation of and feeling in movement.

Incoordination of movement, rather than its loss, is a problem for some with cerebral palsy (CP). Thought and attention towards movement can be counterproductive, a duel between the person's agency and the chaos of the palsy. Too little intention and nothing happens; too much and the body tenses and the movement degrades as increased muscle tone leads to spasm and pain, as people try to impose intention on limbs moving without volition. A man with a hemiplegic form of CP, meaning that he had normal movement of one side of his body but difficulties imposing meaningful movement onto the other, described his world:

> Every movement of his foot, swing of the stick and grip on the stick handle has to be carefully co-ordinated. Without fluidity, he needs constant planning of every move. He is always aware of his body as an object to be controlled, not really 'part of him'.

The effect of this dichotomy, with one side working automatically, while the other misbehaved automatically, led to a curious asymmetry in experiencing his body and beyond.

> The world, that is, his surrounding environment, appears hostile, something he is part of, but certainly not 'in'. Other people appear as obstacles to be avoided, he fears bumping into them and hurting himself, and them; even a hand offering help an unexpected disruption to his 'walking plan'.
>
> His agency is disrupted, split; his relationship to his left side being entirely different to that with his right. When younger this created a sense of physical bi-polarity, as if he had two bodies to deal with. At times it appeared to him as if he had two selves, each in charge of one side of his body.

Those with CP may have known no other situation, but become aware of their situation. Christy Brown, famous for his book and its subsequent film, had such severe cerebral palsy that it prevented controlled movement of any part of him except for his left foot, through which he learnt to communicate and disclose to the world his remarkable sentience.[1] His narrative begins round the age of 5 when he could neither speak nor sit up and showed no real signs of intelligence. Despite this he 'longed to run about and play with the rest'. This was

---

[1] Cerebral palsy eventually shortened his life, since he choked to death in his 40s, as did another great writer with and on CP, Christopher Nolan.

made possible by a home-made go-cart he could sit in and be pushed. It was his 'chariot [and] throne':

> I can remember the wet wind on my face as they raced me along and sitting in it as my brothers played cards under a street lamp on a dark winter night when the gutters in the road were running with water. People would sometimes stare; I had no idea why. I just wanted to be happy and my brothers saw to it that I was. (Brown, 1990, p. 36)

One day the cart broke:

> Everything changed. I could no longer go out with my brothers; the bottom fell out of my world. Everything was different, as I saw it, as I felt it. I was 10, a boy who couldn't walk, speak, feed or dress himself. Only now did I realise how helpless I was. I could not reason out why I was different, I could only feel it. (*ibid.*, p. 46)

The crisis precipitated a sense of self-awareness:

> Up till then I had never thought about myself... Now it was different. I saw everything, not through the eyes of a little boy eager for fun, but through those of a cripple who has just discovered his affliction... (*ibid.*, pp. 48ff.)

After a few weeks he was given a proper invalid chair, with padded seat and rubber tyres. His brothers took him to watch a game of football, 'just like old times, all the "gang" around me, telling jokes.' But it was too late; Christy was aware: 'In a few weeks I had become as different in mind as I now knew I was in body.' He threw himself into painting, then writing to express himself in a way he could never do through word and action. But he was, from then on, always aware of his difference and the impossibility of being like others. He watched as his brothers and sisters moved away to make their own lives, and as that happened he wrote:

> [I] saw and felt the limitations, boredom and narrowness of my condition. All around were signs of activity, effort and growth. Everyone had something to do, to occupy them and keep their hands and minds busy. I had only my left foot.
> I felt myself coming to a greater, more persistent awareness of my own needs, and that in itself was pain enough... no matter how I might overcome my physical limitations, my inner emotion life would never, could never, really be normal... bottled up, suppressed instead of expressed. (*ibid.*, pp. 78ff.)

His book is rich in describing his dawning awareness of difference, but is also of interest for what it does not contain; much analysis of how he moved his left foot, or much attention towards his body. Without controlled movement, without effective agency, his attention was towards his thoughts and mind, and towards his defects and difference, rather than to his body.

This has striking parallel with a young girl with another congenital condition, Möbius syndrome, which is characterized by the absence of movements of facial expression and outward, divergent, movement of the eyes. People with Möbius cannot shut their eyes or mouths, show facial expressions, or move their eyes from side to side. Additional problems include swallowing and speech problems and abnormalities of foot and hand development. 'Celia' was able to reflect on her childhood from womanhood with rare clarity:

> I did not do ballet or horse riding; I did hospitals and operations. I had the eye doctor and the foot doctor and a speech therapist, and a face doctor. My limitations were a fact of life. Not being able to see the blackboard, or not being able to see someone over there. I never thought I was a person; I used to think I was a collection of bits. I thought I had all these different doctors to look after all the different bits.
>
> 'Celia' was not there; that was a name people called the collection of bits. I did not like my feet; I liked my spirit, I liked my brain. I loved reading and read very early on. I could think and dream and imagine. I always knew there was something strong inside that I had a mental dialogue with, but it was not the physical body; it was very separate from the physical. (Cole and Spalding, 2008, pp. 40ff.)

Like Brown, she felt she was her mind, her thoughts; her embodied being was problematic; she was, more than anyone perhaps, a Cartesian child. Though such a failure to inhabit and be embodied in our flesh does not appear to be experienced by all with Möbius, its presence in some is still of interest and importance as we tease away at how agency and movement—usually given and automatic—when reduced or absent, can lead to profound alterations in our sense of self.

Another problem for some with CP and with Möbius is the attitudes of others who make assumptions about their cognitive and other abilities. Children with Möbius have been considered, in the past, to have falsely high incidences of learning difficulties and autism. Since they can be late in speech, cannot make easy eye contact, and have difficulties in interpersonal relatedness for a number of reasons, more pervasive developmental problems have been assumed. Though we may think we see others for what they are, sometimes we see physical difficulties and presume a person has cognitive problems too, because they live within a somatic straightjacket, prevented by their bodies from communicating as they would like. For some with CP it takes years before they can reveal their sentience and even then there can be problems. One woman with CP and severe speech difficulties wrote of how her speech, and thought, was sometimes limited by the words she could say.

Thus far we have concentrated on problems in movement. But just as we take it for granted that our bodies will move when commanded, so we take sensation as a given, a characteristic of what bodies are. In spinal cord injury the focus is naturally on paralysis, but people are as lacking in sensation as they are in movement. Wittgenstein asked if the absence of a feeling was a feeling (Wittgenstein, 1980). What remains when sensation is removed and what are the consequences for any sense of inhabiting one's body? One person with complete paralysis, and lack of sensation from the neck down, tried to describe what was left:

> This is not the same sort of sensation somehow [as after dental anaesthesia]... That numbness is sort of a void, as though something has been taken away and is no longer there. My loss of sensation is not quite like that; it is a different feeling... (Cole, 2004, p. 31)

Try to imagine feeling only one's head and neck, with paralysis and insentience from the neck down, and then, after a several months, being placed in a wheelchair for the first time:

> I felt like a balloon being wafted around. It is a sensation of nothing. I immediately compared with before. To me it was a sensation. You're saying that if you can't feel you can have a sensation. But it wasn't numbness. It was nothingness. It was a sensation because you can sense nothing. It was a definite sensation. My head floating... I've never been up in a balloon, but that's how I imagine it would be... Because you had no sensation it was like sitting in a wheelchair and floating. This was extraordinary. Initially you could sit *in* it, but it felt like you were just a head and so felt unstable and convinced you were going to fall out. (*ibid.*, pp. 31 and 18)

This was 30 years ago, and since then he has adapted:

> It took weeks and months to feel OK. Now I can almost kid myself that I can feel something when I sit in a chair, even though I know I cannot. It feels exactly the same sitting in a chair now to before I was injured. It can't but it does. My mind tells me so. My mind makes me think I am like you over there. It learns what is the norm for this body. It tells me there is nothing wrong, so I feel comfortable and correct. (*ibid.*, pp. 18-21)

Another man, 'Tony', described a similar perception:

> I still view my body as whole; it's just motionless. I'm not a head on a bag of potatoes. I still know it's there, I still like it; I like to see it, it is still me, and I am still it, totally. I am still me, body and all. (*ibid.*, p. 183)

This long term reconstitution of a feeling of normality despite continued loss of sensation, paralysis, and incontinence is extraordinary. And this was not described in terms of the results of training, effort, or rehabilitation. It was as though the coming to terms with a new radically different normal body state occurred on its own, slowly, in the background, as other aspects of life carried on.

Others with tetraplegia describe another, surprising, way in which embodiment can be experienced despite lack of sensation. Most people with SCI live with chronic pain, a form of phantom body pain. Tony's is usually not severe unless he becomes constipated, or something else is wrong. He also has a lower level, background pain he describes thus, 'I would not want to be without some buzzing. It gives me a sense of identity in my body.' Despite not feeling his body, the pain appears to be arising in it, and this allows some feeling of embodiment. Another man made the same observation more forcefully, and only a few months after his injury:

> My physical pain is in the hands and down the legs and in the feet. The pain does not come on; it is there, the whole time, 24 hours per day every day, every day of the year. When it's angry it gets me so down, it's almost unbearable.

Fortunately the pain is not always severe, and since it is felt in the body, it allows some feeling of having a body again. Then, 'It is almost comfortable, almost my friend. I know it's there; it puts me in touch with my body. The pain is the connection — my friend the pain.' (*ibid.*, p. 67)

Ian Waterman has lived for over 40 years without the sensations of touch and movement/position sense (proprioception) below the neck. His perceptions of pain and temperature remain and his motor and movement nerves were unaffected, though he was initially unable to move without the non-conscious proprioceptive feedback that Head discussed. Very early on, lying on a bed, unable to feel or see his body, he felt disembodied, as did Oliver Sacks' case Christina (Sacks, 1985). But this disappeared immediately after he was able to see his body. He learnt to move again, over months and years, by substituting mental thought and effort towards movement and using vision for supervisory feedback. As a consequence, with this necessarily heightened attention towards movement and his body, he is — if anything — more aware of his body than others, cognitively and continuously.

> I was disembodied at the start, lost, and in the early stages needed everything done for me. Even now, I wake up and think, 'Where I am in space?' Each day I need to gradually re-associate myself before moving. So, rather than being disembodied, I am completely, totally, embodied. If I was not I would not know where I am. I re-associate and reconnect

constantly. Everything is through vision. I think about it, before I do it; I think it out and I think it back. (Cole, 2016, p. 119)

Ian lives with background lights on by day and night. This enables him to see in order to move, but it also allows him continually updated reinforcement of his body — sensations from pain and temperature are not enough. During a functional imaging experiment he had to spend two hours on his back, in a scanner, without vision of his body. Someone was with him the whole time reassuring him that his body and legs were OK. But, for Ian, not seeing his body for that length of time was profoundly disquieting. 'It's very difficult to explain. I got in panic mode, not knowing where I was, or if I was safe.' Ian is usually so adept at deflecting attention from the precariousness of his condition, and the experiment had not been designed to reveal how necessary vision was, but at its end — for that brief period — without feeling from his body and without visual reassurance, he revealed once more his initial state of disembodied anxiety.

In the introduction we, very simply, separated action into instrumental, locomotor, and expressive. These, however, overlap and our actions affect how others view us and also how we feel, in terms not only of tactile perceptions, but also in relation to emotional experience. People with Möbius lose not only instrumental actions of eating, drinking, blinking normally, they also never express through the face. This lack of embodied expression and problematic communication led Celia to suggest that, as a child:

> I did not express emotion. I am not sure that I felt emotion, as a defined concept. At my birthday parties I did not get excited. I don't think I was happy, or even had the concept of happiness as a child. (Cole and Spalding, 2007, p. 43)

She had no gesture, why would she move part of the body she felt no real connection with? Though not all people with Möbius have such an experience, Celia was not alone. James, a priest in his forties, described how he had a more dispassionate, cognitive form of emotion:

> I have a notion which has stayed with me over much of my life — that it is possible to live in your head, entirely in my head... I think I get trapped in my mind or my head. I sort of *think* happy or I *think* sad, not really saying or recognising actually feeling happy or feeling sad. Perhaps I have had a difficulty in recognising that which I'm putting a name to is not a thought but a feeling, maybe I have to intellectualise mood. I have to say this thought is a happy thought and therefore I am happy. (Cole, 1998, pp. 118-27)

When he met his wife, 'I think initially was I thinking I was in love with her. It was some time later when I realised that I really felt in love.'

Perhaps more interesting, and certainly more mysterious, is how Celia was able to develop emotional experience as a young woman. At university, escaping family and friends' benevolent but restrictive protectiveness, she aped the clothes of her peers, immersed herself in their talk, and imitated their gesture, though without feeling anything as she did. Then she took a job in Spain teaching English and in that highly expressive, gesture-rich society her expressive movements became more than imitation, she began to feel the emotional experiences within the gesture and speech; she had bootstrapped her feelings into bodily expression, though how remains tantalizingly unclear. Another woman with a similar emotionless childhood described the same process at Music School; as she had to rehearse opera and *Lieder* she began to feel the emotions expressed in the music. The mechanisms for this remain unclear, as feelings become felt as they are expressed, through the body, shared in a rich cultural environment.

Such experiences, though rare, suggest that emotional experience requires embodiment, expression, and communication and conversation. Celia agrees and remarks how much more difficult this is with Möbius, even as a mature and successful woman. For despite being able to gesture,

> [a]ll my gesture is voluntary, even now aged 40. Everything I do, I think about... All the things I am doing, whether turning my head or moving my hands, is all self-taught.
>
> With Möbius [and without facial expression] you have to be more wordy and articulate and this requires intelligence and can be hard and tiring. For me the word is stronger than facial expression. Without the word, how to express the feeling? I am interested now in non-facial aspects; gesture and tone of voice. Gesture is part of language, is a language and people with Möbius do not always learn it; they must be taught. Without gesture, thought is impoverished, as is language without gesture and thought without language.
>
> As a child I did not know I was missing out. To reach out to others I needed embodied expression—feeling. As you grow up the social feedback from others has far more meaning than as a child. A meaningful smile from you triggers an emotional response from me. (Cole and Spalding, 2007, pp. 190; 198–9)

Even now, in her 40s:

> When I express being happy, that has to be vocal and intellectual... There is an element of artificiality of the expression but not the feeling. I have two sorts of happy, intellectual happy to express, and happy happy. I can be happy, not think happy, but to express I have to think. (*ibid.*, p. 192)

Celia had to remember to gesture, and habitually monitors very carefully how successful she is being in conversations with others, as she gestures and employs her voice to compensate, beautifully, for her absence of facial movement. The movements she makes are initiated voluntarily but then take place, like most movements, without need for thought or attention towards them. This was mentioned by Bell, in his 1833 book:

> In truth we stand by so fine an exercise of this power, and the muscles are, from habit, directed with so much precision and with an effort so slight, *that we do not know how we stand*. But if we attempt to walk on a narrow ledge, or stand in a situation where we are in danger of falling we become subject to apprehension; the actions of the muscles are magnified and demonstrative to the degree in which they are excited. (Bell, 1833, p. 197)

That we do not know how we stand was recently revealed via an unlikely source. The eminent UK modern choreographer Siobhan Davies' work from 2014, *Table of Contents*, is a live performance created by her with five others, comprising several performance pieces, as they meditate on the relation between remembering and performing movements, between thought, memory, and action. The pieces take place in a shared space between performers and audiences, with people able to come in and out as they like over the several hours it unfolds.

There are two pieces which are directly traceable to Ian Waterman's influence. In one, *To Hand*, Matthias Sperling moves round the floor spread-eagled and supported on upturned clear plastic pots. Every movement involves risk, as he balances, and thought as to the next movement and how he can manipulate his body in new ways. For some this is bewildering; a man balancing and moving on small pots for several hours; his attention focused not on a goal—say reaching the other end of the room—but more on the constituent parts of each movement. But as a live illustration of what Ian and many people with physical conditions go through each day, it is compelling.

The second piece, *Manual*, is simple yet startling. One of the troupe lies on the ground and invites a member of the audience to instruct them—through words—to stand. The performer moves exactly as told to; no more, no less. We can stand from lying easily, without thought. But it is extraordinarily difficult to analyse exactly what we do and to put that in words to instruct someone else to do it. Siobhan related what happens:

> After the first nervous requests the member of the audience becomes intrigued by the difficulty of the task and notices that language doesn't help as much as they thought. There is a measure of frustration, curiosity and pleasure, and it normally takes at least 15 minutes to complete the task. People learn to attend to the simplest of actions,

common to us all. No one can describe the enormous ensemble of actions, and counter actions, needed to do this manoeuvre.

They were inviting people to reach down into how intention and action are joined, into areas we are not normally conscious of at all. Ian Waterman's dilemma is even more unusual. Without movement/ position sense and cutaneous touch, his brain had none of the unconscious peripheral feedback it needs to coordinate movements; skill and automaticity in action lost forever. From then on he has had to think about any movement he makes and use visual supervision to make sure that actions he orders have come to pass. Over the two years after the illness, he learned to stand and then walk, to dress and to eat, and became completely independent. Only over the last few years over forty years on, with a bad back and being less demanding of himself, has he begun to live more safely from a wheelchair; previously he thought/walked his way everywhere. There have been many accounts of this, so here I will focus on the level of his thoughts towards actions as he moves. Any controlled movement requires thought, and if he is tired, or has a drink of alcohol, his attentional capacity reduces and his movements degrade. If he has a head cold he retires to bed until it's over.

I asked Ian what he actually had to think about to pick up a cup of tea from a table:

> I am initially being aware of my body position to hang it all off. Sitting down on a chair my legs are in a tripod, and I have a mental image of this. Having one arm resting on the table is a good triangle position, and I know I can reach the cup. Once the framework is safe and that I can monitor hand out and in then I can begin. I can see it all except what the fingers are doing behind the cup, but I have learnt to do this [by grasping without using the handle]. I don't know how heavy the cup is so pick it up and monitor it visually. I need to see my arm up to my face, but then I can feel the cup with my face. I may not need to see the cup return until it reaches the table. This is all very controlled and involved.
>
> In the early days every finger was monitored. Now… when I move the hand it is almost automatic that the fingers open, though I keep 3 fingers out of way. To make any reaching movement first I tighten up the back muscles to make the start safe, and then move the arm out, less hand open but watch it to make sure it is doing what I want. (Cole, 2016, p. 128)

All this takes mental effort and so is tiring, so he has to really want the cup of tea. His attention to movement also reaches down further into the body's movement capacities than usual. He has learnt some extraordinary facets of the physicality of the body. He knows for instance that if he relaxes his wrist on the forearm, then the hand will come to rest in a mid-position between flexion and extension. His mental capacities towards action are also huge.

He cannot enjoy a movement for its own sake, cannot be lost in the moment or he might fall. He also has to assume when he asks for a movement that it will occur. So there is always an element of risk and of trust as he transforms intention into action.

> When a bird learns to fly it leaves the nest and goes sit on a branch, and its expectation is that the branch won't break; it trusts the branch. To become mobile and live independently I have learned to manage a range of movements, and I trust my ability to choreograph them. That's the payoff for all the hours, weeks and months of training, an implicit self-belief; that my branch won't break. (*ibid.*, p. 129)

As a result of this attention to movement, especially navigation in the world, Ian has a prodigious memory for places at a minute level. Even now, thirty years on, Ian can still remember almost every hotel he has audited (he works as a disability access auditor for hotels, banks, hospital, etc.). It is second nature for him to absorb the details of accessibility.

Ian kindly collaborated with Siobhan Davies in *Manual*, with Ian instructing Matthias Sperling how to stand. Matthias soon realized the differences between Ian and naïve subjects.

> Immediately and wonderfully apparent for me was the marked difference of working with someone who already *has* their own highly developed, readily accessible personal 'manuals' for consciously accomplishing deconstructed everyday movements, rather than someone searching them out for the first time. The second thing was simply how good his 'manuals' were; how efficient, robust, effective, well-planned, streamlined, architecturally sound... (*ibid.*, p. 130)

'Architecturally sound'; in a way Ian is an architect of his movements, planning and watching them unfold rather than being immersed in them like the rest of us, with a rich knowledge of the physics of the body in movement.

Conclusion

Impairments associated with severe neurological conditions such as paralysis, sensory loss, incoordination in cerebral palsy, and loss of facial animation, present to a person in various ways but each person has to come to terms with their new physicality as they construct as full a life as is possible in their situation. So, with impairment arise creative ways of adapting to new situations. In this those with impairments might be compared with athletes, having to perform at their maxima each day. But, perhaps, some might also be compared with creative artists as they search for new ways of being and feeling. Oliver Sacks wrote of the resourcefulness of people when faced with disease and the requirement when investigating this to go beyond the external and objective:

> ...[N]ature's richness is to be studied in the phenomena of health and disease, in the endless forms of individual adaptation by which human organisms, people, adapt and reconstruct themselves faced with the challenges and vicissitudes of live. (Sacks, 1995, pp. xii–xiv)

He continued by pleading for an intersubjective approach:

> The study of disease, for the physician, demands the study of identity, the inner worlds that patients, under the spur of illness, create. But... these worlds cannot be comprehended wholly from the observation of behaviour, from the outside. In addition to the objective approach of the scientist, we must employ an intersubjective approach, to see the world with the eyes of the patient himself. (*ibid.*, pp. xii–xiv)

This may not only allow us to see the world through another's eyes. Their experiences may open out new ways of seeing, not only disease and how people cope, but also our own experiences of the perception of action and sensation which, in Bell's words, may have been lost by long familiarity. The ways in which people experience and adapt to their losses may awaken us to our habitual automaticity in movement.

References

Bell, C. (1833) *The Hand: Its Mechanism and Vital Endowments as Evincing Design*, Bridgwater Lecture series, reprinted (1979) New York: Pilgrims Press.

Brown, C. (1990) *My Left Foot*, New York: Vintage.

Cole, J. (1998) *About Face*, Cambridge, MA: MIT Press.

Cole, J. (2004) *Still Lives: Narratives of Spinal Cord Injury*, Cambridge, MA: MIT Press.

Cole, J. (2016) *Losing Touch: A Man Without His Body*, Oxford: Oxford University Press.

Cole, J. & Spalding, H. (2007) *The Invisible Smile*, Oxford: Oxford University Press.

Decety, J. & Cacioppo, S. (2012) The speed of morality: A high-density electrical neuroimaging study, *Journal of Neurophysiology*, **108**, pp. 3068–3072.

Haggard, P. (2008) Human volition: Towards a neuroscience of will, *Nature Reviews Neuroscience*, **9** (12), pp. 934–946.

Head, H. (1920) *Studies in Neurology*, Oxford: Oxford University Press.

Kahneman, D. (2011) *Thinking, Fast and Slow*, London: Macmillan.

Libet, B., Gleason, C.A., Wright, E.W. & Pearl, D.K. (1983) Time of conscious intention to act in relation to onset of cerebral activity (readiness-potential) — the unconscious initiation of a freely voluntary act, *Brain*, **106**, pp. 623–642.

Murphy, R. (1987) *The Body Silent*, New York: Henry Holt.

Sacks, O. (1985) *The Man Who Mistook His Wife for a Hat*, London: Duckworth.

Sacks, O. (1995) *An Anthropologist on Mars*, London: Picador.

Wittgenstein, L. (1980) *Remarks on the Philosophy of Psychology*, Oxford: Blackwell.

Chris D. Frith

An Epilogue

> *Before the birth of consciousness,*
> *When all went well,*
> *None suffered sickness, love, or loss*
> *None knew regret, starved hope, or heart burnings.*
>   —Thomas Hardy, 'Before Life and After' (1909)

As Joe LeDoux says in the preface to this volume, brains were unconscious long before they were conscious. From such an assumption it follows that consciousness has evolved and that humans have much richer conscious experience than other animals. But why did consciousness evolve? To answer this question we need to know what can be achieved by the unconscious and what was added when consciousness emerged. The problem is that we have no direct access to our unconscious mental life. We can only make inferences about unconscious operations through observing behaviour and measuring neural activity.

Most critically, we must be able to distinguish between effects due to conscious rather than unconscious cognition. As Newell (this volume) points out, this can be extremely difficult. In particular it can be hard to assess whether or not there was truly an absence of awareness for the critical components of the task being performed. This problem is exacerbated by the demonstration from Cleeremans (this volume) that you can get different answers depending on how you measure awareness. It was, therefore, to be expected that the major concern of this book is to characterize the differences and relationships between conscious and unconscious operations.

There is still much debate about how much unconscious cognition contributes to our behaviour and experience (see Cleeremans, this volume). Few people go quite as far as T.H. Huxley, who declared that consciousness was merely an epiphenomenon, and 'as completely without any power of modifying that working [of the body] as the steam-whistle which accompanies the work of a locomotive engine is without influence upon its machinery' (Huxley, 1874). Nevertheless, there remains a wide range of opinions. At one extreme, unconscious cognition is considered to be restricted to low-level, automatic habits.

At the other extreme, unconscious cognition is thought to be important for making high-level, complex decisions. Indeed, Rosenthal seems to give everything to the unconscious, concluding that it is 'unlikely that there is much, if any, (added) utility to mental states being conscious'. Opinions varying across this whole range can be found in this volume. Here I shall consider three aspects of the relations between unconscious and conscious cognition.

1. Contrasting Unconscious and Conscious Cognition

As a starting point we might assume that conscious operations simply improve on what can already be achieved by unconscious cognition. This assumption implies, first, that the differences between unconscious and conscious cognition are quantitative rather than qualitative and, second, that there is nothing that can be achieved better by unconscious cognition. The consensus of the contributions to this volume is clear. These assumptions are wrong.

1.1. When unconscious cognition is better

There are many situations in which conscious cognition causes performance to deteriorate. As Cappuccio (this volume) points out, there is evidence that conscious control is highly detrimental to skilled athletic performance. The fluency of automatic motor routines is not disrupted by cognitive load or distraction, but can be severely disrupted by conscious monitoring of performance. The opposite of this coin is starkly revealed in Cole's account (this volume) of what happens when, as a result of injury or nerve damage, unconscious and automatic control of movement is no longer possible. Even the simplest of actions, such as picking up a mug, requires conscious deliberation and becomes anxiety provoking and mentally exhausting. At the very least the normal unconscious and automatic control of movement allows us to lead much easier lives.

Likewise conscious deliberation is not always good for decision making. While the claim that complex decisions are better made without consciousness (Dijksterhuis *et al.*, 2006) remains controversial, Cleeremans (this volume) has shown that conscious deliberation can impair decisions through deterioration of initial representations of the problem. This reminds me of work on face recognition. Participants who describe a face they have just seen perform less well on a subsequent recognition test than control participants who do not engage in memory verbalization (Schooler and Engstler-Schooler, 1990). It seems that the conscious deliberation required to verbalize a memory image produces long-lasting degradation of the memory image.

To understand these results it might be useful to consider different kinds of conscious systems. Nick Shea and I (Shea and Frith, 2016) have proposed that it is important to distinguish between representations and processes in cognition. In the type zero system (the unconscious) neither representations nor processes are conscious. In decision making without deliberation (type 1) representations of the problem are conscious, but the processes by which the decisions are made are not conscious. Deliberation brings type 2 into play in which the processes of decision are also conscious.

One explanation for impaired performance with conscious deliberation is that this system cannot handle complex, multidimensional representations. This problem is overcome by reducing dimensionality. As a result of this reduction, in many cases, rather than being weighted appropriately, stimulus dimensions are weighted variably and inconsistently (Levine, Halberstadt and Goldstone, 1996). Related to this reduction of dimensionality is the reduction of ambiguity observed when words become conscious. Both meanings of an ambiguous word (e.g. palm; tree/hand) are available when the word is presented subliminally, while only one meaning is available after supraliminal presentation (Marcel, 1980).

1.2. When conscious cognition is better

In other situations, however, the limitations caused by conscious deliberation can become advantages. For example, Prinz (this volume) suggests that unconscious vision cannot be modulated by attention and concludes that (top-down) attention is one process that cannot take place unconsciously. One mechanism likely to underlie the top-down control of attention is 'dimensional-weighting' (Found and Muller, 1996; Memelink and Hommel, 2013). One possible mechanism for achieving selective attention would be to increase the weight applied to a particular dimension or feature of the environment, as when the participant in an experiment is instructed to respond to the 'red objects' or the 'words in the left ear'. Instructions for deliberate selectivity override the dimensional weightings that the unconscious has learned through past experience. In these cases conscious operations enable us to rapidly and flexibly modify habits that are inappropriate in the current context.

2. The Interplay between the Unconscious and the Conscious

These conclusions about the relative merits of unconscious and conscious cognition lead naturally to the second theme of this book. This concerns the interplay between the two systems.

This interplay runs in both directions. I have already mentioned how conscious processes can modulate unconscious cognition, either for ill, in the case of compulsive monitoring of skilful activities, or for good, in the case of selective attention. This point is addressed by Bernacer and Murillo (this volume), who suggest that skills are integrators of conscious and non-conscious behaviour. Playing music, for example, involves a hierarchy of habits from low-level motor control to high-level concerns with interpretation and phrasing. All of these levels are open to conscious modulation. Radman (this volume) goes as far as to suggest that 'whatever we do consciously acts back on the non-conscious'.

At a higher level of cognition, the ability to think by analogy might be expected to require conscious representations (e.g. Holyoak, 2012). However, there is evidence that analogical thinking can occur without awareness and deliberation (see Hristova, this volume). Hristova concludes that the same basic mechanisms are used for both implicit and explicit analogy making, but suggests that instructions for making and verifying analogies engages modification of these mechanisms by deliberate and effortful conscious cognition. This is another example of conscious cognition influencing the workings of the unconscious.

Another high-level cognitive process for which conscious representations are thought to be necessary is multisensory integration (Singer, 1998; Tononi, 2008). Nevertheless, recent studies (Faivre *et al.*, 2014) have shown that sensory signals in different modalities can be integrated without participants being aware of either signal. However, this integration of unconscious representations only occurred when participants had undergone previous training on stimuli of the same kind presented consciously. Prior conscious experience thereby facilitated information integration. This is another case where conscious experience can modify the functioning of automatic, unconscious cognition.

But this interplay between conscious and unconscious cognition can also run the other way. In addition to our external senses, such as vision and hearing, we also experience bodily sensations from limbs, heart, and stomach. These signals associated with unconscious bodily processes can have effects on conscious processing without us necessarily being aware of them. Gallagher (this volume) calls these prenoetic effects and quotes a study demonstrating that hunger can distort cognitive cognition. The reasoning of judges is modified by their bodily state so that legal decisions are, on average, less favourable just before lunch when the judges are hungry (Danziger, Levav and Avnaim-Pesso, 2011).

However, as LeDoux points out in his introduction, we can also consciously interpret these bodily signals as feelings, such as tiredness, or anxiety, or fullness. Berthoz (this volume) suggests that sympathy, or emotional contagion, is an automatic, unconscious process through which we have the same emotional responses as the person we are observing. Empathy, in contrast, requires a conscious process through which we can inhibit our egocentric perspective and recognize that someone else is having the emotions. This is an example of a bodily signal being reinterpreted in consciousness.

In addition to these experiences associated with internal, bodily states, we also have access to signals reflecting aspects of automatic, unconscious processing. Examples include perceptual fluency—the awareness of the speed with which a sensory signal is interpreted—and action selection fluency—awareness of the speed with which an action is selected. These are examples of metacognitive representations since they are concerned with properties of cognitive processes (Shea et al., 2014).

These signals can be used in explicit, conscious judgments. For example, people base their liking of pictures on their feeling of the fluency with which the picture is perceived (Forster, Leder and Ansorge, 2012). Likewise, people base their judgment of motor control on the basis of their feeling of fluency when selecting an action (Chambon and Haggard, 2012).

However, as Rosenthal (this volume) points out, these metacognitive signals can sometimes be in error (misrepresentations). Perceptual fluency is a marker of familiarity, i.e. that the picture has been seen before, and therefore need not be a marker of liking (Kunst-Wilson and Zajonc, 1980) and action fluency is a consequence of response priming that can also occur in cases of reduced control when performance is impaired. This leads to the experience of feeling more in control when this is not actually the case (Chambon and Haggard, 2012). This fallibility in the interpretation of conscious experiences is relevant to the last aspect of consciousness that I want to consider.

3. Explicit Metacognition and the Social Function of Consciousness

The most obvious difference between conscious and unconscious cognition is that we can report aspects of conscious cognition, while unconscious cognition is, by definition, hidden from us. Thus, although we cannot directly experience the conscious mental lives of others, we can talk to each other about our experiences.

Furthermore such discussions can have effects on both conscious and unconscious cognition. For example, persuading people that free will does not exist makes them behave more selfishly (Baumeister, Masicampo and DeWall, 2009) and reduces post-error slowing in reaction time tasks (Rigoni *et al.*, 2013). These effects probably occur because the discussion changes beliefs about how the mind works, in this case causing a reduction in expectations about the efficacy of top-down, inhibitory control (Rigoni *et al.*, 2012).

Discussions can also alter our interpretation of metacognitive signals about the functioning of unconscious cognition. We all feel the after-effect of an extended period of mental exertion, but is this feeling a sign of tiredness or of being energized? It is widely believed that we are 'tired' after mental exertion (ego depletion; Baumeister *et al.*, 1998). However, a recent study has shown that behaviour after mental exertion can be modified by instructions. If participants are told that 'working on a strenuous mental task can make you feel energized for further challenging activities', then their performance on the Stroop task is better, rather than worse, after mental exertion (Job, Dweck and Walton, 2010).

This result shows how the various metacognitive signals that we experience are open to interpretations that can be altered by discussions with others. Presumably is it precisely the fallibility of these experiences, pointed out by Rosenthal, that enables their reinterpretation. These considerations give another example of the two-way interplay between conscious and unconscious cognition that I discussed in the last section. Signals reflecting unconscious cognition become part of our conscious experience and responses to these conscious signals alter the functioning of the unconscious cognition. However, there is a similar two-way interplay at the next level of the hierarchy where the person interacts with culture (i.e. other people). We describe our experiences to other people, potentially changing their interpretations of their experiences, while what others tell us can alter our interpretations, which, in turn, alter our unconscious processing.

This two-way process between the person and culture requires explicit metacognition and two kinds of transformation (Shea *et al.*, 2014). First, signals arising from unconscious cognitive processes must be converted into representations that can be communicated. A mental representation is made public. Second, communications from others are converted into signals that can modulate unconscious cognitive processes. A mental version of a public representation is internalized.

According to Sperber (1996) such processes are the necessary basis for cultural transmission and the creation of cumulative culture. I have suggested that this is a function for this aspect of consciousness (Frith, 2014b; Frith and Metzinger, 2016).

These observations are entirely compatible with Huxley's metaphor about the effects of the steam-whistle on the workings of the locomotive. The steam-whistle can certainly have an effect on another locomotive. It might cause the train to apply the brakes and stop. The steam-whistle enables steam engines to interact with other steam engines and the outside world more generally.

A nice example of a mental version of an internalized public representation is the inner speech that is such a striking feature of our consciousness. Wilkinson and Fernyhough (this volume) follow Vygotsky in suggesting that this inner speech originates as overt speech in young children that is shaped by the social dialogues from which it derives and gradually gets internalized. One consequence of this social shaping of inner speech is that we will find it easier to share this aspect of conscious experience with others from the same cultural background.

This normative effect of our ability to share our experiences can be seen in behaviour more generally. While our experience of action seems to have rather little effect at the time the action is selected, it has a large effect when we are called upon to describe and justify the action we have just performed. People are even willing to justify choices that they have not actually made (Johansson *et al.*, 2005). These post-hoc explanations and justifications of actions (e.g. I didn't do it deliberately) are essentially discussions about the appropriate causes of action and lead to a cultural consensus as to when certain behaviour is acceptable and when it is not (Frith, 2014a).

These discussions about agency are also an important feature of child development. Young children rapidly learn to claim, 'it was an accident'. Through such discussions they are learning folk psychological theories about the various mental states that cause actions and about the importance of prior intentions. This learning makes it easier to talk to others about our experiences of action, but it also makes our behaviour easier to interpret since it will tend to be constrained by the common beliefs we have acquired about the causes of action (McGeer, 2007).

As both Cleeremans and Newell (this volume) point out, the beliefs and expectations of the experimenter can affect the way subjects behave. I believe these observations may be explained by the mechanisms, discussed above, by which beliefs can be acquired from others and can affect our behaviour. On the one hand, these explicit metacognitive processes have critical implications for the way we conduct and interpret our experiments. We must pay much more attention to the beliefs that experimenters and participants bring to the experiment. On the other hand, this is an exciting development in our understanding of the roles of conscious and unconscious cognition.

Acknowledgments

I am grateful to Uta Frith and Nick Shea for their comments. Funding: this work was supported by the Arts and Humanities Research Council [grant number AH/M005933/], the John Templeton Foundation, the Wellcome Trust and the Institute of Philosophy, School of Advanced Study, University of London.

References

Baumeister, R.F., Bratslavsky, E., Muraven, M. & Tice, D.M. (1998) Ego depletion: Is the active self a limited resource?, *Journal of Personality and Social Psychology*, **74** (5), pp. 1252–1265.

Baumeister, R.F., Masicampo, E.J. & DeWall, C.N. (2009) Prosocial benefits of feeling free: Disbelief in free will increases aggression and reduces helpfulness, *Personality and Social Psychology Bulletin*, **35** (2), pp. 260–268.

Chambon, V. & Haggard, P. (2012) Sense of control depends on fluency of action selection, not motor performance, *Cognition*, **125** (3), pp. 441–451.

Danziger, S., Levav, J. & Avnaim-Pesso, L. (2011) Extraneous factors in judicial decisions, *Proceedings of the National Academy of Science USA*, **108** (17), pp. 6889–6892.

Dijksterhuis, A., Bos, M.W., Nordgren, L.F. & van Baaren, R.B. (2006) On making the right choice: The deliberation-without-attention effect, *Science*, **311** (5763), pp. 1005–1007.

Faivre, N., Mudrik, L., Schwartz, N. & Koch, C. (2014) Multisensory integration in complete unawareness: Evidence from audiovisual congruency priming, *Psychological Science*, **25** (11), pp. 2006–2016.

Forster, M., Leder, H. & Ansorge, U. (2012) It felt fluent, and I liked it: Subjective feeling of fluency rather than objective fluency determines liking, *Emotion*, **13** (2), pp. 280–289.

Found, A. & Muller, H.J. (1996) Searching for unknown feature targets on more than one dimension: Investigating a 'dimension-weighting' account, *Percept Psychophysics*, **58** (1), pp. 88–101.

Frith, C.D. (2014a) Action, agency and responsibility, *Neuropsychologia*, **55**, pp. 137–142.

Frith, C.D. (2014b) How the brain creates culture, *Nova Acta Leopoldina*, **120** (405), pp. 3–26.

Frith, C.D. & Metzinger, T. (2016) What's the use of consciousness?, in Engel, A.K., Friston, K. & Kragic, D. (eds.) *Where's the Action? The Pragmatic Turn in Cognitive Science*, vol. 18, Cambridge, MA: MIT Press.

Holyoak, K.J. (2012) Analogy and relational reasoning, in Holyoak, K.J. & Morrison, R.G. (eds.) *The Oxford Handbook of Thinking and Reasoning*, New York: Oxford University Press.

Huxley, T.H. (1874) On the hypothesis that animals are automata, and its history, *Nature*, **10**, pp. 362–366.

Job, V., Dweck, C.S. & Walton, G.M. (2010) Ego depletion—is it all in your head? Implicit theories about willpower affect self-regulation, *Psychological Science*, **21** (11), pp. 1686–1693.

Johansson, P., Hall, L., Sikstrom, S. & Olsson, A. (2005) Failure to detect mismatches between intention and outcome in a simple decision task, *Science*, **310** (5745), pp. 116–119.

Kunst-Wilson, W.R. & Zajonc, R.B. (1980) Affective discrimination of stimuli that cannot be recognized, *Science*, **207** (4430), pp. 557–558.

Levine, G.M., Halberstadt, J.B. & Goldstone, R.L. (1996) Reasoning and the weighting of attributes in attitude judgments, *Journal of Personality and Social Psychology*, **70** (2), pp. 230–240.

Marcel, A.J. (1980) Conscious and preconscious recognition of polysemous words: Locating the selective effects of prior verbal context, in Nickerson, R.S. (ed.) *Attention and Performance VIII*, Hillsdale, NJ: Erlbaum.

McGeer, V. (2007) The regulative dimension of folk psychology, in Hutto, D. & Ratcliffe, M. (eds.) *Folk Psychology Re-Assessed*, New York: Springer.

Memelink, J. & Hommel, B. (2013) Intentional weighting: A basic principle in cognitive control, *Psychological Research-Psychologische Forschung*, **77** (3), pp. 249–259.

Rigoni, D., Kuhn, S., Gaudino, G., Sartori, G. & Brass, M. (2012) Reducing self-control by weakening belief in free will, *Consciousness and Cognition*, **21** (3), pp. 1482–1490.

Rigoni, D., Wilquin, H., Brass, M. & Burle, B. (2013) When errors do not matter: Weakening belief in intentional control impairs cognitive reaction to errors, *Cognition*, **127** (2), pp. 264–269.

Schooler, J.W. & Engstler-Schooler, T.Y. (1990) Verbal overshadowing of visual memories: Some things are better left unsaid, *Cognitive Psychology*, **22** (1), pp. 36–71.

Shea, N.J., Boldt, A., Bang, D., Yeung, N., Heyes, C. & Frith, C.D. (2014) Supra-personal cognitive control and metacognition, *Trends in Cognitive Science*, **18** (4), pp. 186–193.

Shea, N. & Frith, C.D. (2016) Dual-process theories and consciousness: The case for 'Type Zero' cognition, *Neuroscience of Consciousness*, **2016** (1), niw005.

Singer, W. (1998) Consciousness and the structure of neuronal representations, *Philosophical Transactions of the Royal Society of London, Series B, Biological Sciences*, **353** (1377), pp. 1829–1840.

Sperber, D. (1996) *Explaining Culture: A Naturalistic Approach*, Oxford: Wiley-Blackwell.

Tononi, G. (2008) Consciousness as integrated information: A provisional manifesto, *Biological Bulletin*, **215** (3), pp. 216–242.

Notes on Contributors

Javier BERNACER is researcher in the Mind-Brain Group of the Institute for Culture and Society (ICS, University of Navarra), and associate professor in the Faculty of Education and Psychology of the University of Navarra. He has a PhD in Neuroscience and a Master's Degree in Philosophy, and has published empirical and theoretical research in high impact journals within the fields of neuroscience (*American Journal of Psychiatry*) and psychology (*Behavioral and Brain Sciences*). His current research interest is to elucidate the brain correlates of habit acquisition and consolidation, understanding habits from an Aristotelian perspective.

Alain BERTHOZ is Emeritus Professor at the Collège de France (Chair of Physiology) and former founder and Director of the Perception and Action Lab (CNRS Collège de France). He is a specialist of the physiology of multisensory integration, spatial orientation, the vestibular system, the oculomotor system, locomotion, and spatial memory. He has contributed to the understanding of the neural mechanisms of eye movements, posture and locomotion, multisensory integration, and spatial memory. He pioneered use of intracellular recordings in animals, functional MRI in humans, and intracranial neural recordings in patients. He has worked on sensory-motor and cognitive pathologies in children and adults, and cognitive functions including psychiatric diseases, and the neural basis of empathy and sympathy. He cooperates with robotics groups in Japan and Italy for neuro-inspired robotics and humanoids. He is a member of the French Academy of sciences and the Academy of Technologies, the Academia Europae, American Academy of Arts and Sciences, and others. Recently he has obtained the Castang Prize of the European Society of Child Neurology (2015), the Hallpike-Nylén Prize for patho-physiology of the vestibular system (2016); he has also been elected at the Royal Belgium Academy of Sciences and Humanities (2016) and the Honoris Causa title of University of Salerno (2016). He is the author of more than 300 papers in international journals.

Massimiliano L. CAPPUCCIO is Associate Professor of cognitive science and philosophy of mind at UAE University, the national university of the United Arab Emirates, where he directs the Interdisciplinary Cognitive Science Lab. His research combines empirical methods and phenomenological investigation, and applies various elements of embodied cognition theory to issues in sport and performance psychology and social robotics, focusing on empathy, unreflective action, joint attention, gestures, and the origins of symbolic culture. He is the editor of the MIT Press *Handbook of Embodied Cognition and Sport Psychology*.

Axel CLEEREMANS is a Research Director with the Fonds de la Recherche Scientifique (F.R.S.-FNRS) and a professor of cognitive psychology at the Université libre de Bruxelles (ULB), where he heads the Consciousness, Cognition and Computation (CO3) Group and presides the ULB Neurosciences Institute. His research is essentially dedicated to the differences between information processing with and without consciousness, particularly in the domain of learning and memory, and more recently in the domains of perception, social cognition, and cognitive control. He has acted as president of the European Society for Cognitive Psychology and is a board member of the Association for the Scientific Study of Consciousness. He has recently edited, together with Tim Bayne and Patrick Wilken, *The Oxford Companion to Consciousness*. He is Field Editor-in-Chief of *Frontiers in Psychology* and is also a member of the Royal Academy of Belgium.

Jonathan COLE is a consultant in clinical neurophysiology at Poole Hospital and a professor at Bournemouth University. His research has been in motor control and sensory loss, in chronic pain and in facial perception. He has also written books on the first-person experience of neurological impairments. *About Face* (MIT Press, 1998) and *The Invisible Smile* (OUP, 2007) are on the face and facial immobility; *Still Lives* (MIT Press, 2004) on living with spinal cord injury; and *Pride and a Daily Marathon* (MIT Press, 1995) and his latest, *Losing Touch* (OUP, 2016), on the experience and neuroscience of loss of touch and proprioception.

Donish CUSHING is a Masters student at San Francisco State University. He is interested in how actions are rendered adaptive through conscious processing and through feedback from various neural systems.

Charles FERNYHOUGH is a psychologist and writer. He is a Professor in Psychology at Durham University and principal investigator of Hearing the Voice, an interdisciplinary project on voice-hearing funded by the Wellcome Trust. His research focuses on how ideas from developmental psychology can be utilized to enhance our understanding of adult inner experience and psychopathology.

Chris D. FRITH is Emeritus Professor of Neuropsychology at the Wellcome Trust Centre for Neuroimaging at University College London and Honorary Research Fellow at the the Institute of Philosophy, School for Advanced Studies, University of London. He studies the function of consciousness, the mechanisms of social interaction, and the advantages and disadvantages of group decision making. Some recent publications: Bahrami, B., Olsen, K., Latham, P.E., Roepstorff, A., Rees, G. & Frith, C.D. (2010) 'Optimally Interacting Minds', *Science*, 329 (5995); Frith, C.D. (2014) 'Action, Agency and Responsibility', *Neuropsychologia*, 55; Shea, N., Boldt, A., Bang, D., Yeung, N., Heyes, C. & Frith, C.D. (2014) 'Supra-personal Cognitive Control and Metacognition', *Trends in Cognitive Science*, 18 (4); Friston, K.J. & Frith, C.D. (2015) 'Active Inference, Communication and Hermeneutics', *Cortex*, 68; Frith, C.D. & Metzinger, T. (2016) 'What's the Use of Consciousness? How the Stab of Conscience Made Us Really Conscious', *The Pragmatic Turn: Toward Action-Oriented Views in Cognitive Science*, ed. A.K. Engel, K.J. Friston & D. Kragic, Strüngmann Forum Reports, vol. 18, MIT Press; Shea, N. & Frith, C.D. (2016) 'Dual-Process Theories and Consciousness: The Case for 'Type Zero' Cognition', *Neuroscience of Consciousness*, 1.

Penka HRISTOVA is an assistant professor of cognitive psychology at New Bulgarian University, and a manager of the Experimental Psychology Laboratory at the Department of Cognitive Science and Psychology at New Bulgarian University. She received her PhD in psychology from the New Bulgarian University in 2014. Her research focuses on the mechanisms that underpin analogies with and without intention and awareness, as well as on the influence of emotions on analogy-making. Her research interests include constructive memory, executive functions, judgment, and decision making.

Shaun GALLAGHER holds the Moss Chair of Excellence in Philosophy at the University of Memphis. He is also Professorial Fellow at the University of Wollongong. He has held visiting positions at the Cognition and Brain Sciences Unit, Cambridge, the Ecole Normale Supériure, Lyon, the Centre de Recherche en Epistémelogie Appliquée

(CREA), Paris, the Humboldt University, Berlin, and most recently, Keble College, Oxford University. His recent books include *How the Body Shapes the Mind* (OUP, 2005); *The Phenomenological Mind* (with Dan Zahavi, Routledge, 2012 — 2nd ed.), and his forthcoming monograph: *Enactivist Interventions: Rethinking the Mind*. He is editor of *The Oxford Handbook of the Self* (2011) and editor-in-chief of the journal *Phenomenology and the Cognitive Sciences*.

Reza D. GHAFUR is an undergraduate student at San Francisco State University who is interested in both cognitive psychology and social cognition.

Joseph LeDOUX is the Henry and Lucy Moses Professor of Science at NYU in the Center for Neural Science, and he directs the Emotional Brain Institute of NYU and the Nathan Kline Institute. He is also a Professor of Psychiatry and Child and Adolescent Psychiatry at NYU Langone Medical School. His work is focused on the brain mechanisms of memory and emotion and he is the author of *The Emotional Brain*, *Synaptic Self*, and *Anxious*. LeDoux has received a number of awards, including the Karl Spencer Lashley Award from the American Philosophical Society, the Fyssen International Prize in Cognitive Science, Jean Louis Signoret Prize of the IPSEN Foundation, the Santiago Grisolia Prize, the American Psychological Association Distinguished Scientific Contributions Award, and the American Psychological Association Donald O. Hebb Award. His book *Anxious* received the 2016 William James Book Award from the American Psychological Association. LeDoux is a Fellow of the American Academy of Arts and Sciences, the New York Academy of Sciences, and the American Association for the Advancement of Science, and a member of the National Academy of Sciences. He is also the lead singer and songwriter in the rock band, *The Amygdaloids* and performs with Colin Dempsey as the acoustic duo *So We Are*.

Ezequiel MORSELLA is a theoretician and experimentalist, he has devoted his entire career to investigating the differences in the brain between the conscious and unconscious circuits in the control of human action. He is the lead author of *Oxford Handbook of Human Action*. The conclusions from his investigations have appeared in journals such as *Psychological Review* and *Behavioral and Brain Sciences*. After his undergraduate studies at the University of Miami, he carried out his doctoral and postdoctoral studies at Columbia University and Yale University, respectively.

José Ignacio MURILLO is director of the Mind-Brain Group of the Institute for Culture and Society (ICS, University of Navarra), and Professor in the Faculty of Philosophy of the University of Navarra, teaching Philosophical Anthropology. He is author of several books, book chapters, and papers in philosophy, but has also published his research in journals within the fields of psychology (*Behavioral and Brain Sciences, Frontiers in Psychology*) and neuroscience (*Frontiers in Human Neuroscience*). He is interested in interdisciplinary research on human action, including decision making, agency, habit formation, and free will.

Ben R. NEWELL is Professor of Cognitive Psychology and Deputy Head of the School of Psychology at the University of New South Wales in Sydney, Australia. His research focuses on the cognitive processes underlying judgment, choice, and decision making, and the application of this knowledge to environmental, medical, financial, and forensic contexts. He has an enduring interest in the role that implicit or unconscious processes play in higher-level cognition (learning, categorization, decision making) and has published several papers that question and critique the explanatory burden placed on the 'powers' of unconscious thinking (e.g. Newell, B.R. & Shanks, D.R. (2014) 'Unconscious Influences on Decision Making: A Critical Review', *Behavioral and Brain Sciences*, 37). He also co-authored (with D. Lagnado and D. Shanks) the book *Straight Choices: The Psychology of Decision Making* (Psychology Press, 2015).

Jesse J. PRINZ is a Distinguished Professor of Philosophy at the City University of New York. He works primarily in the philosophy of psychology and has produced books and articles on emotion, moral psychology, aesthetics, and consciousness. He is the author of *The Conscious Brain* (OUP, 2012), *Beyond Human Nature: How Culture and Experience Shape Our Lives* (Allen Lane, 2012), *The Emotional Construction of Morals* (OUP, 2009), *Gut Reactions: A Perceptual Theory of Emotion* (OUP, 2006), and *Furnishing the Mind: Concepts and Their Perceptual Basis* (MIT Press, 2004). He is currently Einstein Fellow at the School of Mind and Brain, Humboldt University, Berlin.

Zdravko RADMAN is a Senior Researcher at the Institute of Philosophy, Zagreb, and teaches philosophy at the University of Split, Croatia. As an Alexander von Humboldt and a William J. Fulbright Fellow he was affiliated with the University of Konstanz and the University of California, Berkeley; as a visiting scholar he conducted research at the ANU, Canberra, and IMéRA, Marseille, among others.

He has published in the philosophy of mind and consciousness, aesthetics, philosophy of language, and theory of knowledge. He is the author of *Metaphors: Figures of the Mind* (Kluwer, 1997/Springer, 2010) and editor of collected volumes: *The Hand, an Organ of the Mind: What the Manual Tells the Mental* (MIT Press, 2013), *Knowing without Thinking: Mind, Action, Cognition, and the Phenomenon of the Background* (Palgrave Macmillan, 2012), *Horizons of Humanity* (Peter Lang, 1997), and *From a Metaphorical Point of View* (W. de Gruyter, 1995).

Thomas REBER is a postdoc whose research is funded by the Swiss National Science Foundation (SNSF) currently working at the Department of Epileptology at the University of Bonn Medical Center. In his research he investigates neural correlates and mechanisms of memory, perception, and awareness in the human medial temporal lobe. To this aim, he compares effects of visual stimuli presented below, near, and above the threshold of conscious perception on behaviour and measures of neural activity such as intracranial EEG, human single units, and fMRI. Major publications include (together with R. Luechinger, P. Boesiger & K. Henke) (2012) "Unconscious Relational Inference Recruits the Hippocampus', *Journal of Neuroscience*, 32 (18), and 'Detecting Analogies Unconsciously', *Frontiers in Behavioral Neuroscience*, 8.

David ROSENTHAL is professor of philosophy at the Graduate Center of the City University of New York, and Coordinator of their Interdisciplinary Concentration in Cognitive Science. He works mainly in empirically informed philosophy of mind and philosophy of psychology, focusing on consciousness, the qualitative character of experience, intentionality, the self, emotions, metacognition, and the relation of thought to speech. His 2005 book, *Consciousness and Mind* (Clarendon Press), addresses many of those topics, as do recent articles and book chapters such as 'Awareness and Identification of Self' (*Consciousness and the Self*, Liu and Perry, eds., CUP, 2011), 'Quality Spaces and Sensory Modalities' (*Phenomenal Qualities*, Coates and Coleman, eds., OUP, 2015), 'How to Think about Mental Qualities' (*Philosophical Issues*, 20, 2010), and 'Consciousness and Its Function' (*Neuropsychologia*, 46, 2008).

Sam WILKINSON is a Research Fellow in Philosophy at Durham University. He works on Hearing the Voice—an interdisciplinary project on voice-hearing funded by the Wellcome Trust. Before that, he completed his PhD on psychiatric delusions at the University of Edinburgh. He has published numerous papers on delusions, auditory verbal hallucinations, and the nature of explanation in psychiatry in journals like *Mind and Language, Philosophical Psychology, Consciousness and Cognition, The Review of Philosophy and Psychology, The Journal of Consciousness Studies,* and *Frontiers in Psychiatry*. He has worked closely with psychologists and psychiatrists.

Index

Action, vi, ix, 4, 5, 8,10-12, 14-18, 34, 41, 50-53, 53-63, 65, 67, 70-72, 85, 93, 95, 99, 112, 143, 147, 153, 157-59, 163, 165, 168, 175, 225-26, 228-42, 248-51, 253-61, 263-74, 286, 290, 293, 295-98, 304-08, 312, 314-17, 320, 323, 325
- automatic, 249, 256, 260, 261, 305, 315
- bodily, 286, 296
- control of, 10, 53, 109, 239
- deliberative, 158-59
- goal directed, 55, 228-29, 232, 234, 238, 240-41, 269, 273, 275
- immanent, 235, 237-38
- selection of, 56, 59-63, 66-67, 71-72, 323
- simulation of, 8, 16
- skilled, 257, 267, 320
- unreflective, 249

Agency, v, 14, 41-42, 50, 68, 95, 135, 226, 258, 305-09, 325
- sense of, 271-72, 299, 301

Amygdala, vii, 8
Analogy, 206-16, 322
Anticipation, 2, 11-12, 15, 195, 214
Aristotle, 225-26, 235-39, 242
Attention/inattention, x, xii, 1-2, 31, 34, 89-90, 92, 94, 97, 123, 147-48, 151-57, 230, 240, 242, 246, 250-54, 256, 259-62, 263-67, 270-75, 276-77, 293, 304, 307, 308, 311, 314, 316, 321-22

Automatic (automaticity), 1, 17, 38, 41, 50, 51, 54, 64, 66-68, 70, 72, 89, 91-92, 94, 129, 158-59, 207-08, 229, 237, 237, 239-40, 249, 255-56, 266, 272, 274, 304-05, 307, 309, 315, 317, 319-20, 322-23

Autonomic system, 166-68, 305
Awareness, vi, xi, 1, 3, 5, 14, 15, 28, 31-33, 35, 37, 40, 42, 52, 84, 87, 88, 89-98, 106-08, 110-19, 121-25, 129, 132-36, 144, 146-47, 151, 174, 178, 182-87, 192-93, 197, 200, 206-08, 210, 212, 215-16, 228-32, 246-51, 253, 255, 258-61, 263-64, 266-67, 269, 272, 274-75, 304, 308, 319, 322-23
- body, 10
- self-awareness, 10, 85, 93, 298, 304, 308

Blindsight, xi, 90, 117, 132, 145
Block, N., 109-12, 116-17, 121, 124, 128, 257
Body schema, 1, 5-6, 8-11, 17, 163, 168
Bourdieu, P., 17-18

Chalmers, D., vi, 26
Clark, A., vi
Cleeremans, A., 96-98, 319-20, 325
Cognition, vii, 8, 30-31, 36, 38, 87, 159, 206, 207, 226, 237, 240, 248, 250, 257, 260, 267, 274, 285-86, 288-90, 300, 319-24
- high-level (conscious contents), vii, 1, 15, 30-31, 35-36, 50, 63, 67, 72
- unconscious, 325, 326

Cognitive (neuro)science, ix, x, xi, 26, 30, 87, 155, 158, 225-29, 231-32, 234-35, 238, 240, 248-49, 258, 270
Cognitive unconscious, x

Decision making, vii, 1, 34, 38, 39, 41, 84, 97, 159, 174-78, 181, 186-87, 206-07, 254, 256, 320
Default system, 1, 111-12
Descartes, R., 18, 116, 125, 226-29, 230-32, 241
Dewey, J., viii

Dijksterhuis, A., 35, 38, 130, 187

Embodied, 16, 33, 37, 40-41, 99-100, 163, 166-69, 268, 274, 309, 311-13
- cognition, 164-65, 167-69, 226, 248, 250, 254-57, 260, 274
- coping, 3
- intentionality, 226, 272
- knowledge, 260, 268, 274
- mind, 32, 39
- perception, 15
- routines, 266
- skill, 36

Emotion, vii, xii, 1, 7-9, 15-17, 29-30, 38-39, 111, 133, 143, 148, 152-53, 163-64, 233, 248, 252, 257-59, 270, 274, 292, 297, 300, 308, 312-13

Empathy, 15-16, 323

Explicit, 32, 65, 89, 91-92, 94, 118, 133-34, 144, 147, 174, 177, 182, 186-87, 199-200, 212, 214, 250, 255-56, 261, 263, 266
- analogies, 207, 211-12, 216, 322
- attitudes, 183-84, 186
- control, 252, 254, 262
- intention, 271
- judgment, 273, 323
- knowledge, 185
- memory, 197, 202
- self-awareness, 84
- thought, 274

Freud, S., x, 2, 4-5, 17, 143, 148
Freudean unconscious, x
Frith, C., viii, 134, 275, 286

Gallagher, S., viii, 4, 275, 322
Gazzaniga, M., xi

Habits, 18, 225-30, 232-42, 268, 319, 321-22
Haggard, P., 4, 14, 112
Hallucinations, 284-85, 289, 291-92
"Hard problem", 27
Heidegger, M., 32
Hippocampus, 2, 7, 13, 192-94, 196, 198-99, 227

Implicit, x, 30, 32, 40, 91-92, 144, 199, 251, 255-56, 261, 298, 316
- analogies, 207, 214-16, 322
- attitude, 183-84, 186

- cognition, 89, 95
- inference, 199
- knowledge, 250
- learning, 86, 98, 195-96, 198-99, 227-28, 237, 263, 265
- memory, x, 192-200, 227-29

Inference, 87, 116-17, 131, 143, 199, 267, 290-91, 298, 319
- conscious (rational), 130, 249
- explicit, 255
- implicit, 200
- transitive, 198
- unconscious, 67

Inhibition, vii, 11, 13-17, 39, 153, 157, 293-95

Inner speech, 69, 284-89, 292-99, 301, 325

Intentionality, vii, 38, 125, 226, 260
- motor, 268, 272-74

Introspection, ix, 182, 183, 187, 257, 262, 266-67, 273-74, 298

Involuntary, 63-65, 67-68, 232-33, 271

James, W., xiii, 2, 33, 54, 70, 163, 225, 229, 234, 237, 239

Judgment, 33-34, 37, 40, 96, 98, 113-19, 167-68, 174-76, 183, 185-87, 197-98, 206, 211, 270, 323
- explicit, 269, 273, 323
- intuitive, 175, 185, 187, 273
- metacognitive, 113-15, 117-19, 129, 135, 323

Kandel, E., 34
Kant, I., 18, 32, 108, 120, 230
Kouider, S., 84, 93, 197

Learning, vi, 14-15, 18, 31, 38, 59, 83, 85-90, 93, 98, 100, 113, 127-28, 131, 148, 152, 187, 194-96, 198-99, 209, 225-30, 235-41, 248, 263, 265, 309, 325

LeDoux, J., vi, viii, 30, 319, 323
Leibniz, G.W., 230
Libet, B., 4, 35, 66, 112, 304

Merleau-Ponty, M., 5, 260, 268, 272
Memory, vi-x, 2, 4, 7, 13, 15, 31, 38, 54, 57-58, 61, 65, 70, 84, 145-47, 152, 157-58, 194-95, 197-99, 200, 206, 209, 227-28, 284, 292, 314, 316, 320-21
- declarative, x, 194

- episodic, 195, 267, 292-94
- explicit, 193, 195, 200, 227, 235
- implicit, x, 192-198, 200, 227, 229-30, 235
- long-term, 2, 152, 192, 194, 197
- non-declarative, x, 195, 228-29
- short-term, 193
- working, x, xii, 94, 157-58, 207-12, 215, 267, 270

Metacognition, 32, 91, 106, 113-16, 118, 120, 129-30, 323-24
Metarepresentation, 90-92, 99
Misrepresentation, 95, 106-07, 119-24, 126-27, 323
Möbius syndrome, 305, 309
Multisensory integration, 8, 322

Non-conscious, v-viii, ix, xi-xiii, 13, 28-30, 32, 35, 37-38, 41-42, 107-08, 111-12, 119, 122, 163, 167-68, 225-29, 241, 246, 248, 252, 254, 257, 260-61, 269, 311, 322
- neuroscientific perspective, vi

Passive Frame Theory, 50-51, 70, 72
Penfield, W., 9
Perception, vii, 3, 6, 9, 12, 15, 30-31, 38, 54, 63, 69, 72, 87, 90-91, 95, 97, 99, 107-09, 111, 120-21, 124, 126, 132, 134, 143, 147-49, 152, 155, 163-68, 175, 193, 197, 200, 206, 215, 229, 231, 248, 250, 255, 258, 268-69, 275, 290, 292, 295, 300-01, 304-05, 311-12, 317
Petit, J.-L., 5
Predictive processing (coding), 87-88, 99, 167, 267, 284-85, 288-89, 290, 291-93, 296-98
Prenoetic, 163-165, 167-68, 170, 249, 323
Proprioception, 6-7, 257, 311

Rationality, 29, 130-31
Radman, Z., 1, 3, 18, 160, 322
Ricoeur, P., 4, 226, 230, 232-35, 237, 241

Self, v, vi, 3-6, 8-9, 16-17, 29, 71, 85, 232, 234, 239, 241, 259, 305, 309
- conscious, viii, 41, 232
- construction of, 6
- distributed, 4
- organismic, 3, 37-39, 42

- phenomenological, 4
- narrative, 4

Simplicity, 1, 11
Skill, vi, 1, 18, 32, 36-37, 42, 89-91, 193, 195, 227, 241, 246, 248-49, 250-54, 257, 260-63, 267-68, 270-76, 274, 315, 320, 322
Sport, 246-52, 257-60, 262-63, 266-67, 270, 274

Unconscious, v, vi, vii, x, xiii, 1-7, 11-17, 26, 28, 29-32, 34-42, 50-56, 58-59, 61, 66-68, 72, 83, 88, 92-99, 142-43, 174-77, 181-87, 192, 197, 200-06, 210, 212-16, 226, 228-35, 239, 230-39, 241, 246, 256-57, 261, 284, 319-24
- analogy, 207, 212-16
- brain, viii, 1-2, 6, 12, 88, 99, 319
- cognition, 95, 319-20, 322-24, 326
- decision-making, 98, 174-76, 181, 186-87
- inferences, 67
- judgment, 175
- knowledge, 6, 92
- learning, 99
- mind, vi, 5, 149
- perception (vision), 142-43, 145, 147-49, 151-53, 157, 159
- thinking, vii, 35

Vision, 2, 6-7, 13, 31, 33, 35, 60, 67, 121-22, 142-46, 148-55, 156, 167, 311-12, 322
- unconscious, 149-54, 158-59, 321
Visual cortex, x-xi, 6, 7, 10, 67, 109-10, 122-24, 143, 145, 168
Voluntary, 67, 70, 134, 232-33, 258, 271-72, 313
- action, 50, 58, 61, 72, 271-72, 258
- behaviour, 60, 134, 233
- control, 258, 271-72
- decision, 272
von Helmholtz, H., 229

Watson, J. (behaviourism), ix
Wittgenstein, L., 28, 36-37, 310
Wundt, W., iv

Zeki, S., 2

www.ingramcontent.com/pod-product-compliance
Lightning Source LLC
Chambersburg PA
CBHW071228230426
43668CB00011B/1347